Bioactive Compounds in Agricultural Soils

Lech Wojciech Szajdak
Editor

Bioactive Compounds in Agricultural Soils

Springer

Editor
Lech Wojciech Szajdak
Institute for Agricultural and Forest Environment
Polish Academy of Sciences
Poznan, Poland

ISBN 978-3-319-43106-2 ISBN 978-3-319-43107-9 (eBook)
DOI 10.1007/978-3-319-43107-9

Library of Congress Control Number: 2016953334

Printed on acid-free paper

This Springer imprint is published by Springer Nature
The registered company is Springer International Publishing AG Switzerland

Contents

Chapter 1
Introduction: Biologically Active Compounds

Lech Wojciech Szajdak

Abstract Each chemical compound possesses numerous biological activities. Biological activity spectrum of a compound should be predicted on the basis of the structure-activity-relationships. The biological activity spectrum of a compound shows its all actions and its participation in the biological, physiological and metabolically pathways despite the difference in the experimental conditions. The biological activity spectrum of a compound shows compound's all actions and the participation in the biological, physiological and metabolically pathways despite the difference in the experimental conditions. If the differences in species, sex, age, dose, and the participation in the metabolic processes and pathways etc. are neglected, the biological activity may be identified only qualitatively. Thus the biological activity spectrum is defined as the "intrinsic" property of a substance depending only on its structure and physicochemical characteristics. Structure-activity-relationship (SAR) is an approach to finding the relationships between the chemical structure (or structural-related properties) and the biological activity of studied compounds. It links the chemical structure to a chemical property (e.g., water solubility) or the biological activity including toxicity.

Keywords Chemical compounds • Biological potential • Biological activity spectrum • Structure-activity-relationship

Each chemical compound possesses numerous biological activities. However, its activity always depends on the object, dose, and participation in the chemical conversions or biochemical pathways. On the other hand, the biological potential of the substance may be discovered under the specific experimental conditions.

As per the Oxford Dictionary, "Biochemistry refers the biological activity of a substance demonstrated in living organisms." The biologically active substances are often from biological origin. Biological activity characterizes the biological effectiveness of a substance and describes the changes caused by biological material like enzymes, hormones, vitamins, etc., in a tissue or in an organ under constant condi-

L.W. Szajdak (✉)
Institute for Agricultural and Forest Environment, Polish Academy of Sciences,
Str. Bukowska 19, 60-809 Poznań, Poland
e-mail: lech.szajdak@isrl.poznan.pl

L.W. Szajdak (ed.), *Bioactive Compounds in Agricultural Soils*,
DOI 10.1007/978-3-319-43107-9_1

tions as compared with certain international standard. Certain biological activity is measurable in quantity, and the international unit (IU) corresponds to a certain amount of a biologically active substance which produces a certain biological response. However, another measure is the IC_{50}—value corresponding to that amount of an effective substance, which inhibits the biochemical reaction in target organism by 50 % effectiveness. Concentration in this context means the concentration of a given substance in the target tissue—often in mg/kg or g/kg (Uosukainen and Pihlaja 2006).

Some scientists call the *biological potential* the *biological activity spectrum*. Biological activity spectrum of the compound should be predicted based on the structure–activity relationships. The biological activity spectrum of a compound shows compound's all actions and the participation in the biological, physiological, and metabolic pathways despite the difference in the experimental conditions. If the differences in species, sex, age, dose, participation in the metabolic processes and pathways, etc., are neglected, the biological activity may be identified only qualitatively. Thus, "the biological activity spectrum" is defined as the "intrinsic" property of a substance depending only on its structure and physicochemical characteristics (Filimonov and Poroikov 1996; Filimonov et al. 1999).

The molecular structure of an organic compound determines its properties (Jurs et al. 1988).

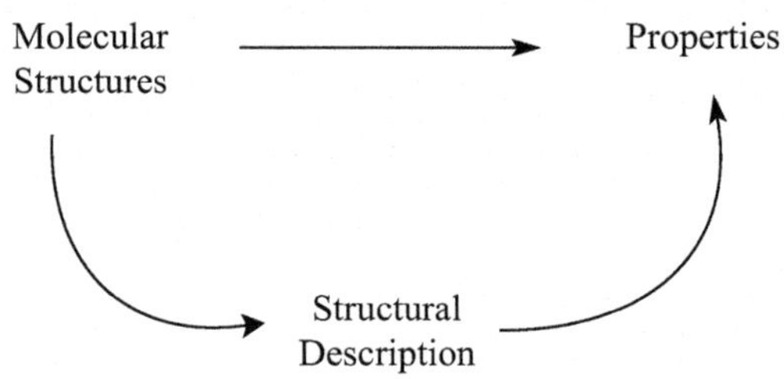

The term *active substance* has been used in this chapter to refer to raw material (organic matter, mineral and organic soil, peat, moorsh, sapropel) substrates and growing media for agriculture and horticulture. Relationship between the molecular structure and biological activity or molecular structure and physical properties can be investigated for most organic compounds using different analytical methods. High-performance liquid chromatography (HPLC), gas chromatography, infrared analysis, UV-VIS spectrophotometry, spectrofluorimetry, colorimetric test, thermal analysis, electron paramagnetic resonance, enzyme activity, etc., are recommended for the quantitative and qualitative determination of chemical and biochemical compounds, their metabolites, and degradation products.

Specification for the active substances should include the tests for (i) appearance, (ii) identification, (iii) content/assay, (iv) impurities (residual solvents, ash, heavy metals, and related substances, metabolites, and the products of the degradation), and (v) other parameters relevant to the individual substances (water content, loss on drying, the presence or ratio of isomers, optical rotation, melting point and the clarity, color and pH of solutions).

Structure–activity relationship (SAR) is an approach to find the relationships between chemical structure (or structural-related properties) and its biological activity of studied compounds. It links the chemical structure to a chemical property (e.g., water solubility) or biological activity including toxicity. Qualitative SARs and quantitative SARs are collectively named (Q)SARs. Qualitative relationships are derived from non–continuous data (e.g., yes or no data), while quantitative relationships are derived from continuous data (e.g., toxic potency data). The approach is not new [AFA Cros in 1863 noted in "Action de l'alcool amylique sur l'organisme," the relationship between the toxicity of primary aliphatic alcohols and their water solubility (Chirico and Gramatica 2012)]. The concept of structure biodegradability relationships (SBR) explains the variability in persistence of organic chemicals in the environment.

The main concept of SAR is that the structure of compound is responsible for the activity. Therefore, similar molecules should reveal similar biological activities. Thus, the SAR approach assumes that the structure of a molecule (e.g., its geometric, electronic properties) contains the features responsible for its physical, chemical, and biological properties (Nantasenamat et al. 2009, 2010; Thompson et al. 2006).

Biological activity (e.g., mutagenicity, cancerogenicity, teratogenicity, inhibition of enzyme activity, toxicity) of substances is governed by their properties, which in turn are determined by their chemical structure. The objectives of SAR are twofolds:

(i) To accurately determine the limits of variation in the structure of a chemical that are consistent with the production of a specific effect (e.g., can a chemical elicit a specific toxic endpoint)
(ii) To define the ways, which alters the structure and thereby the overall properties of a compound to influence the endpoint potency (Patani and LaVoie 1996; Brown 2012)

The action of biologically active substances in soil depends upon the environmental conditions (such as temperature, light, water, oxygen tension, and availability of nutrients). Biologically active substances exert their greatest effects on increasing the plant yield, when conditions are suboptimal for plant growth.

Chemical compounds may be classified according to several criteria. One common method is based on the specific elements present (oxides, hydrides, halides, halogens). Organic compounds are those compounds with a backbone of carbon atoms, and all the remaining compounds are classified as inorganic. As the name suggests, organometallic compounds are organic compounds bonded to metal atoms. Another classification scheme for chemical compounds is based on the types of bonds that the compound contains. *Ionic compounds* contain ions and are held together by the attractive forces among the oppositely charged ions. Common *salt* (sodium chloride) is one of the best-known ionic compounds.

However, molecular compounds contain discrete molecules, which are held together by sharing electrons (covalent bonding), e.g., water (contains H_2O molecules), methane (contains CH_4 molecules), and hydrogen fluoride (contains HF molecules). Furthermore, another classification system is based on reactivity—specifically, the types of chemical and biochemical reactions, conversions, and pathways that the compounds are likely to undergo.

Some biologically active substances act at low concentrations. Phytohormones can be also considered in the same category as antagonisms as they affect the plant growth. Vitamins from B group are normal constituents of fertile soils. Their presence is due to release from organic residues, liberation from plant roots, and synthesis by soil microorganisms. Phenolic acids derived from the lignin decay by microorganisms influence the activity of peroxidase in soils (Stevenson 1982).

The following organic compounds occur in soil under continuous cropping, and intensive rotations are ecotoxic and allelopathic (Smyk 1992; Szajdak and Życzyńska-Bałoniak 1994):

1. Aliphatic and aromatic amines (primary, secondary, tertiary) as the precursors of nitrozoamines
2. Amides, imides, esters, derivatives of carboxylic and hydroxycarboxylic acids, peptides (Fig. 1.1)
3. Phenols: hydroxyphenolic acids (p-hydroxybenzoic, p-coumaric, ferulic, salicylic, syringic, vanillic, protocatechuic), catechol, quinone, hydroquinone, resorcine, derivatives of o,m,p, p-aminophenols and nitrophenols (Table 1.1)
4. Coumarins and their derivatives (Table 1.2) (Fig. 1.2 and 1.3)
5. Terpenoids (saponins, diterpenoids)
6. Flavonoids as per IUPAC nomenclature (Table 1.3):

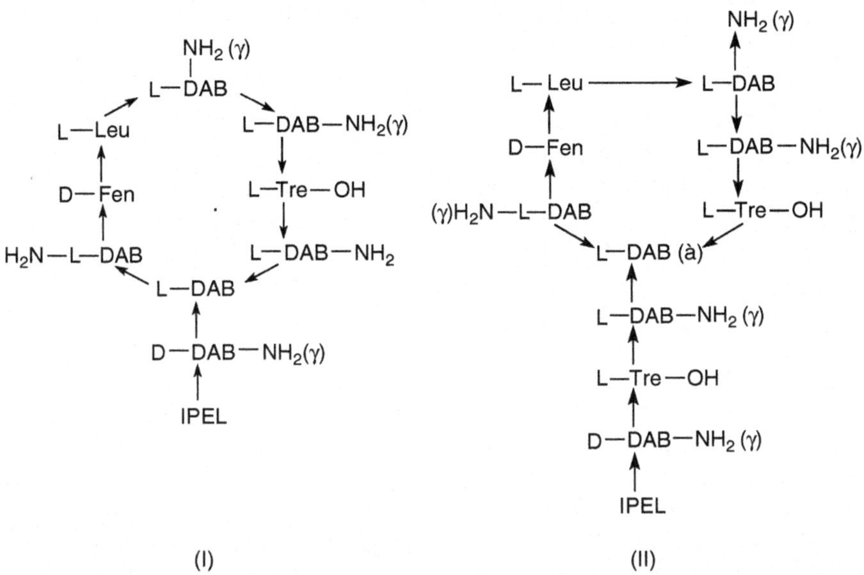

Fig. 1.1 Polymyxin B—the product of *Bacillus polymyxa* strains. Structure of polymyxin B from eight (I) or seven (II) of amino acids (Szajdak 2011)
Bond CO–NH, *Tre* threonine, *DAB* L-α,γ-diaminobutyric acid, *Fen* phenylalanine, *Leu* leucine, *IPEL* isopelargonic acid

Table 1.1 Phenolic acids in soils

Chemical compound	Structure
Benzoic acid	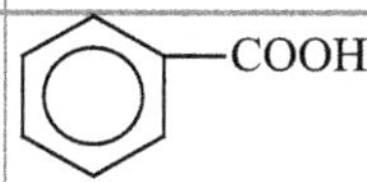
Cinnamic acid	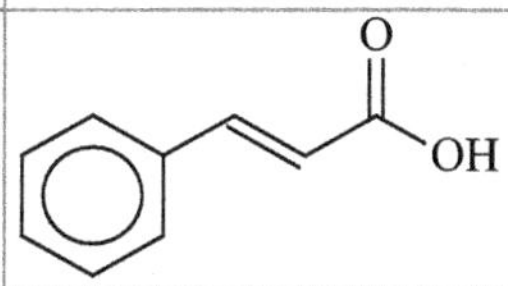
o-Hydroxycinnamic acid	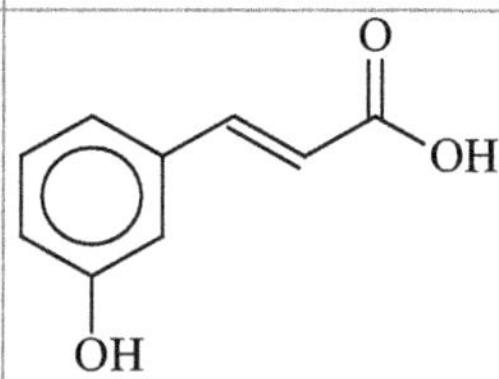
Vanillic acid	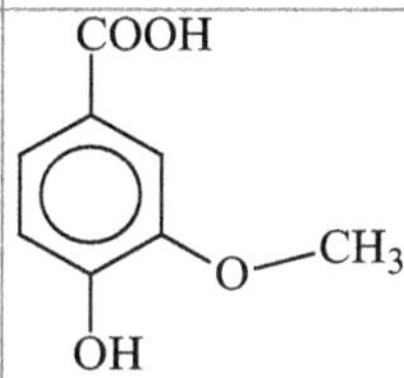
p-Coumaric acid	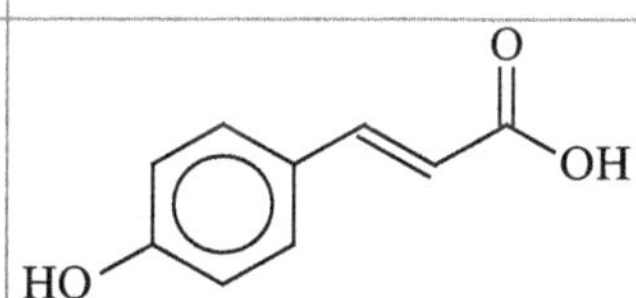
Ferulic acid	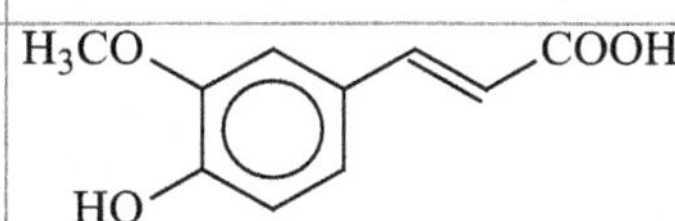
Caffeic acid	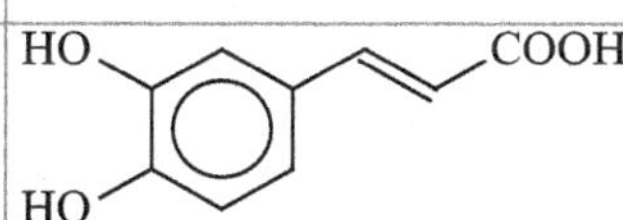

Table 1.2 Derivatives of coumarin in soils

Compounds	Structure
Dicumarol	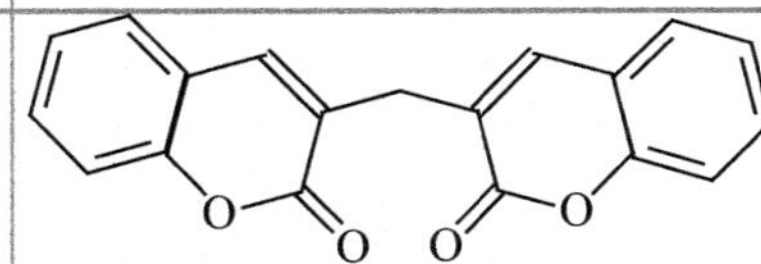
7-Hydroxycoumarin	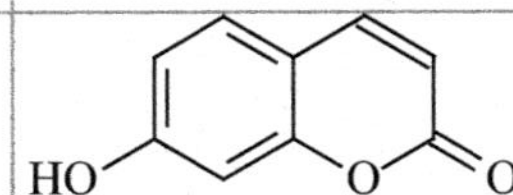
Psoralen	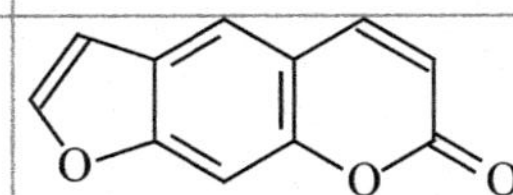
Angelicin	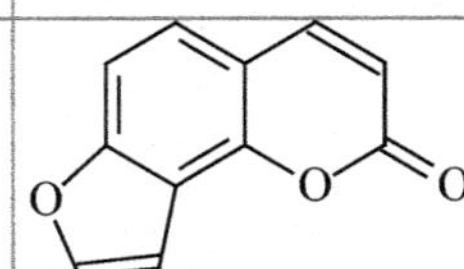
Warfarin	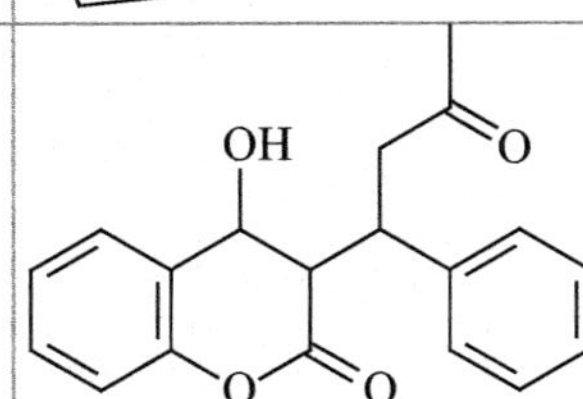

Novobiocin	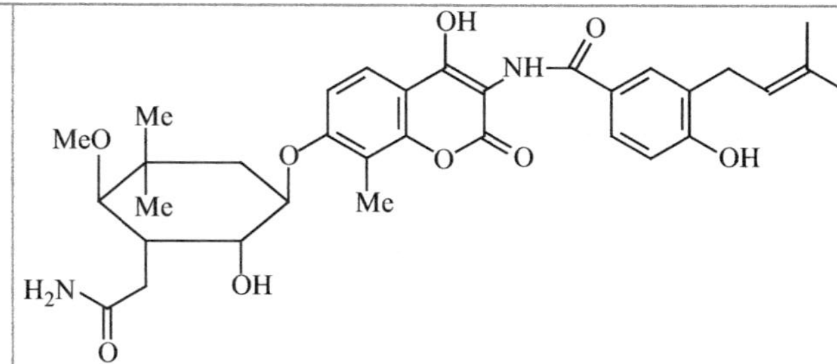
Coumermycin	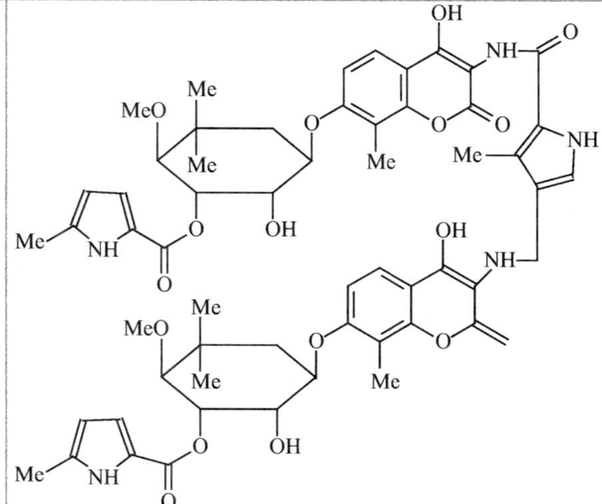

(continued)

Table 1.2 (continued)

Compounds	Structure
Clorobiocin	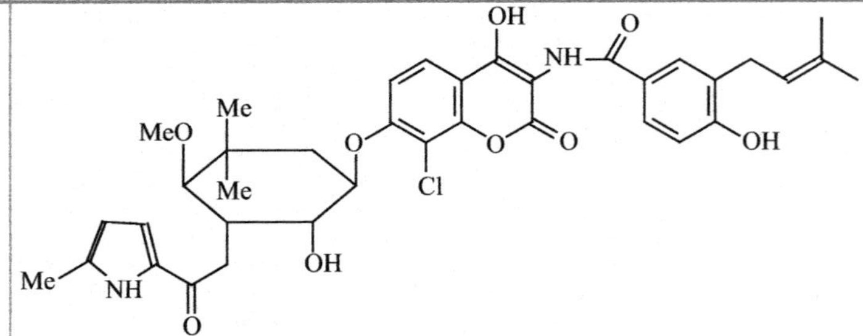

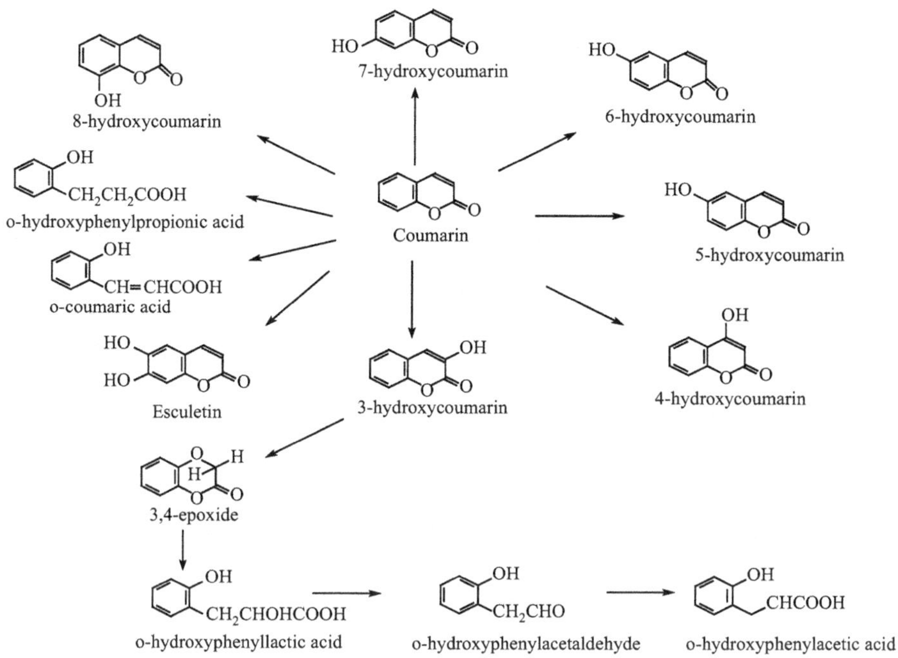

Fig. 1.2 Conversion of coumarin. All biotransformations are possible (Lacy and O'Kennedy 2004)

(i) Flavones, derivatives of 2-phenylchromen-4-one (2-phenyl-1,4-benzopyrone) (examples: quercetin, rutin)
(ii) Isoflavonoids, derivatives of 3-phenylchromen-4-one (3-phenyl-1,4-benzopyrone)
(iii) Neoflavonoids, derivatives of 4-phenylcoumarin (4-phenyl-1,2-benzopyrone)

7. Nitrozoamines and nitrozoamides
8. Mycotoxins
9. Polycyclic aromatic hydrocarbons: anthracene, chrysene, corannulene, naphthalene, phenanthrene, triphenylene, benzo[α]pyrene, coronene, tetracene, pentacene, pyrene, ovalene (Table 1.4)
10. Heterocyclic compounds (Table 1.5)
11. Glycosides (Table 1.6)
12. Alkaloids—ergot alkaloids and others (Table 1.7)
13. Phytoalexins (gossypol, capsidiol, camalexin, pisatin, etc.) (Table 1.8)

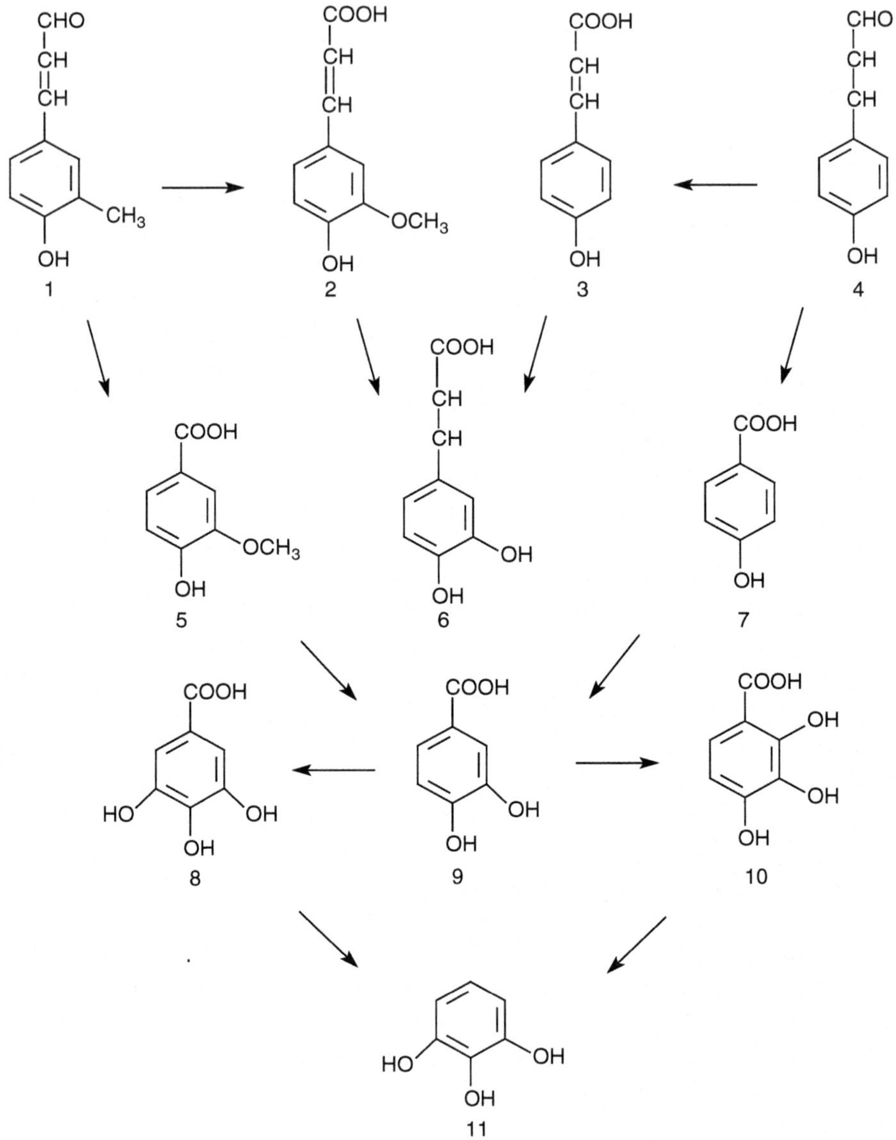

Fig. 1.3 The mechanism of the conversion of methoxyphenyl acids in fungi *Epicoccus nigrum* (Haider and Martin 1967), (*1*) coniferyl aldehyde, (*2*) ferulic acid, (*3*) p-hydroxycinnamic acid, (*4*) p-hydroxycinnamic aldehyde, (*5*) vanillic acid, (*6*) caffeic acid, (*7*) p-hydroxybenzoic acid, (*8*) gallic acid, (*9*) protocatechuic acid, (*10*) 2,3,4-trihydroxybenzoic acid, (*11*) pyrogallol

Table 1.3 Flavonoids in soils

Compound	Structure
Flavone	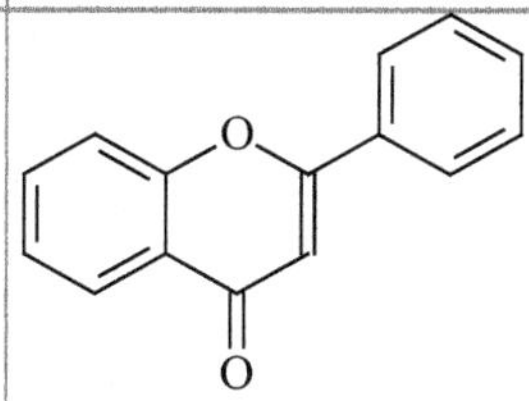
Isoflavone	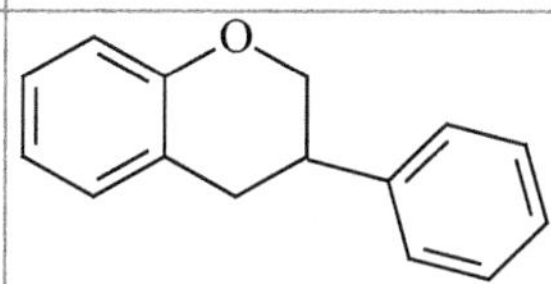
Neoflavonoids	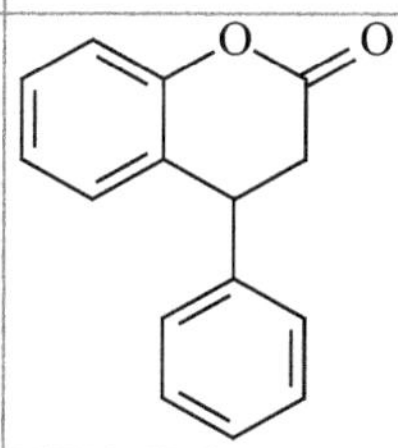
Quercetin	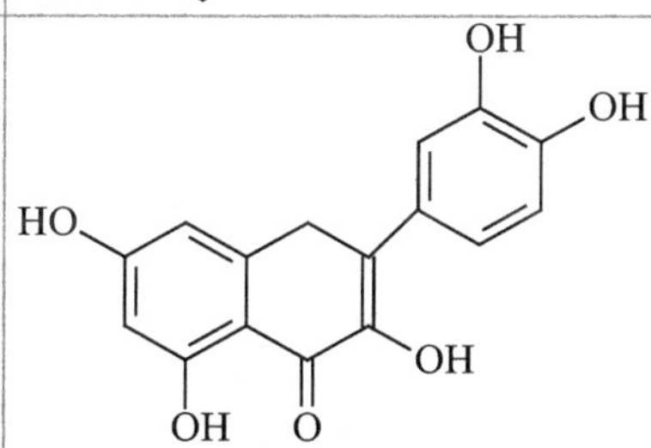
Epicatechin	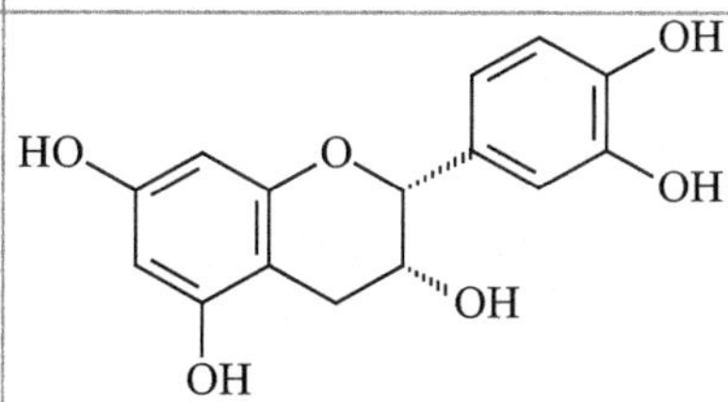
Rutin	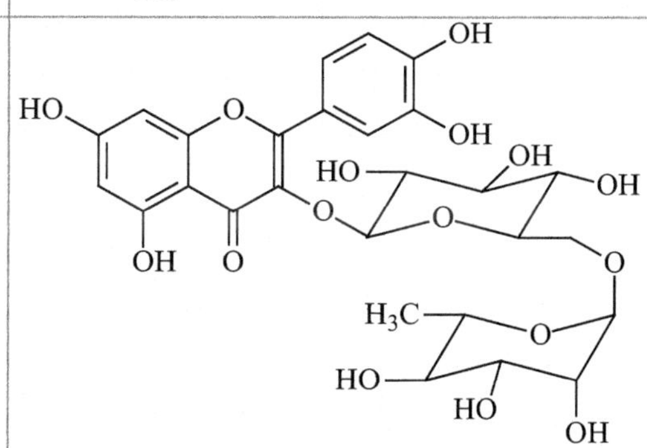

Table 1.4 Polycyclic aromatic hydrocarbons in soils

Compound	Structure
Anthracene	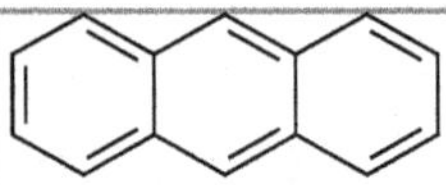
Chrysene	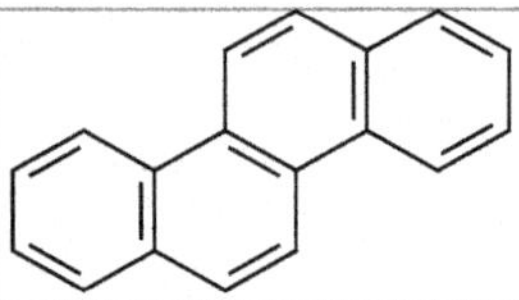
Naphthalene	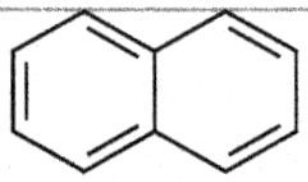
Phenanthrene	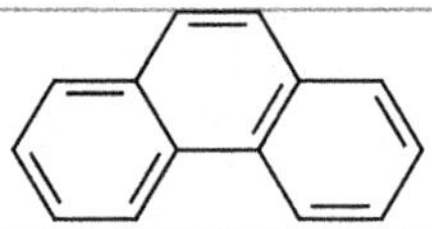
Triphenylene	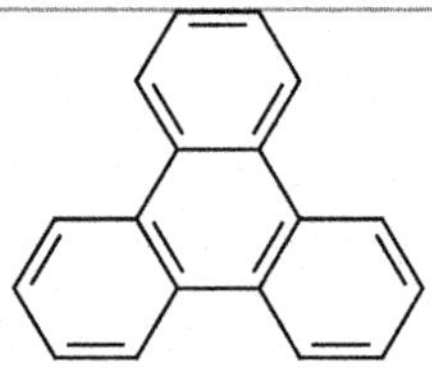
Benzo[α]pyrene	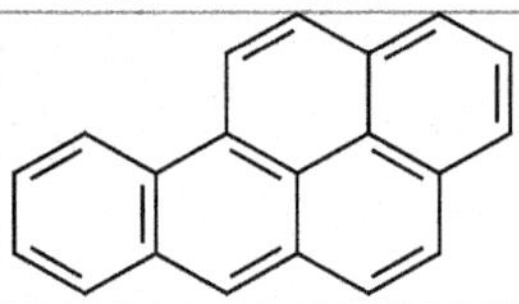
Coronene	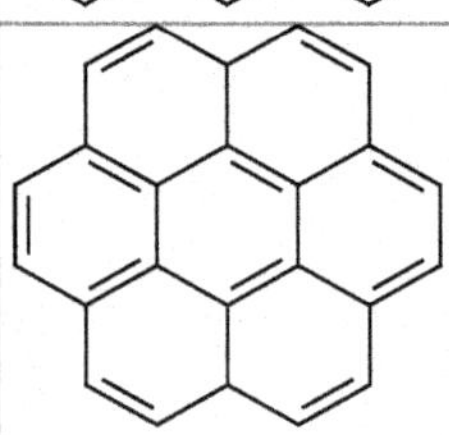
Tetracene	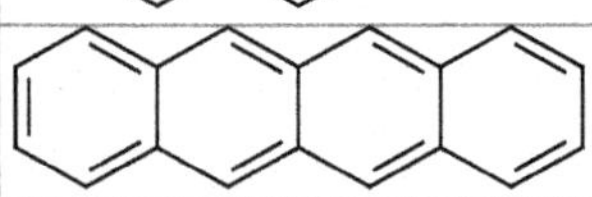
Pentacene	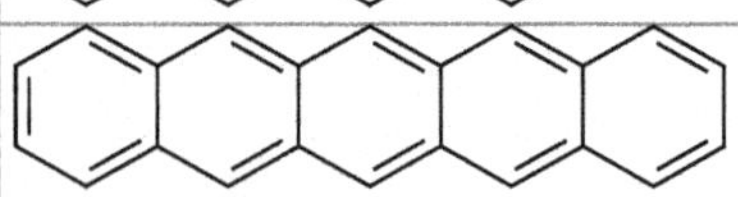
Pyrene	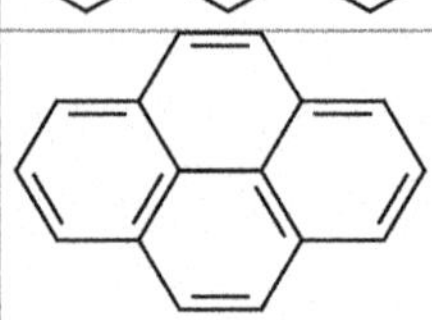

(continued)

Table 1.4 (continued)

Compound	Structure
Ovalene	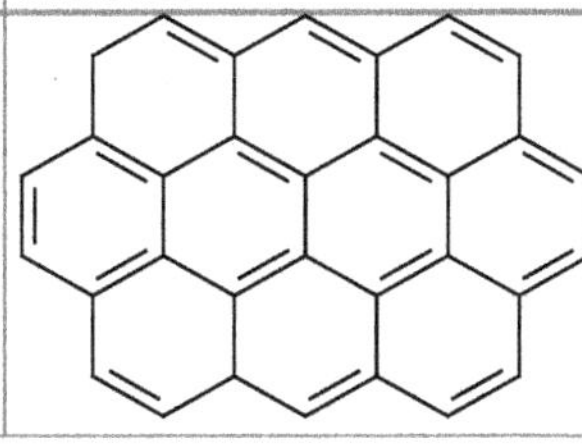

Table 1.5 Heterocyclic compounds in soils

Compound	Structure
Indole	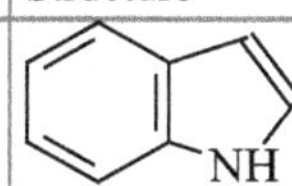
Anthraquinone	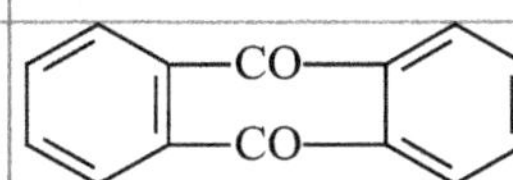
Fluorene	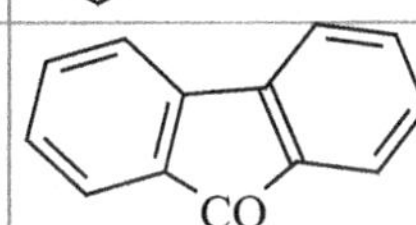
Quinoline	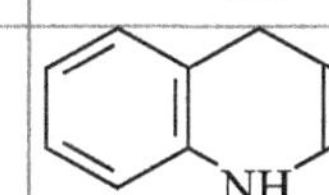
Carbazole	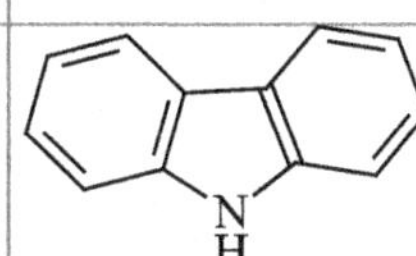
Furan	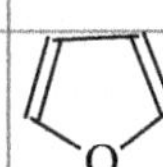
Thiophene	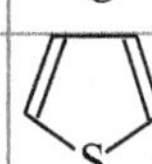
Pirolle	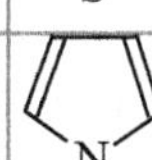
Benzofuran	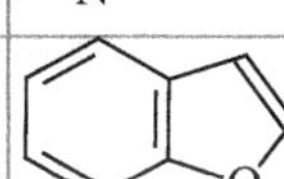
Benzothiophene	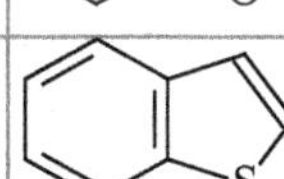

(continued)

Table 1.5 (continued)

Compound	Structure
Benzopirol	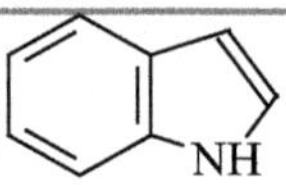
Xanthol	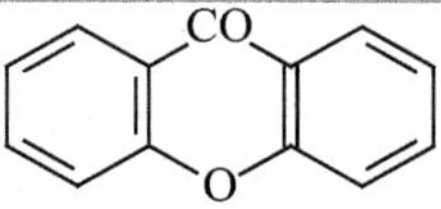
Purine	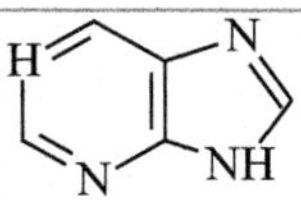
Pyrimidine	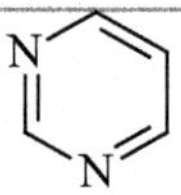

Table 1.6 Glycosides in soils

Compound	Structure
Solanine	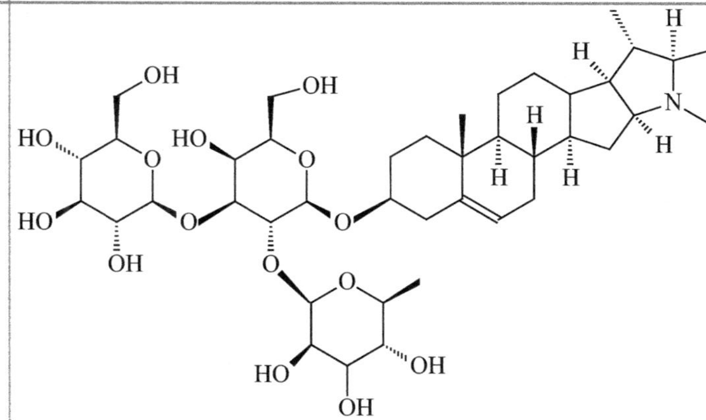

(continued)

Table 1.6 (continued)

Compound	Structure
Tomatine	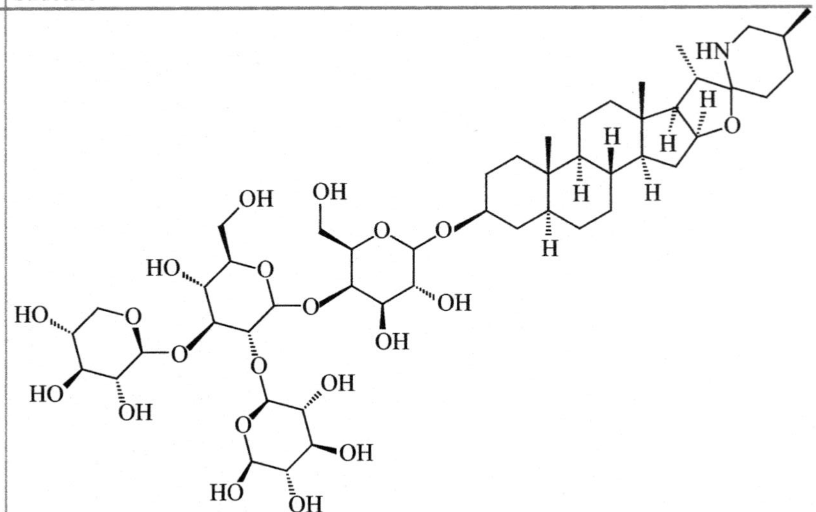

Quercetin glycoside	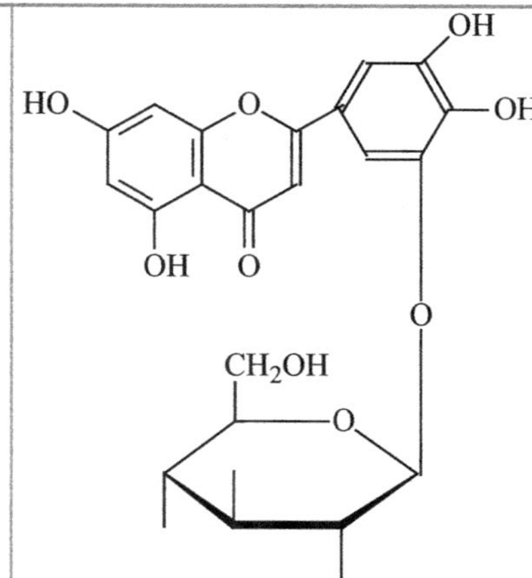
Amygdalin	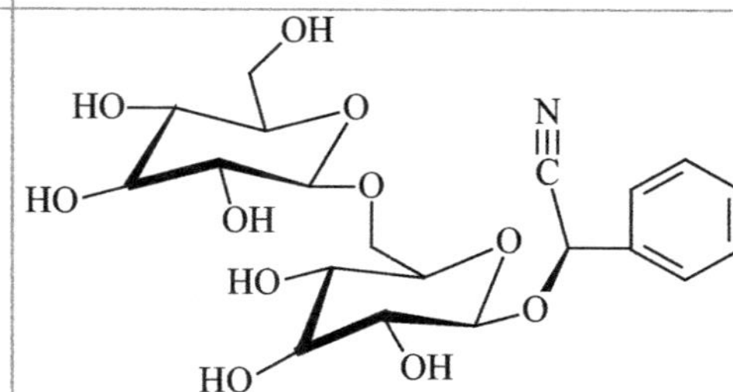

(continued)

Table 1.6 (continued)

Compound	Structure
Lotaustralin	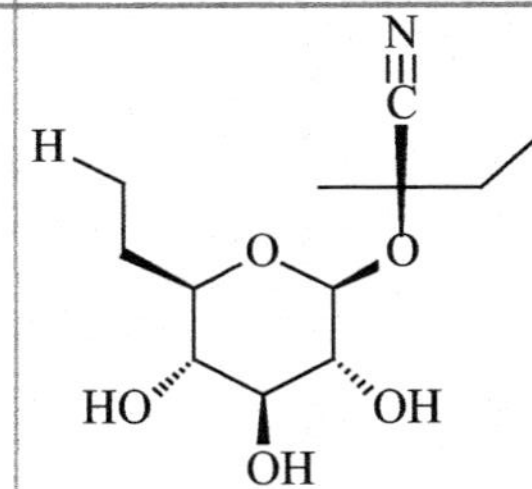
Linamarin	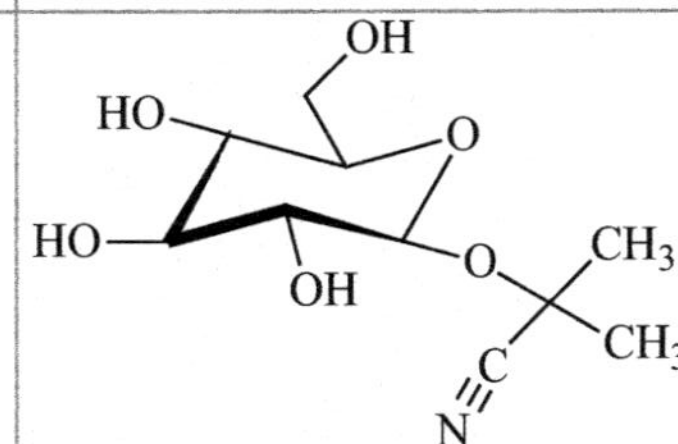

Table 1.7 Alkaloids in soils

Compounds	Structure
Mitomycin	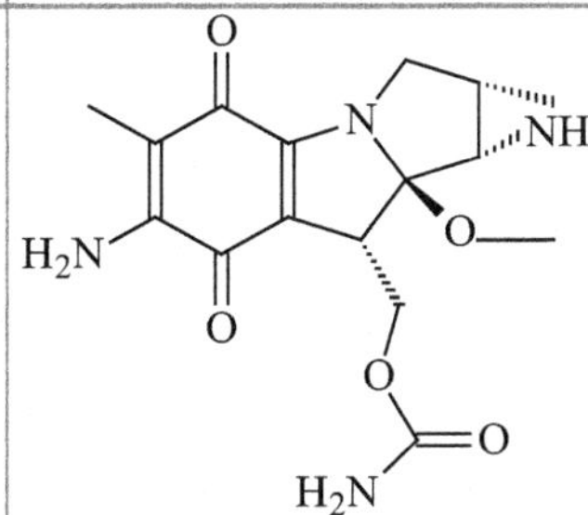
Ergotamine	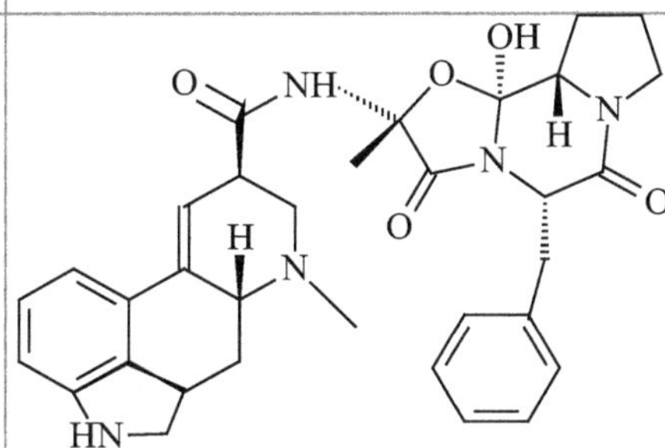
Ergocristine	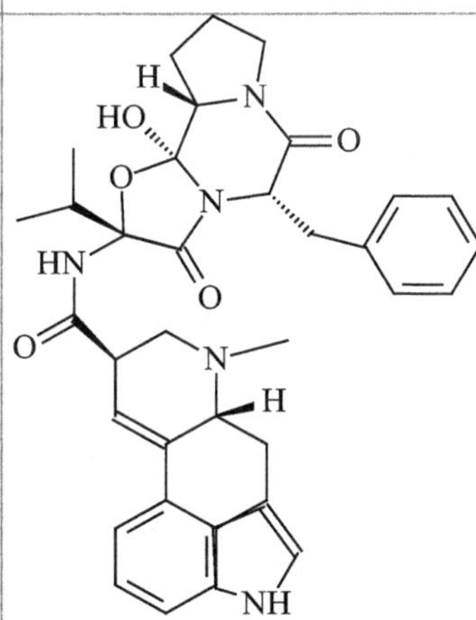
Ergometrine	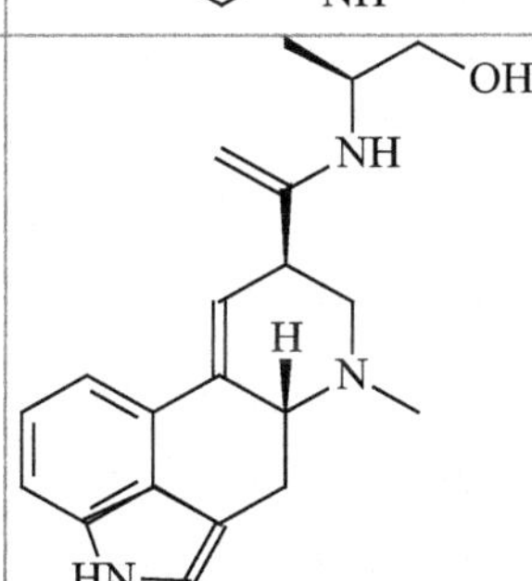

(continued)

Table 1.7 (continued)

Compounds	Structure
Ergocryptine	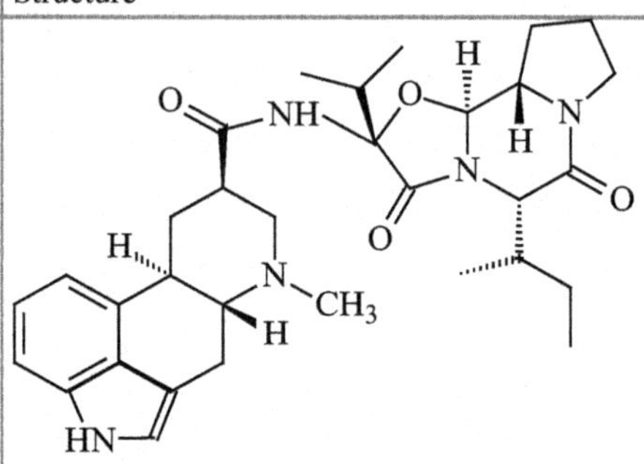
Solanidine	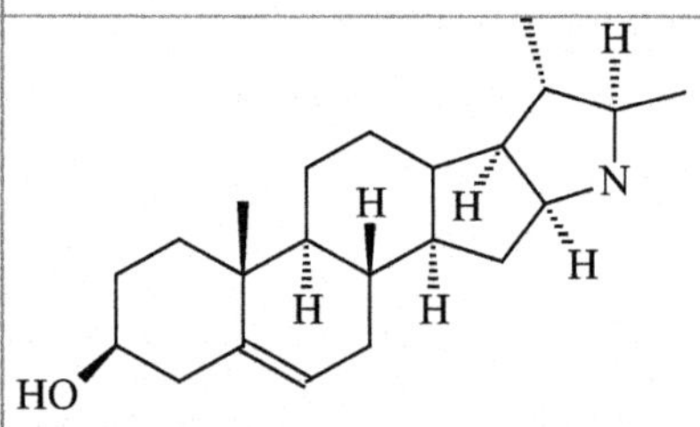
Sparteine	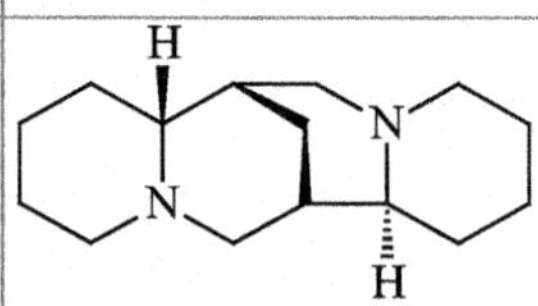

Table 1.8 Phytoalexins in soils

Compound	Structure
Pisatin	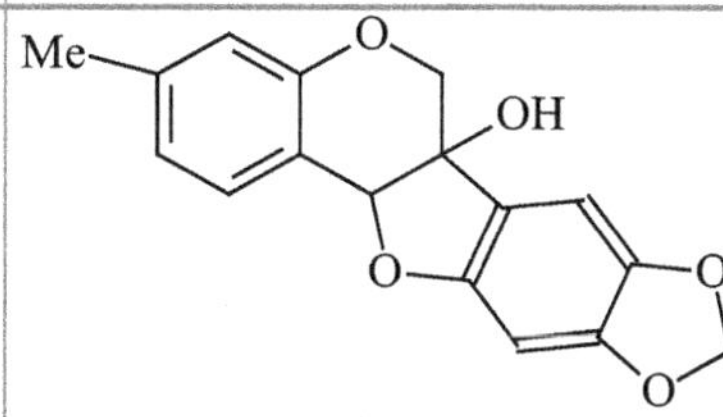
Camalexin	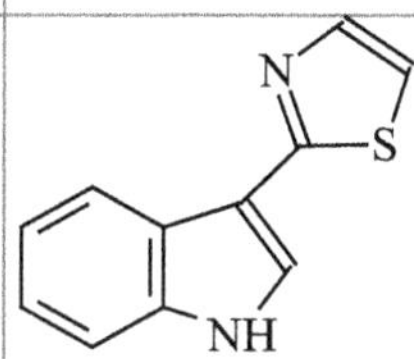
Gossypol	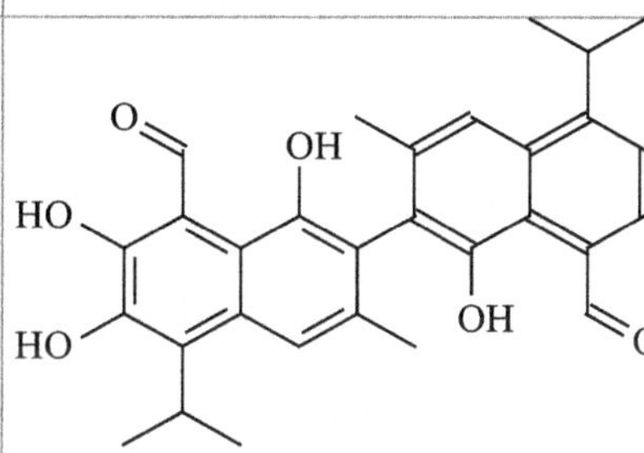
Capsidiol	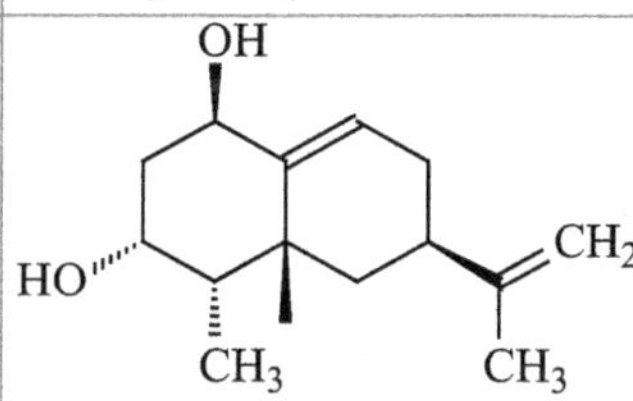

References

Brown N (2012) Bioisosteres in medicinal chemistry. Wiley-VCH, Weinheim, p 237. ISBN 978-3-527-33015-7

Chirico N, Gramatica P (2012) Real external predictivity of QSAR models. Part 2. New intercomparable thresholds for different validation criteria and the need for scatter plot inspection. J Chem Inf Model 52:2044–2058. doi:10.1021/ci300084j.PMID22721530

Filimonov DA, Poroikov VV (1996) PASS: computerized prediction on biologically active spectra for chemical substances. In: Bioactive compound design. Possibilities for industrial use. Bios Scientific Publishers, Oxford, pp 47–56

Filimonov DA, Poroikov VV, Borodina Y, Gloriozova T (1999) Chemical similarity assessment through multilevel neighborhoods of atoms: definition and comparison with the other descriptors. J Chem Inf Comput Sci 39:666–670

Haider K, Martin PJ (1967) Synthesis and transformation of phenolic compounds by *Epiccocum nigrum* in relation to humic acids formation. Proc Soil Sci Soc Am 31:766–772

Jurs PC, Hasan MN, Rohrbaugh RH (1988) Prediction of physicochemical properties of organic compounds from molecular structure. In: Jochun C, Hicks MG, Sunkel J (eds) Physical property prediction in organic chemi- physicochemical properties of organic compounds from molecular structure. Springer, Berlin, pp 209–233

Lacy A, O'Kennedy R (2004) Studies on coumarin- related compounds to determine their therapeutic role in the treatment of cancer. Curr Pharm Des 10:3797–3811

Nantasenamat C, Isarankura-Na-Ayudhya C, Naenna T, Prachayasittikul V (2009) A practical overview of quantitative structure-activity relationship. Exp Clin Sci 8:74–88

Nantasenamat C, Isarankura-Na-Ayudhya C, Prachayasittikul V (2010) Advances in computational methods to predict the biological activity of compounds. Expert Orin Drug Discovery 5:633–654

Patani GA, LaVoie EJ (1996) Bioisosterism: a rational approach in drug design. Chem Rev 96:3147–3176

Smyk B (1992) The soil fatigue as potential threat to natural environments of terrestrial ecosystems. Acta Acad Agric Tech Olst Agric 55:17–31

Stevenson FJ (1982) Humus chemistry, genesis, composition, reactions. Wiley, New York, pp 172–194

Szajdak LW (2011) Shelterbelts: efficient element of landscape. LAP Lambert Academic Publishing, Saarbrücken, p 90

Szajdak L, Życzyńska-Bałoniak I (1994) Phenolic acids in brown soils under continuous cropping system of rye and crop rotation. Pol J Soil Sci 27:113–121

Thompson SJ, Hattotuwagama CK, Holliday JD, Flower DR (2006) On the hydrophobicity of peptides: comparing empirical predictions of peptide log P values. Bioinformation 1(7):237–241

Uosukainen H, Pihlaja K (2006) Peat in balneology and therapy. Terraviva Oy, Kempele, pp 1–96

Part I
Biologically Active Substances in Cropping Systems

Chapter 2
Free, Bounded, and Included in Humic Acids Amino Acids: Thermal Properties of Humic Acids from Cropping Systems

Lech Wojciech Szajdak, Irena Życzyńska-Bałoniak, and Kazimiera Wegner

Abstract Almost all nitrogen in surface soil horizons is in organic form (0–90 %). Nevertheless, the chemical composition of nitrogen in organic soil fraction is not completely understood, and little is known of the factors affecting the distribution of organic nitrogen forms in soils.

The continuous cropping and crop rotation influenced the amino acids and properties of humic acids (HA) in soil; these influenced the crop yields. The results indicated that the composition of bound amino acids depends on cropping system and on the availability of nitrogen, phosphorus, and potassium (NPK) from fertilizers. In the soils under continuous cropping, the NPK fertilizer strongly affected the bound amino acid content than manure 38 % vs. 25%, respectively, while the contents in soils under crop rotation were 41 and 27 %, respectively.

Negative effects of continuous cropping on the content of total bound amino acids were decreased by NPK fertilization, but the manure application in continuous cropping of rye was less effective. Thus, NPK was the main driver causing changes in the total amounts of bound amino acids in HA. Crop yields of rye increase with an increase in organic N in bound amino acids and nitrogen in humic acids.

The humic acids extracted from the soils under crop rotation and fertilized with NPK showed highest aliphatic properties, but the humic acids from soils fertilized with manure were most aromatic. Aromaticities of humic acids from soils under continuous cropping of rye fertilized with manure were higher than from crop rotation fertilized with NPK.

L.W. Szajdak (✉) • I. Życzyńska-Bałoniak
Institute for Agricultural and Forest Environment, Polish Academy of Sciences,
Str. Bukowska 19, 60-809 Poznań, Poland
e-mail: lech.szajdak@isrl.poznan.pl

K. Wegner
Department of Chemistry, Agriculture University,
Str. Bernardyńska 6, 85-290 Bydgoszcz, Poland
e-mail: kwegner@utp.edu.pl

L.W. Szajdak (ed.), *Bioactive Compounds in Agricultural Soils*,
DOI 10.1007/978-3-319-43107-9_2

The evidence gathered suggests strong linkages between the soil and fertilizer and manures' quality affects the productivity. Achieving a balance between the agricultural production and conservation of natural resources is necessary to develop sustainable agricultural systems.

Keywords Free and bunded amino acids • Humic acids • Thermal properties

2.1 Introduction

The individual crop is affected by the cropping system in which it is grown. The ecological conditions are determined by many factors (soil, atmospheric environment, and cropping systems). The diversification of cropping systems leads to higher crop yields by influencing plant diseases, weeds, nutrient availability, and root distribution. Diversified crop rotations alter the pattern and degree of nutrient removal from soil. Increasing the amounts of crop residues to soil increases the soil organic carbon pool and the nutrient availability with time. The nutrient availability increases by reduced tillage including the legumes in crop rotations. The crop rotation influences the nutrient uptake by crops, e.g., phosphorus application to preceding crop increases the activity of vesicular–arbuscular mycorrhiza (VAM), which increases the absorption zone of immobile phosphorus (Honermeier 2007).

Multidisciplinary studies of ecological effects on continuous cereal cropping (*decline effect*—decreased crop yields on fields in which the same plant species has been grown continuously) have shown numerous differences in physical, chemical, and biological composition and biochemical properties of soils under rotations and continuous cropping of rye (Cox 1965; Steineck and Ruckenbauer 1976; Wasilewska 1979; Niewiadomski et al. 1980; Truszkowska et al. 1980; Pimentel and Hall 1984; Ryszkowski 1986; Ketcheson 1980; Myśków et al. 1986; Schönhammer and Fischbeck 1987;·Wicke and Urban 1988; Łoginow et al. 1990; Kaszubiak et al. 1990; Ryszkowski et al. 1990, 1998; Ryszkowski and Karg 1990, 1992; Crookston et al. 1991; Sieling and Hanus 1992; Johnston et al. 1992; Garz and Stumpe 1992; Copeland et al. 1993; Szajdak et al. 1998; Blecharczyk 1999) (Table 2.1).

2.2 Continuous Cropping

In soils of long-term continuous cropping vs. crop rotation, the lower diversity in metabolites of microbes and products of plant biomass decay have been determined. These substances can create stress conditions for many organisms, including cultivated plants, which may lead to higher susceptibility to pathogens and pests and impair their growth. Under these conditions, pests (weeds, pathogens, insects) develop leading to the outbreaks of crop pests (Jelinowski and Mróz 1979;

Table 2.1 Chemical properties of soils under crop rotation and continuous cropping of rye fertilized with NPK or with manure

		$C_{(organic)}$	$N_{(total)}$	$P_{\cdot(available)}$	$K_{(available)}$	$Mg_{(available)}$
Treatments	$pH_{(1\ M.\ KCl)}$	mg/100 g				
CR control	5.8	680± 31.3	72.7± 3.3	7.3± 0.3	5.9± 0.3	2.3± 0.1
CR manure	6.2	1180±54.3	110.7± 5.1	15.1± 0.7	17.4± 0.4	5.7± 0.2
CR NPK	5.9	648± 29.8	69.6± 3.2	10.9± 0.5	6.8± 0.3	1.9± 0.1
CC control	5.7	723± 33.2	72.7± 3.3	6.0± 0.3	3.6± 0.2	2.7± 0.1
CC manure	6.2	1205± 55.4	110.7± 5.0	13.2± 0.6	14.9± 0.7	5.5± 0.3
CC NPK	5.8	696± 32.0	69.6± 3.2	9.9± 0.4	9.1± 0.4	2.5± 0.1

Szajdak et al. (2000)
Where *CR-control* crop rotation – control; *CR-manure* crop rotation fertilized with manure; *CR-NPK* crop rotation fertilized with NPK, *CC-control* continuous cropping of rye – control, *CC-manure* continuous cropping of rye fertilized with manure, *CC-NPK* continuous cropping of rye fertilized with NPK. Where $x \pm \Delta x$ — mean values with their confidence limit Δx at $\alpha = 0.05$

Table 2.2 Total primary production ($g{\cdot}dw{\cdot}m^{-2}$) of rye fields from different crop rotation patterns

	Wielichowo			Jelcz-Laskowice	
Production element	Continuous cropping of rye	Rye after potatoes	Rye after field pea	Continuous cropping of rye	Rye after small beans
Yield: grain	314	489	410	196	307
Straw + rye regrowth	759	794	754	968	1182
Weeds	83	41	51	64	27
Dead shoots + shed leaves	374	262	261	635	497
Roots	160	165	169	172	175
Total	1690	1750	1645	2035	2191
$g{\cdot}dw{\cdot}m^{-2}$	742	708	632	871	700
Input of plant	44	40	38	42	32
Debris into soil (%)					

Łapiński and Ryszkowski (1986) and Ryszkowski and Bernacki (1990)

Wasilewska 1979; Blake et al. 1980; Durska et al. 1986; Witkowski and Zamszyn 1986; Wicke and Urban 1988; Adamiak and Zawiślak 1990; Shcherba 1994; Ryszkowski et al. 1998) (Tables 2.2 and 2.3).

Besides, the soils' incorporation of plant biomass in continuous cropping of rye affects the soils microorganism communities, which leads to the production of secondary metabolites, e.g., phenolic compounds. The soils under continuous cropping of rye accumulate higher quantity of phenolic acids, phenols, and their combinations with amino acids and amino sugars, which inhibit the seed germination, root growth,

Table 2.3 Biomass (B) of the soil invertebrates ($mg \cdot dw \cdot m^{-2}$) in continuous cropping of rye and in energetic cost of maintenance ($kJ \cdot m^{-2}$) (M)

	Wielichowo (1983–1984)				Jelcz-Laskowice (1986–1989)			
	Continuous cropping of rye		Norfolk rotation[a]		Continuous cropping of rye		Norfolk rotation	
Group	B	M	B	M	B	M	B	M
Protozoa	321.4	845.3	373.1	896.1	548.8	1447.4	708.3	1906.9
Nematoda	252.0	163.8	217.4	147.6	179.3	101.2	180.7	102.0
Enchytraeidae	24.4	11.9	15.6	7.6	–	–	–	–
Lumbricidae	85.7	7.4	200.4	15.6	285.0	22.7	3811.0	296.8
Acarina	4.6	0.89	6.9	0.6	11.4	34.6	20.6	55.6
Collembola	26.0	4.7	41.2	7.9	63.5	95.6	45.4	63.5
Winged insect larvae	394.5	8.0	472.6	17.2	1177.6	76.8	1736.0	91.9
Total	1108.6	1041.9	1327.2	1092.6	2265.6	1808.3	6502.0	2516.7

Karg et al. (1986), (1990) and Witkowski and Zamszyn (1986)

[a]Mean values for rye after potatoes and rye after field pea in Norfolk rotation

Table 2.4 Phenolic acids in soils under continuous cropping of rye and crop rotation after harvest in mg kg[1]

Type of cultivation	Phenolic acids					Total amount
	p-Hydroxy benzoic	Vanillic	p-Coumaric	Syringic	Ferulic	
Continuous cropping of rye	7.48±0.28	14.24±0.54	5.17±0.20	12.49±0.48	11.49±0.44	50.87
Crop rotation	3.67±0.14	3.77±0.14	0.66±0.03	2.98±0.11	1.62±0.06	12.70

Szajdak and Życzyńska-Bałoniak (1994)
Where $x \pm \Delta x$—mean values with their confidence limit Δx at $\alpha = 0.05$

and development of cultivated plants (Guenzi and McCalla 1966; Bu'Lock 1980; Hruszka 1982; Wójcík-Wojtkowiak et al. 1990; Szajdak and Życzyńska-Bałoniak 1994) (Table 2.4).

All negative effects occur in soil under continuous cropping than in crop rotation, and these decreases the crop yields (Gawrońska-Kulesza 1966; Hageman and Schrader 1979; Gonet and Wegner 1993; Gonet et al. 1993; Blecharczyk and Pudełko 1997) (Table 2.5).

2.2.1 Soil Organic Matter

Organic matter in soils consists of a mixture of plant and animal products in various stages of decomposition and substances synthesized during the breakdown of these compounds (Flaig 1971; Stevenson 1986; Paul and Clark 1989). Soil organic matter or humus consists of two major types of compounds: (i) unhumified substances

Table 2.5 The comparison of crop yields (t/ha) growing continuously and in rotation

Authors	Period of research	Crop rotation	Continuous cropping	Rel. (%)
Winter wheat				
Huet and Boyeldiu (1976)	1971–1975	5.52	4.49	73
Kübler (1977)	1966–1975	4.75	3.25	68
Hanley and Ridgman (1978)	1963–1974	4.07	2.78	68
Kürten and Range (1980)	1971–1978	4.26	2.95	69
Heyn and Brüne (1981)	1965–1981	5.74	4.71	82
Ridgman et al. (1985)	1979–1982	4.57	4.33	95
Schönhammer and Fischbeck (1987)	1970–1984	5.04	4.03	80
Claupein and Zoschke (1987)	1971–1985	5.13	4.49	88
Feinstkorn and Kreuz (1988)	1977–1986	6.80	5.68	84
Wicke and Urban (1988)	1976–1985	6.86	5.50	80
Zawiślak et al. (1991)	1963–1989	4.52	3.26	72
Panse et al. (1994)	1980–1992	6.91	6.06	88
Winter barley				
Kürten and Range (1980)	1971–1978	4.84	4.99	103
Buss and Zoschke (1984)	1975–1981	5.84	4.11	70
Wicke and Urban (1988)	1976–1985	6.92	6.19	89
Baltruschat and Dehne (1989)	1971–1982	4.94	5.19	104
Christen and Sieling (1993)	1987–1991	7.60	7.29	96
Panse et al. (1994)	1980–1992	5.43	5.35	99
Winter rye				
Steinbeck and Ruckenbauer (1976)	1960–1975	3.17	2.77	87
Schönhammer and Fischbeck (1987)	1970–1984	4.70	4.21	90
Wicke and Urban (1988)	1976–1985	6.25	5.62	90
Mercik (1989)	1976–1986	4.20	3.73	89
Zawiślak et al. (1991)	1957–1989	4.19	3.45	82
Garz and Stumpe (1992)	1962–1988	4.17	2.94	71
Panse et al. (1994)	1980–1992	6.85	6.75	99
Blecharczyk and Skrzypczak (1994)	1986–1992	6.23	5.76	92
Spring barley				
Johnston and Mattingly (1976)	1970–1975	5.68	5.45	96
Kübler (1977)	1966–1975	4.25	3.55	84
Schönhammer and Fischbeck (1987)	1970–1984	4.09	3.81	93
Krejĉir (1987)	1973–1983	5.93	5.14	87
Feistkorn and Kreuz (1988)	1977–1986	6.40	5.80	91
Wicke and Urban (1988)	1976–1985	4.26	3.62	85
Zawiślak et al. (1991)	1957–1989	4.63	3.63	78
Rous (1992)	1986–1990	6.24	6.22	100
Blecharczyk et al. (1995)	1957–1992	3.86	3.14	81

Blecharczyk and Pudełko (1997)

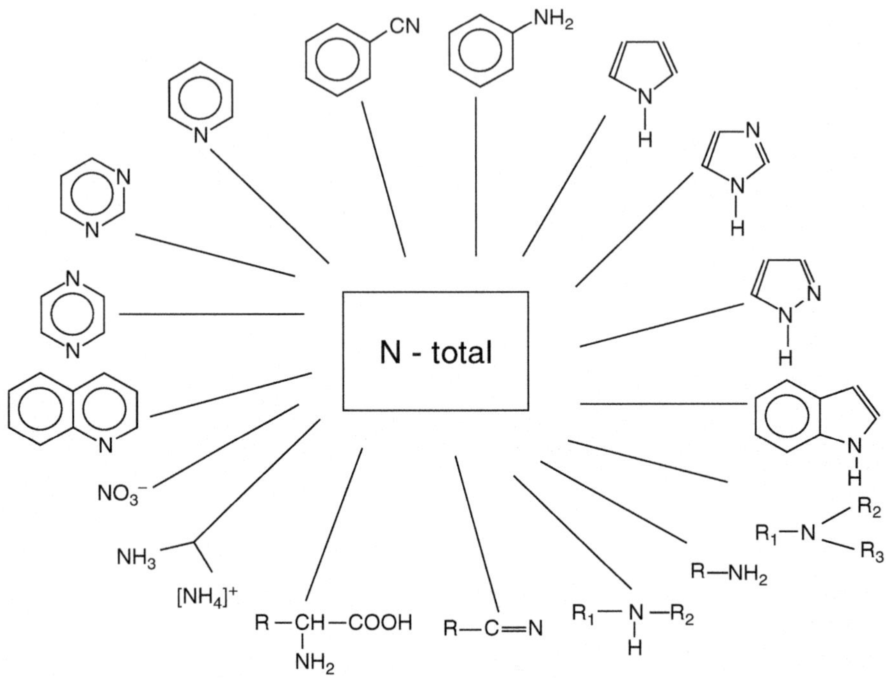

Fig. 2.1 N-derivatives of soil organic matter included in total nitrogen (Schulten and Schnitzer 1998; completed by Szajdak and Życzyńska-Bałoniak 2002)

(ii) and humified remains of plant and animal tissues, which effects on the availability of nutrients for plant growth (Stevenson 1986; Jones et al. 2008).

Almost all nitrogen in surface soil horizons is in organic form (0–90 %) (Fig. 2.1). Nevertheless, the chemical composition of nitrogen in organic soil fraction is not completely understood, and little is known about the factors affecting the distribution of organic nitrogen forms in soils.

Crop rotations, fertilization, and microbiological activity affect the nitrogen levels in soils (Stevenson 1985; Campbell et al. 1991; Szajdak and Österberg 1996; Szajdak and Sokolov 1997; Szajdak et al. 2003).

Humus is composed from 20–60 % of humic acids (HA). The nitrogen (20–40 %) in HA consists of amino acids or peptides, the main unit of protein, which are connected to the central core by hydrogen bonds (Harworth 1971). Amino acids influence the plant growth and thus organic matter increases the soil productivity (Szajdak and Österberg 1996; Szajdak et al. 2003). Amino acids are amphoteric due to the presence of both carboxyl and amino groups in their molecules. Little is known about the variation in amino acid contents in HA in soil under long-time cultivation or with different fertilizer treatments (NPK or manure).

The availability of different forms of nitrogen in soil influences the net primary productivity and vegetation succession gradients in soils (Trojanowski 1973; Vance and Chapin 2001; Jones and Kielland 2002). The cycling of diverse N-containing substances in agriculture soils is still poor. In addition, the relative contribution and function of dissolved organic and inorganic nitrogen compounds in plants and microbial nutrition remains controversial (Owen and Jones 2001; Schimel and Bennett 2004). Some plants are capable of bypassing the mineralization process of nitrogen cycle by directly taking up low molecular weight compounds (up to 1000 Da, viz., amino acids, amino sugars, amines, alkaloids, amides, peptides, etc.). These substances are microbially changed into NH_4^+ and NO_3^- (Chapin et al. 1993; Kielland 1994; Jones et al. 2005a, b; Kielland et al. 2006; Hill et al. 2011).

The size of amino acid pools in soil is small (1–50 μM), but the flux from this pool is very fast (Kielland 1995; Jones and Kielland 2002; Jones et al. 2009). The half-life of amino acid pool in soil is 1–6 h, indicating that free amino acid pool turns over hundreds of times annually in soils (Kielland et al. 2007). Thus, the impact of this quick cycling through the low molecular weight compounds to NH_4^+ and NO_3^- formation remains largely unknown. From the amino acids, there is a very quick formation and excretion of NH_4^+ into the soil with the release of CO_2 (Jones and Kielland 2002). Therefore, they postulated that the soil microbial community might be using free amino acids not as a source of nitrogen but as source of energy. Besides, they (Jones et al. 2005b, 2009) postulated that catabolism and anabolism simultaneously used the 30–40 % of amino acids in respiration and the remaining in cell biomass formation.

The continuous cropping and crop rotation influenced the amino acids and properties of humic acids in soil; these influenced the crop yields. The surface soil condition strongly influenced the qualitative and quantitative forms of amino acids in soil. The adoption of modern, industrialized agricultural production systems has caused degradation of soils (i.e., loss of soil biodiversity, poor soil tilth and imbalanced elemental composition). Environmental consequences of agricultural industrialization were not immediately apparent, but are now being recognized due to (1) fossil-fuel-based mechanization leading to soil erosion in vast areas, (2) animal rearing polluting the nearby water resources with excessive fecal-borne pathogens and nutrient loads, and (3) liberal use of pesticides and fertilizer application having threatened the soil and water quality (Franzluebbers 2008).

2.2.2 *Amino Acids*

There are three soil quality indicators: (i) chemical, (ii) physical, and (iii) biological. In each of these classes, many soil properties or processes can be selected to indicate soil functional capabilities. Root exudates release considerable amounts of organic substances in soil including the amino acids (Vancura 1967; Claudius and Merhotka 1973; Smyth 1976). In soils, amino acids are formed during the decomposition of plant biomass and from the transamination of keto acids. Despite the high input of amino acids into the soil, their actual concentrations are low because microbes rapidly degrade them. Most of the amino acids in soils occur in bound form in humin–peptide fraction. The binding of amino acids and peptides to humic

Table 2.6 Free amino acids in soil under Norfolk crop rotation in $\mu g \cdot kg^{-1}$

	Date				
Amino acids	March 9	May 3	May 30	July 5	August 8
Cysteic acid	255.3 ± 9.7	343.9 ± 13.1	110.5 ± 4.2	125.0 ± 4.8	61.1 ± 2.3
Taurine	67.3 ± 2.6	86.2 ± 3.3	30.3 ± 1.2	55.2 ± 2.1	2.4 ± 0.9
Proline	29.6 ± 1.1	33.0 ± 1.3	–	62.9 ± 2.4	13.8 ± 0.5
Alanine	37.8 ± 1.4	42.4 ± 1.6	14.5 ± 0.6	26.9 ± 1.0	20.8 ± 0.8
Citrulline	202.2 ± 7.6	276.2 ± 10.5	88.9 ± 3.4	171.6 ± 6.5	120.4 ± 4.6
Valine	–	68.9 ± 2.6	13.4 ± 0.5	30.0 ± 1.1	20.0 ± 0.8
Cysteine	59.1 ± 2.3	10.8 ± 0.4	–	–	9.7 ± 0.4
Methionine	15.1 ± 0.6	3.6 ± 0.1	–	3.7 ± 0.1	7.4 ± 0.3
Leucine	7.5 ± 0.3	135.8 ± 5.2	14.7 ± 0.6	53.0 ± 2.0	42.1 ± 1.6
Tyrosine	46.7 ± 1.8	35.3 ± 1.3	24.4 ± 0.9	22.6 ± 0.9	5.9 ± 0.2
Phenylalanine	88.1 ± 3.4	139.1 ± 5.3	39.3 ± 1.5	41.2 ± 1.6	25.0 ± 1.0
β-Alanine	39.3 ± 1.5	45.4 ± 1.7	16.2 ± 0.6	–	–
γ-Aminobutyric acid	9.5 ± 0.4	45.9 ± 1.7	17.2 ± 0.7	70.3 ± 2.7	25.4 ± 1.0
Arginine	–	–	–	–	5.6 ± 0.2
Cystathionine	31.3 ± 1.2	38.8 ± 1.5	7.9 ± 0.3	3.7 ± 0.1	3.4 ± 0.1
Glycine	11.8 ± 0.5	13.2 ± 0.5	6.3 ± 0.2	8.1 ± 0.3	11.2 ± 0.4
Total amount	900.6	1318.5	383.6	674.2	396.7

Życzyńska-Bałoniak and Szajdak (1992)
Where $x \pm \Delta x$—mean values with their confidence limit Δx at $\alpha = 0.05$

acids protects them from rapid degradation (Bremner 1967; Sörensen 1967). In soils under crop rotation, the mean concentration of total free amino acids was 796.0.7 μg kg^{-1} and was 734.7 μg kg^{-1} under continuous cropping of rye (Życzyńska-Bałoniak and Szajdak 1992). There is great variability in free forms of amino acids than bound ones (Tables 2.6, 2.7, 2.8, and 2.9); there are higher turnover rates in crop rotation than in continuous rye cropping.

In 1957, experiments were started on podzolic soil where similar agricultural practices were used, viz. plowing before sowing, sowing, harvesting, and plowing first followed by harrowing. Pesticides were not used in these rotations. The treatments were (i) continuous cropping of rye and (ii) 7-year crop rotation (potato, spring barley, alfalfa, oil seed rape, winter rye, and winter rye). There were three fertilizer treatments: (i) control (soils without any fertilizers), (ii) NPK (N-nitrogen 90 kg/ha, P_2O_5 60 kg/ha, K_2O 120 kg/ha), and (iii) manure (30 t/ha). The experimental field soils were coarse sand (organic carbon: 0.83 %) (Swift 1996). Investigations were made in the 49th year of cultivation. Winter rye had been grown continuously since 1957. The soil pH was 6.02 under continuous cropping of rye and 6.27 under crop rotation. In soil having fertilizer with NPK and fertilizer with manure, the carbon content was 648–796 mg/hg and 1180–1205 mg/hg, and the nitrogen content was 69.7–72.7 mg/hg and 110.7 mg/hg, respectively. The rye culture was fertilized with 270 kg/ha NPK (nitrogen 90 kg/ha, phosphorus 60 kg/ha, and potassium 120 kg/ha).

Table 2.7 Free amino acids in soil under continuous cropping of rye in μg·kg^{-1}

	Date				
Amino acids	March 9	May 3	May 30	July 5	August 8
Cysteic acid	268.1 ± 10.2	395.2 ± 15.0	123.1 ± 4.7	84.2 ± 3.2	106.7 ± 4.1
Taurine	72.8 ± 2.8	108.2 ± 4.1	35.2 ± 1.3	40.4 ± 1.5	45.6 ± 1.7
Proline	212.9 ± 8.1	45.3 ± 1.7	–	71.4 ± 2.8	28.8 ± 1.1
Alanine	30.9 ± 1.2	56.6 ± 2.2	16.2 ± 0.6	18.6 ± 0.7	22.8 ± 0.9
Citrulline	151.0 ± 5.7	374.9 ± 14.3	91.1 ± 3.5	107.5 ± 4.1	143.7 ± 5.5
Valine	–	90.4 ± 3.4	17.1 ± 0.7	28.9 ± 1.1	28.5 ± 1.1
Cysteine	51.0 ± 1.9	5.5 ± 0.2	–	–	3.8 ± 0.1
Methionine	14.1 ± 0.5	4.0 ± 0.2	–	3.0 ± 0.1	5.1 ± 0.1
Leucine	55.6 ± 2.1	164.2 ± 6.2	8.2 ± 0.3	38.0 ± 1.4	49.1 ± 1.9
Tyrosine	44.6 ± 1.7	69.4 ± 2.6	20.2 ± 0.8	21.9 ± 0.8	8.1 ± 0.3
Phenylalanine	65.7 ± 2.5	131.8 ± 5.0	32.7 ± 1.3	39.1 ± 1.5	34.1 ± 1.3
β-Alanine	11.4 ± 0.4	13.2 ± 0.5	–	9.4 ± 0.4	–
γ-Aminobutyric acid	5.9 ± 0.2	40.0 ± 1.5	18.1 ± 0.7	26.5 ± 1.0	2.1 ± 0.1
Cystathionine	26.3 ± 1.0	51.8 ± 2.0	8.2 ± 0.3	–	15.8 ± 0.6
Glycine	24.7 ± 0.9	18.8 ± 0.7	4.9 ± 0.2	7.8 ± 0.3	10.0 ± 0.4
Total amount	1035.0	1569.	347.9	496.7	504.2

Życzyńska-Bałoniak and Szajdak (1992)
Where $x \pm \Delta x$—mean values with their confidence limit at $\alpha = 0.05$

Proline The dominating amino acids in both cropping types were cysteic acid, citrulline, proline, phenylalanine, and taurine. In soils under continuous cropping of rye, the proline concentration was 14–109 % higher than in crop rotation (Tables 2.6 and 2.7). An accumulation of proline under continuous cropping system was reported by Życzyńska-Bałoniak et al. (1986). The accumulation of proline under continuous cropping of rye conditions may be due to its heterocyclic amino acid structure, which could be more resistant to microbial degradation than other amino acids. The accumulation of proline in soils is considered as negative effect. This amino acid is a secondary amine; in the presence of nitrites, proline may form the N-nitrosamines (Fig. 2.2). Proline is a potent toxin, with carcinogenic and mutagenic effects (Pla 1980; Kofoed et al. 1981; Von Hofe et al. 1987; Larsson et al. 1990; Pesci 1992).

Citrulline High concentrations of citrulline were found in soils of both cropping systems. Citrulline is an amino acid, which does not occur in proteins but is an intermediate of the urea cycle. It is a basic, diamine monocarboxylic acid and is readily available source of nitrogen for plants and microorganisms.

β-Alanine The β-alanine content was higher in soils under continuous cropping of rye than in crop rotation. In soil samples taken from fields under crop rotation in March and at the beginning of May, the β-alanine content was threefolds higher than under continuous cropping of rye. Since β-alanine is a constituent of bacterial cell walls, the higher concentration of this amino acid in soils under crop rotation indi-

Table 2.8 Bound amino acids in soil under crop rotation in mg·kg^{-1}

	Date				
Amino acids	Marc 9	May 3	May 30	July 5	August 8
Cysteic acid	9.80±0.40	13.82±0.52	17.92±0.68	1.12±0.04	13.47±0.51
Taurine	1.20±0.05	4.49±0.17	5.52±0.20	2.46±0.09	4.11±0.16
Proline	3.20±0.12	5.09±0.19	2.25±0.09	10.68±0.41	7.03±0.27
Glycine	0.90±0.03	1.79±0.06	3.71±0.14	38.11±1.45	4.98±0.18
Alanine	34.52±1.31	71.03±2.70	77.84±2.95	67.27±2.60	15.88±0.60
Citrulline	34.52±1.31	114.20±4.33	111.55±4.23	94.16±3.58	127.55±4.85
Methionine	0.76±0.03	2.72±0.1	1.66±0.06	1.25±0.05	2.13±0.08
Valine	10.75±0.04	16.86±0.64	19.49±0.74	14.45±0.55	15.34±0.58
Phenylalanine	4.72±0.18	7.54±0.29	4.63±0.17	5.72±0.22	7.06±0.27
Cysteine	24.38±0.93	33.69±1.28	41.42±1.57	28.26±1.07	31.37±1.19
β-Alanine	2.28±0.09	12.24±0.47	21.42±0.81	3.94±1.15	4.18±0.16
Cystathionine	8.69±0.33	13.03±0.49	6.20±0.23	7.06±0.27	3.25±0.12
β-Aminobutyric acid	–	7.88±0.29	7.72±0.29	4.99±0.19	1.23±0.05
Leucine	33.31±1.31	69.05±2.62	67.92±2.60	60.04±2.28	58.04±2.21
γ-Aminobutyric acid	4.15±0.16	5.89±0.22	7.84±0.29	6.30±0.23	5.72±0.22
Ornithine	10.60±0.40	13.69±0.52	19.51±0.74	11.01±0.41	11.80±0.45
Lysine	30.58±1.16	39.84±1.51	49.65±1.88	33.82±1.29	42.58±1.62
Histidine	14.25±0.54	20.41±0.78	21.41±0.81	15.05±0.57	15.17±0.58
1-Methylhistidine	6.25±0.23	5.27±0.20	14.29±0.54	6.18±0.23	8.15±0.31
3-Methylhistidine	–	3.19±0.12	3.87±0.15	8.21±0.31	2.88±0.11
Arginine	15.06±0.57	21.48±0.82	4.12±0.16	18.84±0.72	20.20±0.76
Total amount	249.92	483.2	509.94	438.91	402.11

Życzyńska-Bałoniak and Szajdak (1993)
Where $x \pm \Delta x$—mean values with their confidence limit Δx at $\alpha = 0.05$

cates a higher bacterial biomass in these soils. This result supports the findings of other researchers, showing that in soils under diversified cropping patterns, bacterial biomass increases (Stevenson 1972; Durska and Kaszubiak 1980). Analyzing the amino acid composition in the soils for sulfuric, basic, aromatic, and neutral groups, it was found in sulfur groups, which were most abundant. The concentrations of sulfuric amino acids during plant growth period (except the harvest) were higher in soils under continuous cropping of rye than under crop rotation. Their increased concentrations indicate that soils under continuous cropping of rye were becoming acidic due to biochemical processes originating in this system of cultivation (Życzyńska-Bałoniak and Szajdak 1992).

Bound Amino Acids From the soil samples, the content of bound amino acids was determined after extraction of free amino acids. The mean content of bound amino acids in soils was 416.8 mg kg^{-1}, under crop rotation, and 371.1 mg kg^{-1} in soils under continuous cropping of rye (Tables 2.8 and 2.9).

Table 2.9 Bound amino acids in soil under continuous cropping of rye in mg·kg^{-1}

	Date				
Amino acids	March 9	May 3	May 30	July 5	August 8
Cysteic acid	12.48±0.47	14,06±0.53	11.76±0.45	14.29±0.29	18.27±0.69
Taurine	2.50±0.10	4.89±0.19	6.40±0.24	1.45±0.06	5.36±0.20
Proline	7.50±0.29	9.08±0.35	9.16±0.35	6.57±0.25	20.77±0.79
Glycine	2.95±0.11	4,20±0.16	3.20±0.12	40.19±1.52	2.32±0.09
Alanine	28.15±1.07	69.48±2.64	53.11±2.02	57.48±2.18	10.46±0.40
Citrulline	25.20±0.96	102.19±3.9	76.98±2.92	93.67±3.57	86.50±3.29
Methionine	0.98±0.04	–	1.20±0.05	1.63±0.06	0.53±0.02
Valine	10.04±0.38	16.18±0.61	12.74±0.48	14.57±0.55	16.49±0.63
Phenylalanine	3.80±0.14	5,26±0.20	3.79±0.14	3.62±0.14	3.65±0.13
Cysteine	40.63±1.54	39.78±1.51	24.29±0.92	28.08±1.07	31.25±11.18
β-Alanine	1.74±0.07	2.12±0.08	2.88±0.10	3.85±0.15	3.48±0.13
Cystathionine	5.95±0.23	12.12±0.46	4.64±0.18	8.01±0.30	6.29±0.22
β-Aminobutyric acid	3.88±0.15	6.43±0.24	6.40±0.24	1.72±0.07	2.20±0.08
Leucine	56.33±2.14	67.31±2.56	31.67±1.20	53.98±2.05	42.65±1.62
γ-Aminobutyric acid	6.63±0.25	7.97±0.30	1.57±0.06	6.40±0.24	5.97±0.23
Ornithine	13.96±0.53	10.12±0.38	11.27±0.43	13.70±0.52	12.63±0.48
Lysine	39.28±1.49	28.62±1.09	28.84±1.10	30.44±1.17	36.56±1.39
Histidine	19.13±0.72	18.46±0.70	12.78±0.49	13.10±0.50	17.64±0.67
1-Methylhistidine	10.90±0.40	1.83±0.07	10.21±0.38	9.53±0.36	13.13±0.50
3-Methylhistidine	3.40±0.12	3.50±0.13	8.92±0.34	9.17±0.35	11.54±0.43
Arginine	18.49±0.70	16.20±0.62	15.20±0.58	16.71±0.63	18.64±0.71
Total amount	313.92	439.8	337.01	428.16	366.33

Życzyńska-Bałoniak and Szajdak (1993)
Where $x \pm \Delta x$—mean values with their confidence limit Δx at $\alpha = 0.05$

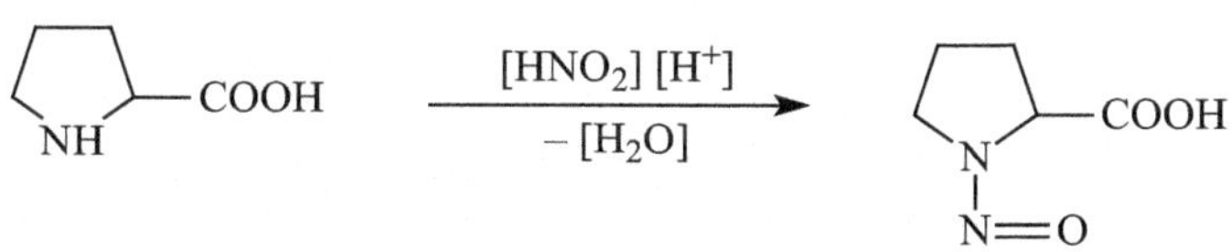

Fig. 2.2 Formation of N-nitrosoproline from proline

The most noticeable differences in the amount of bound amino acid present in the soils occurred during the period of intensive plant growth, i.e., in May. At this time, the concentration of bound amino acids in soils under crop rotation was 51 % higher than in soil samples from continuous cropping of rye. Predominantly bound amino acids in the two systems of cultivation were citrulline, cysteine, alanine lysine, and leucine. The citrulline concentration (20 %) was always higher in soils under crop rotation over the continuous rye cropping soil. Therefore, soils under continuous cropping of rye showed lower nitrogen biological storage capacity,

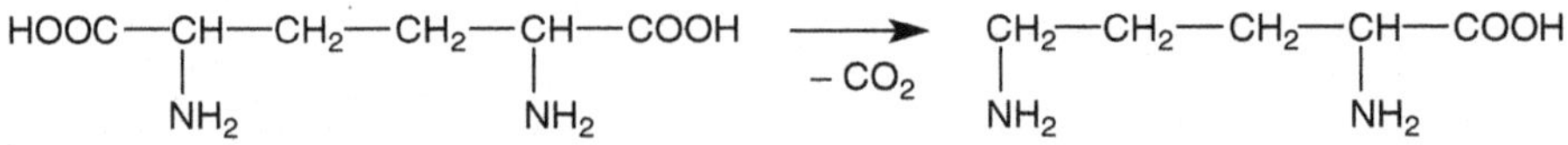

Fig. 2.3 Decarboxylation of α, ε-diaminopimelic acid with the creation of lysine

which is easily available to microbe and plant forms (citrulline). However, analyses of free amino acids did not show differences in citrulline concentrations between soils under continuous and crop rotation. The rates of formation of this amino acid (rates of biochemical processes) are similar in both cropping systems. But lower concentrations of bound forms of citrulline in soils under continuous cropping of rye indicated that catabolic processes (decomposition rates) exceeded the anabolic ones for this amino acid in this cropping system.

Like the free proline, the content of bound proline was considerably higher (200 %) in soils under continuous cropping of rye than in soils under crop rotation (Tables 2.8 and 2.9). The opposite is true for β-alanine. In soils under rotation, the β-alanine concentrations were about 300 % higher than in soils under continuous cropping of rye. The 20 % higher content of bound lysine in soils under crop rotation showed intensive microbiological activity in these soils. Lysine is formed by decarboxylation of α,ε-diaminopimelic acid (Fig. 2.3).

β-Alanine is a constituent of bacterial cell walls (Stevenson 1972; Durska and Kaszubiak 1980). The highest concentrations (49.7 mg kg^{-1}) of bound lysine were found during the intensive growth period of cereals (in May). This concentration was 70 % higher than in soils under continuous cropping of rye (Tables 2.8 and 2.9). The same observation holds true for β-alanine, which is also present in highest concentrations during the intensive growth period of plant. Thus, higher activity of bacteria could be indicated during the intensive growth of plant under crop rotation conditions. Contrarily, the sulfur containing amino acids showed the lowest and neutral amino acids the highest concentrations in bound amino acids in both types of soils (under continuous cropping of rye and under crop rotation). The low concentrations of basic amino acids could be explained due to their high capacity of reacting with reducing sugars and quinones (Holtzlaw et al. 1980).

The quantitative composition of free and bound amino acids in soil depends on the type of cropping. Soils under crop rotation contained higher amount of bound amino acids. The high concentration of some amino acids in crop rotation indicates that these soils are rich in bacterial biomass than soils under monoculture (single plant species cultivated) for years.

Dynamic soil properties are those properties that can change over short time periods (e.g., months, years, and decades) and are used in soil quality assessment because they change quickly with management. They can indicate whether a farm uses agronomically and ecologically sustainable practices. Changes in soil properties with time are a key component of dynamic soil quality assessment. Sustainable cropping systems will improve the soil quality, often through diverse crop rotations, minimal use of tillage for weed control and seedbed preparation, and addition of organic amendments (animal manures, crop residues, and compost). Management systems which decreased the soil quality indicators with time will lead to low soil quality, often induced by cropping systems with low residue production, intensive tillage, and near monoculture cultivation (Franzluebbers 2008).

2.2.3 *Amino Acids in Humic Acids*

Organic matter in soils consists of a mixture of plant and animal products in various stages of decomposition and substances synthesized during the breakdown of these compounds (Bremner 1967; Flaig 1971; Stevenson 1986; Szajdak and Matuszewska 2000), which may affect the availability of nutrients for plant growth (Stevenson 1986). All nitrogen in surface soil horizons is in organic form. The chemical composition of nitrogen in the organic soil fraction is not understood; little is known about the factors affecting the distribution of organic nitrogen forms in soils. Crop rotations, fertilization, and microbiological activity affect the nitrogen levels in soils (Szajdak and Sokolov 1997; Szajdak and Österberg 1996). Humus is composed of 20–60 % humic acids (HA), and 20–40 % nitrogen in HA consists of amino acids or peptides connected to the central core by hydrogen bonds (Harworth 1971). Little is known about the variability in content of amino acids in HA, in soil under long-time cultivation or applied different fertilization (NPK or manure).

The pH of control soils and under continuous cropping of rye ranged from 5.7 to 5.9 (Table 2.10). The fertilization with NPK or manure decreased the soil acidity to pH 6.2. The organic carbon content in control soils and fertilized with manure ranged from 6.48 to 6.96 and 11.8 to 12.05 $g \cdot kg^{-1}$, respectively. The nitrogen content in control soil and fertilized with NPK or manure was from 6.97 to 7.27 and 11.07 $g \cdot kg^{-1}$, respectively (Ryszkowski et al. 1998).

Humic Acids This investigation revealed that total acidity of HA extracted from soils under continuous cropping of rye in control and fertilized with manure was lower than acidity in HA from crop rotation (Table 2.10).

The total acidity of HA from crop rotation was 6.36–10.02 $meq \cdot g^{-1}$ of HA and was from 7.07 to 9.92 $meq \cdot g^{-1}$ of HA for continuous cropping of rye (Table 2.11). The highest total acidity (10.02 meq/g of HA) was observed in HA extracted from crop rotation fertilized with manure, while the lowest total acidity of HA was from crop rotation fertilized with NPK (6.36 meq/g of HA). The HA extracted from soils under crop rotation contained the highest content of phenolic groups (6.46 meq/g of HA).

The total amounts of bound amino acids in all samples of HA from soil under crop rotation were significantly higher than from soils under continuous cropping of rye (Table 2.3). The total amount of bound amino acids was highest (3093.2 $mg \cdot kg^{-1}$) in crop rotation fertilized with NPK. This was 18 % higher than in HA from soils under continuous cropping of rye fertilized with NPK (2541.4 $mg \cdot kg^{-1}$) and the lowest of 1888.7 $mg \cdot kg^{-1}$ for the controls under crop rotation (Table 2.12).

Table 2.10 Chemical properties of soils under crop rotation and under continuous cropping of rye, fertilized with NPK, manure or nonfertilized (control) and ash content in HA

Treatment	$pH_{(1\ M\ KCl)}$	$C_{(organic)}$ [$g \cdot kg^{-1}$]	$N_{(total)}$	HA Ash content [%]
Crop rotation control	5.8±0.2	6.80±0.2	0.727±0.03	8.5
Crop rotation manure	6.2±0.1	11.80±0.4	1.107±0.04	5.6
Crop rotation NPK	5.9±0.2	6.48±0.2	0.696±0.02	7.9
Continuous cropping of rye control	5.7±0.2	7.23±0.3	0.727±0.03	9.0
Continuous cropping of rye manure	6.2±0.3	12.05±0.4	1.107±0.04	5.1
Continuous cropping of rye NPK	5.8±0.2	6.96±0.2	0.696±0.02	7.5

Szajdak et al. (2004)
Control, no NPK or manure, manure 30 t/ha/year; NPK, 90 kg N, 60 kg P_2O_5, 120 kg K_2O/ha/year

Table 2.11 Functional group analysis of HA from soils under crop rotation and under continuous cropping of rye, fertilized with NPK, manure, or nonfertilized (control)

Treatment	Total acidity [$meq \cdot g^{-1}$ of HA]	Phenolic OH groups [$meq \cdot g^{-1}$ of HA]	Carboxylic COOH groups [$meq \cdot g^{-1}$ of HA]
Crop rotation control	7.49±0.3	5.40±0.2	2.09±0.1
Crop rotation manure	10.02±0.4	6.46±0.3	3.56±0.1
Crop rotation NPK	6.36±0.2	4.85±0.2	1.51±0.1
Continuous cropping of rye control	7.42±0.3	4.96±0.2	2.46±0.1
Continuous cropping of rye manure	9.92±0.4	6.06±0.2	3.86±0.2
Continuous cropping of rye NPK	7.07±0.3	4.89±0.2	2.18±0.1

Szajdak et al. (2004)
Control, no NPK or manure, manure 30 t/ha/year; NPK, 90 kg N, 60 kg P_2O_5, 120 kg K_2O/ha/year, meq: milliequivalent

In the soils under continuous cropping, the NPK fertilizer strongly affected the bound amino acid content than manure 38 % vs. 25 %, respectively, while the contents in soils under crop rotation were 41 and 27 %, respectively. Neutral amino acids were in highest proportion (57–69 %) of the total amino acids. The lowest concentrations of basic amino acids (11–19 %) were due to their capacity to react with reducing sugars and quinones. The acidic net charge was smaller (20–27 %) in all samples and was strongest in soils fertilized with manure. The total acidity was higher for HA extracted from soils under crop rotation and continuous cropping of rye and fertilized with manure. The latter was accompanied by correspondingly higher -COOH and phenolic -OH concentration. HA extracted from these soils also contained phenolic groups (19 %) and more carboxylic groups (44 %) than from those soils fertilized with NPK (Table 2.12). This is owing to the concentrations of lignin in manure. During the lignin degradation, components containing benzene rings (factor of aromaticity), carboxyl and hydroxyl, aldehyde, and methoxyl groups are formed (Shu-Yen et al. 1985; Szajdak and Życzyńska-Bałoniak 1994), which

Table 2.12 Mean concentration (mg·kg^{-1} of dry mass of HA ± 95 % confidence interval) of bound amino acids in HA from soils under crop rotation and under continuous cropping of rye, fertilized with NPK, manure, or nonfertilized (control)

Amino acids	Type of cultivation					
	Continuous cropping of rye, control	Crop rotation, control	Continuous cropping of rye, NPK	Crop rotation, NPK	Continuous cropping of rye, manure	Crop rotation, manure
Acidic						
Cysteic acid	34.1 ± 1.3	11.8 ± 0.5	111.5 ± 4.2	98.6 ± 3.7	15.1 ± 0.6	49.5 ± 1.9
Taurine	9.8 ± 0.4	9.1 ± 0.4	23.8 ± 0.9	15.6 ± 0.6	8.9 ± 0.3	20.3 ± 0.8
Phosphoethanolamine	–	9.2 ± 0.3	3.2 ± 0.1	8.7 ± 0.3	8.2 ± 0.4	22.5 ± 0.9
Aspartic acid	–	20.3 ± 0.8	9.9 ± 0.4	12.3 ± 0.5	22.6 ± 0.9	35.6 ± 1.4
Threonine	115.1 ± 4.4	105.9 ± 4.0	102.3 ± 3.9	81.3 ± 3.1	118.1 ± 4.5	135.7 ± 5.1
Serine	116.1 ± 4.4	113.7 ± 4.3	123.6 ± 4.7	211.3 ± 8.0	138.1 ± 5.2	140.5 ± 5.3
Glutamic acid	125.9 ± 4.8	180.6 ± 6.9	137.5 ± 5.2	288.6 ± 11.0	293.3 ± 11.2	257.1 ± 9.8
Neutral						
Proline	171.2 ± 6.5	67.9 ± 2.6	192.4 ± 7.3	154.6 ± 5.9	110.6 ± 4.2	92.4 ± 3.5
Glycine	420.1 ± 16.0	357.6 ± 13.6	314.1 ± 11.9	485.6 ± 18.4	437.8 ± 16.6	492.5 ± 18.7
Alanine	158.5 ± 6.0	165.5 ± 6.3	195.7 ± 7.4	217.2 ± 8.3	190.7 ± 7.3	270.2 ± 10.3
Valine	221.3 ± 8.4	107.3 ± 4.1	171.4 ± 6.5	205.3 ± 7.8	127.2 ± 4.8	162.9 ± 6.2
Cysteine	80.7 ± 3.1	214.1 ± 8.1	218.6 ± 8.3	199.1 ± 7.6	121.5 ± 4.6	144.7 ± 5.5
Cystathionine	66.2 ± 2.5	41.2 ± 1.6	49.6 ± 1.9	80.6 ± 3.1	51.2 ± 2.0	85.9 ± 3.3
Methionine	33.2 ± 1.3	–	98.2 ± 3.7	102.1 ± 3.9	68.4 ± 2.5	53.9 ± 2.0
Isoleucine	49.6 ± 1.9	85.7 ± 3.3	85.7 ± 3.3	160.3 ± 6.1	112.5 ± 4.3	184.2 ± 7.0
Leucine	28.6 ± 1.1	33.1 ± 1.3	93.9 ± 3.6	100.8 ± 3.8	31.5 ± 1.2	64.5 ± 2.5
Tyrosine	107.3 ± 4.1	85.4 ± 3.2	29.5 ± 1.1	52.1 ± 2.0	–	142.7 ± 5.4
β-Alanine	11.3 ± 0.4	8.0 ± 1.6	52.6 ± 2.0	74.1 ± 2.8	16.8 ± 0.6	39.8 ± 1.5
γ-Aminobutyric acid	5.2 ± 0.2	3.6 ± 0.1	41.6 ± 1.6	69.8 ± 2.7	25.7 ± 1.0	34.1 ± 1.3

(continued)

Table 2.12 (continued)

Amino acids	Type of cultivation					
	Continuous cropping of rye, control	Crop rotation, control	Continuous cropping of rye, NPK	Crop rotation, NPK	Continuous cropping of rye, manure	Crop rotation, manure
Basic						
Ornithine	38.1 ± 1.5	57.7 ± 2.2	80.6 ± 3.1	70.6 ± 2.7	66.6 ± 2.5	30.9 ± 1.2
Lysine	120.6 ± 4.6	151.8 ± 5.8	186.7 ± 7.1	195.2 ± 7.4	117.4 ± 4.5	164.3 ± 6.2
Histidine	18.5 ± 0.7	29.1 ± 1.1	131.6 ± 5.0	96.8 ± 3.7	127.3 ± 5.0	98.4 ± 3.7
Arginine	33.2 ± 1.3	30.1 ± 1.1	87.4 ± 3.3	112.6 ± 4.3	60.4 ± 2.3	76.4 ± 2.9
Total amount	*1964.6 ± 74.9*	*1888.7 ± 73.7*	*2541.4 ± 96.5*	*3093.2 ± 117.7*	*2269.9 ± 86.5*	*2799.0 ± 106.4*

Szajdak et al. (2004)

Control, no NPK or manure, manure 30 t/ha/year; NPK, 90 kg N, 60 kg P_2O_5, 120 kg K_2O/ha/year

contributed to acidic properties. Our earlier investigations demonstrated different impacts of NPK and manure on the chemical structure of humic acids extracted from soils under continuous cropping of rye and crop rotation. The application of NPK increased the aliphatic structures in humic acids (HA) than manure application in these treatments. The aromaticity of HA from soils under continuous cropping of rye fertilized with manure was higher than from crop rotation fertilized with manure (Szajdak et al. 2000).

The glutamic acid dominated the acid fraction in all HA, its highest concentrations (293.3 mg·kg^{-1}) were in soils under continuous rye cropping fertilized with manure, and it was 12 % higher than in soils under crop rotation. Among neutral amino acids, glycine, alanine, and valine dominated, with glycine predominant. Their contents in soils under crop rotation fertilized with manure were 492.5 mg·kg^{-1}, i.e., 11 % higher than in soils under continuous rye cropping.

Proline The proline concentrations were 60.3 % lower in soils under crop rotation than under continuous rye cropping. The highest proline concentrations (192.4 mg·kg^{-1}) were in soils under continuous cropping of rye fertilized with NPK, i.e., 20 % higher than under crop rotation. Higher concentrations of proline in soils under continuous cropping of rye than crop rotation may explain its heterocyclic structure. This structure protects it from further degradation. The accumulation of proline in soils is harmful because under acidic conditions, the proline in the presence of nitrite ions may form N-nitrosamine potent toxins, with carcinogenic and mutagenic effects (Kofoed et al. 1981; Larsson et al. 1990).

In soils under crop rotation than under continuous rye cropping, soils fertilized with NPK had higher concentrations of β-alanine (46 %) and lysine (16 %) than with manure. A similar phenomenon was observed in the previous study (Życzyńska-Bałoniak and Szajdak 1993; Szajdak and Österberg 1996; Ryszkowski et al. 1998). This indicated a higher microbial biomass in soils under crop rotation, as the β-alanine and lysine are the typical constituents of bacteria than fungus cell walls (Stevenson 1972; Durska and Kaszubiak 1980). In addition, the positive and linear correlation between the activity of rhodanese and the concentrations of free sulfuric amino acids has been found. Enzyme rhodanese is formed by fungus in soils. Higher activity of rhodanese in soils under continuous cropping of rye indicates higher biomass of fungus, as the activity of rhodanese in soil is the measure of the abundance of fungi (Szajdak 1996).

The nitrogen in the bound amino acids was 22 % higher than in HA from crop rotation fertilized with manure. Lowest nitrogen contents were found in controls for both types of cropping. NPK fertilizers supplied larger amounts of nitrogen, phosphorus, and potassium than manure (Table 2.13).

The crops absorbed more nitrogen, phosphorus, and potassium from NPK than from manure. Probably it was due to disappearance of manure nitrogen, presumably by denitrification (Pratt et al. 1973; Svensson et al. 1991; Klemedtsson et al. 1991; Goulding et al. 1993; Rudaz et al. 1999). Organic and inorganic fertilizers in crop rotation and continuous cropping of rye gave different grain yield of winter rye. The highest yield (5940 kg·ha^{-1}) was observed in crop rotations using NPK as fertilizer,

Table 2.13 Uptake and balance of macronutrients (N, P, K, kg·ha^{-1}·year^{-1}) of winter rye grown continuously and in rotation

	Crop	Fertilizers			Uptake			Balance		
Fertilization	sequence[a]	N	P	K	N	P	K	N	P	K
Control	CR	0	0	0	81.5	17.5	44.5	−81.5	−17.5	−44.5
	CC	0	0	0	44.3	9.5	20.6	−44.3	−9.5	−20.6
Manure	CR	150	39.2	174.3	137.3	27.8	85.5	12.7	11.4	88.8
	CC	150	39.2	174.3	97.2	19.7	44.9	52.8	19.5	129.4
NPK	CR	90	26.2	99.6	129.7	26.0	76.0	−39.7	0.2	23.6
	CC	90	26.2	99.6	103.3	21.7	56.5	−13.3	4.5	43.1

Szajdak et al. (2004)
Control, no NPK or manure, manure 30 t/ha/year; NPK, 90 kg N, 60 kg P_2O_5, 120 kg K_2O/ha/year
[a]Average from 1988 to 1996, CR, crop rotation; CC, continuous cropping of rye

Table 2.14 Grain yield of winter rye (kg·ha^{-1})

	Crop rotation			Continuous cropping		
Years	Control	Manure	NPK	Control	Manure	NPK
1993	4430	5930	5350	2480	4730	5310
1994	3580	5220	6060	1960	4420	4610
1995	4380	6050	6420	2980	5620	5980
Average	*4130*	*5730*	*5940*	*2470*	*4920*	*5300*

Szajdak et al. (2004)
Control, no NPK or manure, manure 30 t/ha/year; NPK, 90 kg N, 60 kg P_2O_5, 120 kg K_2O/ha/year

and the lowest yield (2470 kg·ha^{-1}) was in control of continuous cropping of rye. In unfertilized plots in continuous cropping of rye, the yield was above half of the plot with manure. The NPK fertilizer gave 32 % higher aboveground biomass of winter rye and 27 % higher grain yield than manure (Table 2.14).

In summary, the results indicated that the composition of bound amino acids in HA depends on cropping system and on the availability of nitrogen, phosphorus, and potassium from fertilizers. Negative effects of continuous cropping on the content of total bound amino acids were decreased by NPK fertilization, but the manure application in continuous cropping of rye was less effective. Thus, NPK was the main driver causing changes in the total amounts of bound amino acids in HA. Crop yields of rye increase with an increase in organic N in bound amino acids and nitrogen in HA.

2.2.4 *Thermal Evaluation of HA Structure*

There are three soil quality indicators: (1) chemical, (2) physical, and (3) biological. Within each of these classes, many soil properties or processes can be selected to indicate soil functional capabilities. Soil organic matter is a critical component of soil quality. Accumulation of residues and organic matter at the soil surface is beneficial to soil quality, due to their positive effects on conserving water, preserving nutrients, and creating a suitable habitat for soil biological diversity (Franzluebbers 2008). With the start of degradation of plants and animal residues, HA is produced in soils. During the decomposition, HA is formed with carbon from all major plant components either by direct transformation or by "resynthesis" activity of microorganisms. The polymers of HA are synthesized enzymatically within microbial cell by chemical oxidative condensation following cellular autolysis. Therefore, HA are rarely homogenous; however, they represent accumulation of more resistant end products from many reaction taking place under natural conditions, either directly or indirectly through biological processes.

HA originates randomly from the decay of plant tissues or microbial metabolism–catabolism or both; hence, their chemistry is complex and a function of different ecosystems (vegetation, climate, topography, etc.) in which it is formed (Piccolo et al. 2003). HA provides energy to many biochemical processes in soil (Dziadowiec 1979; Gołębiowska et al. 1996) and also regulates the nutrient dynamics and C/N ratio. The aromatic pathway acts as a major resource quality factor at all levels—both by forming enzymatically recalcitrant molecules and also by the direct toxic effects of phenolic or quinonic monomers.

HA are composed of higher molecular weight (10,000–100,000) compounds containing aromatic rings, peptide chain, and nitrogen in cyclic and aromatic forms. These are created by polymerization, polyaddition, and polycondensation of similar but not identical substrates; therefore, no two humic molecules are identical. Pure humic acids contain 57 % carbon and 4 % nitrogen. Besides, the HA contains primarily COOH groups, phenolic OH groups, alcoholic OH, and some ketonic oxygen. The quantity of HA in soil organic matter depends on the balance between primary productivity and the rate of decomposition (Paul and Clark 1989).

The derivatographic method allows the rapid identification of consecutive stages of organic substances in soils and also estimates the changes in energy levels of these compounds (Leinweber and Schulten 1992; Schnitzer et al. 1974; Shurigina 1971).

In long-term experiments, the continuous rye cropping than crop rotation produces undifferentiated biological metabolites of microbes and plants (Ryszkowski et al. 1990). Therefore, this study aimed (i) to improve the interpretation of derivatographic results of HA from soils under continuous cropping of rye and from soils under crop rotation fertilized with NPK or with manure and (ii) to show that long-term continuous cropping of rye changes the HA structure than in crop rotation.

Table 2.15 shows the weight losses of HA samples in different temperature regions. For all samples, the changes in differential thermal analyses (DTA) and thermogravimetric analyses (DTG) are compatible. Thus, each thermal effect recorded on the DTG corresponded to weight losses of HA. Furthermore, HA are characterized by two exothermic effects: (i) in low temperature (below 350 °C—egzoL) and (ii) in high temperature (above 350 °C—egzoH). All samples of HA characterized also small peak on DTA curve below 100 °C (endothermic effect). The endothermic effect is explained by the disappearance of aliphatic structures

Table 2.15 Parameters of thermal decomposition of humic acids from soils under continuous cropping of rye and from crop rotation fertilized with NPK or manure or without any fertilizer

	Maximum temperature of effects recorded on DTA curves (°C)			Loss of weight corresponding with effects on DTA curves			The rations of area under DTA and DTG curves according to isothermal reactions			
				DTG	DTG	DTG				
Samples of HA	endo	$egzo_N$	$egzo_W$	endo	$egzo_N$	$egzo_W$	$\frac{DTA_N}{DTG_N}$	$\frac{DTA_W}{DTG_W}$	$\frac{DTA_{N+W}}{DTG_{N+W}}$	$Z = \frac{DTG_N}{DTG_W}$
CC control	89	298	429	8.04	6.77	9.99	1.26	4.10	3.14	0.68
CC NPK	100	283	408	10.10	5.92	8.26	1.94	4.51	3.14	0.72
CC manure	98	312	432	12.71	5.27	12.44	2.31	5.33	4.13	0.42
CR control	92	280	415	6.72	6.18	5.42	3.32	3.80	3.55	1.14
CR NPK	98	286	389	8.40	6.69	5.16	3.19	3.44	3.30	1.29
CR manure	90	314	460	7.19	7.14	11.03	3.15	5.36	4.50	0.64

Szajdak et al. (2000)

Where *CC-control* continuous cropping of rye control, *CC-NPK* continuous cropping of rye fertilized with NPK, *CC-manure* continuous cropping of rye fertilized with manure, *CR-control* crop rotation control, *CR-NPK* crop rotation fertilized with NPK, *CR-manure* crop rotation fertilized with manure

in HA (Rakovskiy and Filimonov 1967). HA from both kinds of cultivation and fertilized with NPK have lower value of egzoH in comparison with samples fertilized with manure (Table 2.14). For HA from soil under continuous cropping of rye and fertilized with NPK, the value of egzoH was 8.26. For soils under crop rotation, the value of the same parameter was 38 % lower. The fertilization with manure moved the egzoH effects in both soils toward high temperature with simultaneous increase in them (432 and 460 °C). For soil under continuous cropping of rye, and fertilized with manure, the egzoH parameter was equal to 12.44, and for soils under crop rotation, 11 % lower at 11.03. The lower values of egzoH parameters of HA from soils fertilized with NPK than with manure demonstrate that this HA from soils fertilized with NPK contained more aliphatic structures than HA from soils fertilized with manure (Gonet 1989).

Estimation of weight losses (DTG) in samples of HA for endothermic and exothermic effects has affirmed more weight losses in the endothermic range. It might be caused more hygroscopicity of HA samples. The higher weight losses (6.18–7.14) of HA were in the exothermic range (below 350 °C) in soils under crop rotation than from soils under continuous cultivation (5.27–6.77). This phenomenon was reversed above 350 °C, where higher weight losses in HA were observed from continuous cropping of rye than from crop rotation.

The weight loss with applied manure in HA was 12.44 from soils under continuous cropping of rye and was 11.03 from soils under crop rotation. The weight loss in HA with NPK fertilizers was 5.16 from soils under crop rotation, and it was 60% higher from soils under continuous cropping of rye. Manure in both types of cultivation caused the highest weight losses in HA in higher temperature region.

The value of DTAL means the heat of combustion of HA. The higher temperature of decomposition of HA shows higher energy of activation and also explains the differences in the structures of HA. These differences are caused by the type of cultivation and the use of fertilizers. All samples of HA in high temperatures (above 350 °C) have higher values of heat of combustion (DTA:DTG) than HA in full range of temperatures (DTAL + H:DTGL + H). These highest values of this parameter characterized HA extracted from soils under both types of cultivation and fertilized with manure. The lower values of heat of combustion might indicate that the compounds include more aliphatic bonds that are easily degraded (Gonet and Wegner 1990, 1993). As a result of breaking off aliphatic bridges between aromatic structural units, aromatic compounds eliminate with the highest rate (Maryganova et al. 1992). The higher values of heat of combustion of HA from soils fertilized with manure than NPK might show that the linkages of HA contain more aromatic groups than aliphatic.

Parameter Z reflects the ratio between thermolabile and thermostable parts of the humic molecules. The lower results of Z parameter (parameter of aliphaticity) indicate higher aromaticity properties of HA. Thus, the fertilization of NPK irrespective of cultivation leads to the increase of aliphatic structures in HA. Comparison of Z parameter with the type of cultivation shows that HA from crop rotation contains more aliphatic properties than from continuous cropping of rye. This is in line with the findings of others (Hruszka 1982; Szajdak and Życzyńska-Bałoniak 1994), who found higher accumulation of phenolic compounds in soils under continuous cropping of rye than under crop rotation.

These results showed that the crop rotation (i) decreases the temperature of combustion in high temperature range, (ii) causes higher weight losses in low temperature range (DTA egzoL), (iii) decreases the heat of combustion in high temperature range (DTAH:DTG), and (iv) increases the Z parameter. It might indicate higher aliphatic properties of this HA. Furthermore, the manure application irrespective of the type of cultivation increases the temperature of combustion in low and high range of temperatures and also increases the heat of combustion. The HA extracted from the soils under crop rotation and fertilized with NPK showed highest aliphatic properties, but the HA from soils fertilized with manure was most aromatic. Aromaticities of HA from soils under continuous cropping of rye fertilized with manure were higher than from crop rotation fertilized with NPK.

The evidence gathered suggests strong linkages between the soil and fertilizer, and manures' quality affects the productivity. Achieving a balance between the agricultural production and conservation of natural resources is necessary to develop sustainable agricultural systems.

References

Adamiak E, Zawiślak K (1990) Changes in the weed communities in continuous cropping of cereals and maize. In: Ryszkowski L, Karg J, Pudełko J (eds) Ekologiczne procesy w monokulturowych uprawach zbóż (Ecological processes in continuous cropping of rye). Wydawnictwo Naukowe UAM, Poznań, pp 45–47 (in Polish)

Baltruschat H, Dehne W (1989) The occurrence of vesicular-arbuscular mycorrhiza in agroecosystems. Influence of nitrogen fertilization and green manure in continuous monoculture and in crop rotation on the inoculum potential of winter barley. Plant Soil 113:251–256

Blake G, Coble H, Barnes G, Walters H (1980) Weed control, insect, diseases and soils physical properties: crop rotation vs monoculture. Crops Soil Mag 32:8–17

Blecharczyk A (1999) Forty-years of fertilizing experiment in Brody with crops grown continuously and in crop rotation. Zesz Probl Postep Nauk Rol 465:261–272

Blecharczyk A, Pudełko J (1997) Przyszłość monokultur w rolnictwie europejskim. The future of continuous cropping in European Agriculture. Acta Acad Agric Tech Olst 64:143–15

Blecharczyk A, Skrzypczak G (1994) Tolerancja żyta ozimego na uprawie w wieloletniej monokulturze (Tolerance of rye in continuous cropping of rye). Zesz Nauk ATR Bydgoszcz Rolnictwo 187:45–52 (in Polish)

Blecharczyk A, Pudełko J, Małecka I (1995) Spring barley tolerance to continuous cropping. Fragm Agron 2(46):238–239

Bremner JM (1967) Nitrogenous compounds. In: McLaren AD, Peterson GH (eds) Biochemistry, vol 1. Marcel Dekker, New York, pp 70–85

Bu'Lock JD (1980) Mycotoxins as secondary metabolites. In: Steyn PS (ed) Biosynthesis of mycotoxins. Academic, New York, pp 1–16

Buss B, Zoschke M (1984) Untersuchungen zum Daueranbau von Wintergerste (*Hordeum vulgare* L.) auf geblechter Parabraunerde (Sandboden). J Agron Crop Sci 153:422–437

Campbell A, Schnitzer M, Lafond GP, Zentner RP, Knipfel JE (1991) Thirty-year crop rotations and management practices effects on soil and amino nitrogen. Soil Sci Soc Am J 55:739–745

Chapin FS, Moilanen L, Kielland K (1993) Preferential use of organic nitrogen for growth by a non-mycorrhizal arctic sedge. Nature 361:150

Christen O, Sielin K (1993) The effect of different preceding crops on the development, growth and yield of winter barley. J Agron Crop Sci 171:114–123

Claudius G, Merhotka R (1973) Root exudates from lentil/*Lens culinaris Madic*/seedlings in relation to wilt disease. Plant Soil 38:315–320

Claupein W, Zoschke M (1987) Einfluss langjähriger Winterweizen-Monokultur auf Ertragsbildung, Krankheitsbefall und Ncmatodenbesatz im Vergleich zum Winterweizenanbau in de Fruchtfolge. J Agron Crop Sci 158:227–235

Copeland P, Allnaras R, Crookston R, Nelson W (1993) Corn-soybean rotation effects on soil water depletion. Agron J 85:203–210

Cox J (1965) Continuous wheat growing and the decline of take-all. Report of Rothamsted Experimental Station, pp 133–134

Crookston R, Kurle J, Copeland P, Ford J, Lueschen W (1991) Rotational cropping sequence affects yield of corn and soybean. Agron J 82:229–232

Durska G, Kaszubiak H (1980) Occurrence of α, ε-diaminopimelic acid in soil. II. Usefulness of α, ε-diaminopimelic acid determination for calculations of the microbial biomass. Pol J Ecol Stud 6:195–199

Durska G, Kaczmarek W, Kaszubiak H, Muszyńska M, Pędziwilk Z (1986) Abundance of bacteria and fungi in soil and their respiration activity under rye in continuous cropping. In: Mysków W, Kuś J, Kamińska M (eds) Ekologiczne skutki monokulturowej uprawy zbóż. Instytut Uprawy Nawożenia i Gleboznawstwa, Puławy, pp 95–106

Dziadowiec H (1979) Zmiany energetyczne towarzyszące humufikacji ściółek leśnych (Changes in energy associated with humification of forest litter). Stud Soc Sci Torun Sectio D 1:11–104 (in Polish)

Feinstkorn M, Kreuz E (1988) Die Effektivität der Bedennutzung bei komplexer Anbeuintensivierung und Fungizidensatz in zehnjährigen Monokulturen von Winterweizen und Sommergerste auf Löß-Schwarzende. Arch Acker Pflanz Bod Berlin 9:611–617

Flaig W (1971) Organic compounds in soil. Soil Sci 111:19–33

Franzluebbers AJ (2008) Linking soil and water quality in conservation agricultural systems. J Integr Biosci 6(1):15–29

Garz J, Stumpe H (1992) Der von Julius Kühn begründete Versuch "Ewiger Roggenbau" in Halle nach 11 Jahrzenten. Kühn-Archich 86:1–8

Gawrońska-Kulesza A (1966) Wpływ nawożenia organiczego i mineralnego stosowanego w zmianowaniu 3- i 4-letnim na niektóre właściwości chemiczne gleby, wysokość i jakość plonów (Impact of mineral and organic fertilization used in crop rotation on some chemical soil properties, and crop yield). Rocz Nauk Rol Ser A 92:3–28 (in Polish)

Gołębiowska D, Ptak W, Wegner K (1996) Correlation between derivatographic and chemiluminescence analysis data in relation to elemental composition of humic acids. Environ Int 22(5):495–500

Gonet SS (1989) Badania kwasów huminowych metodą analizy różnicowej (Investigations of humic acids by the method of differential thermal analysis). Rocz Gleb T 40(1):27–38 (in Polish)

Gonet SS, Wegner K (1990) Wpływ nawożenia na próchnicę gleb (Impact of fertilization on humus of soils). Zesz Nauk AR Wrocław Rol 8(196):128–135

Gonet SS, Wegner K (1993) Wpływ wieloletniego nawożenia mineralnego i organicznego na właściwości kwasów huminowych (Impact of long-term mineral and organic fertilization on properties of humic acids). Zesz Nauk AR Kraków 37(277):51–63 (in Polish)

Gonet SS, Wiśniewski W, Wegner K (1993) Właściwości kwasów huminowych ekstrahowanych z gleb o różnym nawożeniu azotowym (Properties of humic acids extracted from soils of different fertilization). Rocz Gleb 37:277–285 (in Polish)

Goulding KWT, Webster CP, Powlson DS, Poulton PR (1993) Denitrification losses of nitrogen fertilizer applied to winter following ley and arable rotations as estimated by acetylene inhibition and ^{15}N balance. J Soil Sci 44:63–72

Guenzi WD, McCalla TM (1966) Phytotoxic substances extracted from soil. Soil Sci Soc Am Proc 30:214–216
Hageman NR, Schrader WD (1979) Effects of crop sequence and fertiliser level on soil bulk density. Agron J 71:1005–1008
Hanley F, Ridgman J (1978) Long-term effects of growing winter wheat continuously. J Agric Sci UK 90(3):517–521
Harworth RD (1971) The chemical nature of humic acid. Soil Sci 106:188–192
Heyn J, Brüne H (1981) Ergebnisse einess 13jährigen Feldversuches mit Weizen in Fruchtfolge und Monokultur. Landwitsch Forsch Sond (Kongressband) 38:709–721
Hill PW, Farrar J, Roberts P, Farrell M, Grant H, Newsham KK, Hopkins DW, Bardgett RD, Jones DL (2011) Vascular plant success in a warming Antarctic may be due to efficient nitrogen acquisition. Nat Clim Chang 1:50
Holtzlaw KM, Schaumberg GD, LeVasque-Madore CS, Sposito Heick JA, Johnson CT (1980) Analytical properties of soluble, metal-complexing fractions in sludge-soil mixtures. V. Amino acids, hexosamines and other carbohydrates in fulvic acid. Soil Sci Soc Am J 44:736–740
Honermeier B (2007) Diversity in crop production systems. In: Benckiser G, Schnell S (eds) Biodiversity in agricultural production systems. Taylor & Francis/CRC, Boca Raton, pp 1–19
Hruszka M (1982) Studies on toxicity of phenolic acids in continuous cropping. Acta Univ Agric (Brno) 30:79–85
Huet PH, Boyeldieu J (1976) Compertement varietal et conduite du ble tender en monuculture de longue duree. Ann Agron 27(5–6):969–981
Jelinowski S, Mróz A (1979) Studies of influence of crop rotation on appearance of root disease in winter wheat. Zesz Probl Postep Nauk Rol 218:133–139 (in Polish)
Johnston A, Mattingly E (1976) Experiments in the continuous growth of arable crops at Rothamsted and Woburn experimental stations: effects of treatments on crop yields and soil analyses and recent modifications in purpose and design. Ann Agron 27(5–6):927–956
Johnston N, Copeland P, Crookston R, Pfleger F (1992) Mycorrhizae: possible explanation for yield decline with continuous corn and soybean. Agron J 84:387–390
Jones DL, Kielland K (2002) Soil amino acid turnover dominates the nitrogen flux in permafrost-dominated taiga forest soils. Soil Biol Biochem 34:209–219
Jones DL, Healey JR, Willett VB, Farrar JF, Hodge A (2005a) Dissolved organic nitrogen uptake by plants an important N uptake pathway? Soil Biol Biochem 37:413–423
Jones DL, Kemmitt SJ, Wright D, Cuttle SP, Bol R, Edwards AC (2005b) Rapid intrinsic rates of amino acid biodegradation in soils are unaffected by agricultural management strategy. Soil Biol Biochem 37:1267–1275
Jones DL, Hughes LT, Murphy DV, Healey JR (2008) Dissolved organic carbon and nitrogen dynamics in temperate coniferous forest plantations. Eur J Soil Sci 59:1038–1048
Jones DL, Kielland K, Sinclair FL, Dahlgren RA, Newsham KK, Farrar JF, Murphy DV (2009) Soil organic nitrogen mineralization across a global latitudinal gradient. Glob Biogeochem Cycles 23:GB1016
Karg J, Czarnecki A, Paprocki R (1986) Soil fauna under rye in crop rotation and continuous rye cropping. In: Mysków W, Kuś J, Kamińska M (eds) Ekologiczne skutki monokulturowej uprawy zbóż (Ecological effects of continuous cropping of rye). Instytut Uprawy Nawożenia i Gleboznawstwa, Puławy, pp 113–133 (in Polish)
Karg J, Czarnecki A, Witkowski T, Paprocki R (1990) Density and biomass of edaphon in continuous cropping of rye and crop rotation. In: Ryszkowski L, Karg J, Pudełko J (eds) Ekologiczne procesy w monokulturowych uprawach zbóż (Ecological processes in continuous cropping of rye). Zakład Badań Środowiska Rolniczego i Leśnego PAN, Poznań, pp 187–195 (in Polish)
Kaszubiak H, Kaczmarek W, Pędziwilk Z, Sawicka A, Muszyńska M, Durska G (1990) Microbial communities in crops under continuous cropping of rye and crop rotation. In: Ryszkowski L, Karga J, Pudełko J (eds) Ekologiczne procesy w monokulturowych uprawach zbóż. Zakład Badań Środowiska Rolniczego i Leśnego PAN, Poznań, pp 77–90 (in Polish)

Ketcheson JW (1980) Long-range effects of intensive cultivation and monoculture on the quality of southern Ontario soils. Can J Soil Sci 41:403–410
Kielland K (1994) Amino acid absorption by arctic plants implications for plant nutrition and nitrogen cycling. Ecology 75:2373–2383
Kielland K (1995) Landscape patterns of free amino acids in arctic tundra soils. Biogeochemistry 31:85–98
Kielland K, McFarland JW, Olson K (2006) Amino acid uptake in deciduous and coniferous taiga ecosystems. Plant Soil 288:297–307
Kielland K, McFarland JW, Ruess RW, Olson K (2007) Rapid cycling of organic nitrogen in taiga forest ecosystems. Ecosystems 10:360–368
Klemedtsson L, Simkins S, Svensson BH, Johnson H, Rosswall T (1991) Soil denitrification in three cropping systems characterized by differences in nitrogen and carbon supply. II. Water and NO_3^- effects on the denitrification process. Plant Soil 138:273–286
Kofoed D, Nemming O, Brunfeldt K, Nebelin E, Thomsen J (1981) Investigations on the occurrence of nitrosamines in some agricultural products. Acta Agric Scand 31:40–48
Krejĉir J (1987) Plonowanie Jęczmienia jarego w monokulturzez i zmianowaniu. Acta Acad Agric Tech Olst Agricultura 44:23–39 (in Polish)
Kübler E (1977) Auswirkungen von Versuchsdauer und Häufigkeit des Sommergerstenbaus auf den Kornertrag und das Auftreten von Schadfaktoren. J Agron Crop Sci 145:36–50
Kürten P, Range W (1980) Ergebnisse eines achtjährigen Fruchtfolge-Düngungsversuches mit Anbau von Winterweizen und Wintergerste in Getredefolge und Monokultur. Land Forsch 33:385–407
Łapiński S, Ryszkowski L (1986) Primary production of rye in continuous cropping. In: Myskòw W, Kuś J, Kamińska J (eds) Ekologiczne skutki monokulturowej uprawy zbóż (Ecological effects in continuous cropping of rye). Instytut Uprawy Nawożenia i Gleboznawstwa, Puławy, pp 35–42 (in Polish)
Larsson BK, Österdahl BG, Regner S (1990) Polycyclic aromatic hydrocarbons and volatile N-nitrosamines in some dried agricultural products. Swed J Agric Res 20:49–56
Leinweber L, Schulten HR (1992) Differential thermal analysis thermogravimetry and in source pyrolysis-mass spectrometry studies on the formation of soil organic matter. Thermochim Acta 200:151–167
Łoginow W, Andrzejewski J, Wiśniewski W, Kusińska A, Cieścińska B, Karlik B, Jankowiak J (1990) Influence of continuous cropping on transfer of organic matter and nitrogen in soil (in Polish). In: Ryszkowski L, Karg J, Pudelko J (eds) Ekologiczne procesy w monokulturowych uprawach zbóż (Ecological processes in continuous cropping of rye). UAM, Poznań, pp 111–132 (in Polish)
Maryganova VV, Bambalov NN, Lukashenko IM, Kalinkevich GA, Khmelnitskiy RA (1992) Study of peat humus acids by pyrolysis mass-spectroscopy. Soviet Soil Sci (Pochvovedenie) 1:152–155 (in Russian)
Mercik S (1989) Plonowanie żyta, pszenicy i ziemniaków w zależności od wieloletniego zróżnicowanego nawożenia i zmianowania. Cz. I. Żyto. Yield of the rye, wheat and potatoes in long-term continuous cropping. Part I. Rye. Rocz Gleb 40(91):191–201 (in Polish)
Myśków W, Kuś J, Kamińska M (1986) Ekologiczne skutki monokulturowej uprawy zbóż (Ecological processes in continuous cropping of rye). Instytut Uprawy Nawożenia i Gleboznawstwa, Puławy, p 146 (in Polish)
Niewiadomski W, Adamiak J, Zawiślak K (1980) Tolerance of 9 most important species of cultivated plants for continuous cropping. Zesz Nauk ATR Olsztyn Rolnictwo 29:271–281 (in Polish)
Owen AG, Jones DL (2001) Competition for amino acids between wheat roots and rhizosphere microorganisms and the role of amino acids in plant N acquisition. Soil Biol Biochem 33:651–657
Panse A, Maidl X, Dennert J, Brunner H, Fischbeck G (1994) Ertagsbildung von getreidereichen Fruchtfolgen und getreidemonokulturen in einem extensiven un intensiven Anbausystem. J Agron Crop Sci 173:160–171

Paul EA, Clark FE (1989) Soil microbiology and biochemistry. Academic Press, Inc./Harcourt Brace Javonovich Publisher, San Diego, pp 93–116

Pesci P (1992) Effect of light on abscisic acid-induced proline accumulation in leaves: comparison between barley and wheat. Plant Physiol 6:209–214

Piccolo A, Conte P, Cozzolino L, Spaccini R (2003) The conformation structure of humic substances. In: Bendi DK, Nieder R (eds) Handbook of processes and modeling in the soil-plant system. Food Products Press/The Haworth Reference Press, New York, pp 83–120

Pimentel D, Hall C (1984) Food and energy resources. Academic, New York, p 268

Pla GL (1980) Toxic responses of the liver. Toxicology. The basic sciences of poisons. MacMillan Publishing Co. Inc., New York/Toronto/London, pp 216–231

Pratt PF, Broadbent FE, Martin JP (1973) Using organic wastes as nitrogen fertilizers. Calif Agric 27:10–13

Rakovskiy V, Filimonov V (1967) Mechanism of the reactions of thermal destruction of oxygen containing components of fuels. Chem Chem Technol 3(16):302–311, Nedra, Moscow (in Russian)

Ridgman J, Walters D, Wedgwood R, Muller-Wilmes U (1985) Some effect of rate of application of nitrogenous fertilizer to wheat grown continuously compared with wheat in a four course rotation. J Agric Sci 105:389–396

Rous D (1992) The problems of monocultures and high concentrations of cereals. Acta Acad Agric Tech Olst Agric 55:193–200

Rudaz AO, Wälti E, Kyburz G, Lehmann P, Fuhrer J (1999) Temporal variation in N_2O and N_2 fluxes from a permanent pasture in Switzerland in relation to management, soil water content and soil temperature. Agric Ecosyst Environ 73:83–91

Ryszkowski L (1986) Ecological effects of simplified rotations (in Polish). In: Myśków W, Kuś J, Kamińska M (eds) Ekologiczne skutki monokulturowej uprawy zbóż (Ecological effects of continuous cropping of rye). Instytut Uprawy Nawożenia i Gleboznawstwa, Puławy, pp 5–20 (in Polish)

Ryszkowski L, Bernacki Z (1990) Primary production in continuous cropping. In: Ryszkowski L, Karg J, Pudełko J (eds) Ekologiczne procesy w monokulturowych uprawach zbóż (Ecological processes in continuous cropping of rye). Wydawnictwo Naukowe UAM, Poznań, pp 37–45 (in Polish)

Ryszkowski L, Karg J (eds) (1990) Badania monokultur zbożowych (Investigations of continuous rye cropping). SGGW-AR, Warszawa, p 171 (in Polish)

Ryszkowski L, Karg J (1992) Energy flow in rye grown in continuous and Norfolk rotation cultures. Acta Acad Agric Tech Olst Agricultura 55:210–213

Ryszkowski L, Karg J, Pudełko J (eds) (1990) Ekologiczne procesy w monokulturowych uprawach zbóż. Ecological processes in continuous cropping of rye. Wydawnictwo Naukowe UAM, Poznań, p 328 (in Polish)

Ryszkowski L, Szajdak L, Karg J (1998) Effects of continuous cropping of rye on soil biota and biochemistry. Crit Rev Plant Sci 17:225–244

Schimel JP, Bennett J (2004) Nitrogen mineralization: challenges of a changing paradigm. Ecology 85:591–602

Schnitzer M, Turner RC, Hofman JA (1974) A thermogravimetric study of organic matter of representative Canadian soils. Can J Soil Sci 44(1):7–13

Schönhammer A, Fischbeck G (1987) Untersuchungen an getreidereichen Fruchtfolgen und getredemonokulturen. Bay Landwirt Jahr 64(2):175–191

Schulten HR, Schnitzer M (1998) The chemistry of soil organic nitrogen: a review. Biol Fertil Soils 26:1–15

Shcherba VV (1994) Degradation of rye straw and flax scutch lignins by submerged cultures of mycelial fungi. Prikl Biochim Mikrobiol 30:403–409

Shurigina EA (1971) Differential thermal analysis (DTA) and thermogravimetry (TG) of soils humus substances. Geoderma 6:169–177

Shu-Yen L, Freyer AJ, Minard RD, Bellag JM (1985) Enzyme-catalyzed complex-formation of amino acid esters and phenolic humus constituents. Soil Sci Soc Am J 49:337–343
Sieling H, Hanus H (1992) Yield of winter wheat influenced by the interaction between crop management measures and take-all. Eur J Agron 1:201–206
Smyth WH (1976) Character and significance of forest tree root exudates. Ecology 57:324–331
Sörensen JH (1967) Duration of amino acid metabolites formed in soil during decomposition of carbohydrates. Soil Sci 104:204–241
Steineck O, Ruckenbauer P (1976) Results of 70 years long-term rotation and fertilization experiment in the main cereal growing area of Austria. Ann Agron 27(5–6):803–818
Stevenson G (1972) Biology of fungus, microbes and viruses. PWRiL, Warszawa, p 280 (in Polish)
Stevenson FJ (1985) Amino acids. In: Humus chemistry, genesis, composition, reactions. Wiley, New York, pp 65–78
Stevenson FJ (1986) Amino acids. In: Cycles of soils. A Wiley-Interscience Publication/John Wiley and Sons, New York, pp 155–215
Svensson BH, Klemedtsson L, Simkins S, Paustian K, Rosswall T (1991) Soil denitrification in three cropping systems characterized by differences in nitrogen and carbon supply. I. Rate-distributed frequencies, comparison between systems and seasonal N losses. Plant Soil 138:257–271
Swift RS (1996) Organic matter characterization. In: Methods of soil analysis. Part 3. Chemical methods, SSSA book series no. 5. Science Society of America, Madison, pp 1011–1069
Szajdak L (1996) Impact of crop rotation and phenological periods on rhodanese activity and free sulfuric amino acids concentrations in soils under continuous rye cropping and crop rotation. Environ Int 22:563–569
Szajdak L, Matuszewska T (2000) Reaction of woods in changes of nitrogen in two kinds of soil. Pol J Soil Sci 33:9–17
Szajdak L, Österberg R (1996) Amino acids present in humic acids from soils under different cultivations. Environ Int 22:331–334
Szajdak L, Sokolov G (1997) Impact of different fertilizers on the bound amino acids content in soils. Int Peat J 7:29–32
Szajdak L, Życzyńska-Bałoniak I (1994) Phenolic acids in brown soils under continuous cropping of rye and crop rotation. Pol J Soil Sci 27:113–121
Szajdak L, Życzyńska-Bałoniak I (2002) Influence of mid-field afforestation on the changes of organic nitrogen compounds in ground water and soil. Pol J Environ Stud 11(1):91–95
Szajdak L, Ryszkowski L, Życzyńska-Bałoniak I, Matuszewska T (1998) Wpływ wieloletniej uprawy monokulturowej żyta na transformację materii organicznej gleby (Impact of long-term continuous cropping of rye on the transformation of organic matter). Zesz Probl Postep Nauk Rol 460:121–130 (in Polish)
Szajdak L, Wegner K, Matuszewska T (2000) Effects of cropping systems on chemical and thermal properties of humic acids in soil. Allelopath J 7(2):235–242
Szajdak L, Jezierski A, Cabrera MT (2003) Impact of conventional and no-tillage management on soil amino acids, stable and transient radicals and properties of humic and fulvic acids. Org Geochem 34:693–700
Szajdak L, Życzyńska-Bałoniak I, Meysner T, Blecharczyk A (2004) Bound amino acids in humic acids from arable cropping systems. J Plant Nutr Soil Sci 167:562–567
Trojanowski J (1973) Przemiany substancji organicznych w glebie (Transformation of organic substances in soil). PWRiL, Warszawa, p 331 (in Polish)
Truszkowska W, Cieśla J, Dorenda M, Kania T (1980) Studies on black root in winter wheat and rye under conditions of continuous cropping. Rocz Nauk Rol Ser E T 10:119–134 (in Polish)
Vance ED, Chapin FS (2001) Substrate limitations to microbial activity in taiga forest floors. Soil Biol Biochem 33:173–188
Vancura V (1967) Root exudates of plant. III. Effects of temperature and "cold shock" on the exudation of various compounds from seeds and seedlings of maize and cucumber. Plant Soil 27:319–328

Von Hofe E, Schmerold I, Lijiński W, Jeltsch W, Kleihus P (1987) DNA methylation in rat tissues by series of homologous aliphatic nitrosamines ranging from N-nitrosodimethylamine to N-nitrosomethyldodecylamine. Carcinogenesis 8:1337–1341

Wasilewska L (1979) The structure and function of soil nematode communities in natural ecosystems and agrocenoses. Pol Ecol Stud 5:97–145

Wicke H, Urban G (1988) Wirkung der Fruchtfolgegestaltung auf die Getreideertrage 1976 bis 1985 in Kohnzentrationsversuch Seehausen. Tag. Ber., Akad. Landwirtsch. – Wiss. DDR, Berlin 261:91–98

Witkowski T, Zamszyn Z (1986) Occurrence of Nematoda in soils under continuous cropping of rye and crop rotation. In: Myśków W, Kuś J, Kamińska M (eds) Ekologiczne skutki monokulturowej uprawy zbóż (Ecological effects of continuous cropping of rye). Instytut Uprawy Nawożenia i Gleboznawstwa, Puławy, pp 134–146 (in Polish)

Wójcik-Wojtkowiak D, Politycka B, Schneider M, Perkowski J (1990) Phenolic substances as allelopathic agents arising during the degradation of rye (Secale cereale) tissues. Plant Soil 124:143–147

Zawiślak K, Adamiak J, Gawrońska-Kulesza A, Pudełko J, Blecharczyk A (1991) Yield of main cereals and maize in continuous cropping. Plonowanie podstawowych zbóż i kukurydzy w monokulturach. In: Ryszkowski L, Karg J, Pudełko J (eds) Ekologiczne procesy w monokulturowych uprawach zbóż (Ecological processes in continuous cropping of rye). Zakład Badań Środowiska Rolniczego i Leśnego PAN, Poznań, pp 197–231 (in Polish)

Życzyńska-Bałoniak I, Szajdak L (1992) Seasonal changes of free amino acids content in soils under continuous cropping of rye and crop rotation. In: Bałazy S, Ryszkowski L (eds) Produkcja pierwotna, zasoby zwierząt i wymywanie materii organicznej w krajobrazie rolniczym (Primary production, animal resources and leaching of organic matter in agricultural landscape). Zakład Badań Środowiska Rolniczego i Leśnego PAN, Poznań, pp 7–15 (in Polish)

Życzyńska-Bałoniak I, Szajdak L (1993) The content of bound amino acids in soil under rye monoculture and Norfolk crop rotation in different periods of plants development. Pol J Soil Sci 26:111–117

Życzyńska-Bałoniak I, Czemko M, Mendelewski P (1986) Biochemical differentiation of soil in relation to crop rotation pattern. In: Myśków W, Kuś J, Kamińska M (eds) Ekologiczne skutki monokulturowej uprawy zbóż (Ecological effects in continuous cropping of rye). Instytut Uprawy Nawożenia i Gleboznawstwa, Puławy, pp 54–66 (in Polish)

Chapter 3
Conversions and Pathways of Organic Carbon and Organic Nitrogen in Soils

Stanisław Kalembasa and Dorota Kalembasa

Abstract In mineral cultivated soils in the aspect of agricultural utilization, the content of soil organic matter is one of the most important parameters which influences the physical, chemical, and biological properties. The turnover of organic matter in soil and especially its basic elements carbon and nitrogen are the main investigated elements in the aspect of their dynamics. Full understanding of their turnover is possible by the application of sequential fractionation methods in which mobile and active fractions of those elements are separated. In this chapter we present a few groups of organic carbon compounds present in soil with the possible ways of their turnover (immobilization–mineralization process) as well as some chemical methods which are used for the separation of selected biologically active compounds (humic and fulvic acids, polysaccharides, organic acids, phenols) or further degradation methods like chemolysis, susceptibility for oxidation, and fractional separation with the application of different chemical reagents ($KMnO_4$, H_2SO_4, $K_2Cr_2O_7$). The nitrogen element in soils is mainly organic compound in form (95 %) but is available for plants (NH_4^+ , NO_3^-) for only 5 % of the total content. Dynamics and direction of the turnover processes of organic nitrogen compounds in soils on the big scales depend upon the quality and quantity of those compounds. This chapter contains discussion about the turnover of protein, amino acids, and amino sugars, nitrification and denitrification processes, as well as isolation of different groups of organic nitrogen compounds by chemolysis. Also a new method of sequential fractionation of organic matter from mineral soils is proposed in which through fractional separation the amount of carbon and nitrogen can be determined. For the agricultural utilization especially of fertilizers, the possibility of calculating the amount of nitrogen mineralization during vegetation period from so-called easily mineralized forms of organic nitrogen compounds is proposed, which helps to calculate the dose of nitrogen applied as a fertilizer. This step in the equilibrium of nitrogen in soils prevents pollution of the environment.

Keywords Carbon • Nitrogen • Turnover • Bioactive forms • Compounds • Fractionation • Available forms

S. Kalembasa (✉) • D. Kalembasa
Siedlce University of Natural Science and Humanities,
Str. B. Prusa 14, 08-110 Siedlce, Poland
e-mail: europrof171@googlemail.com

L.W. Szajdak (ed.), *Bioactive Compounds in Agricultural Soils*,
DOI 10.1007/978-3-319-43107-9_3

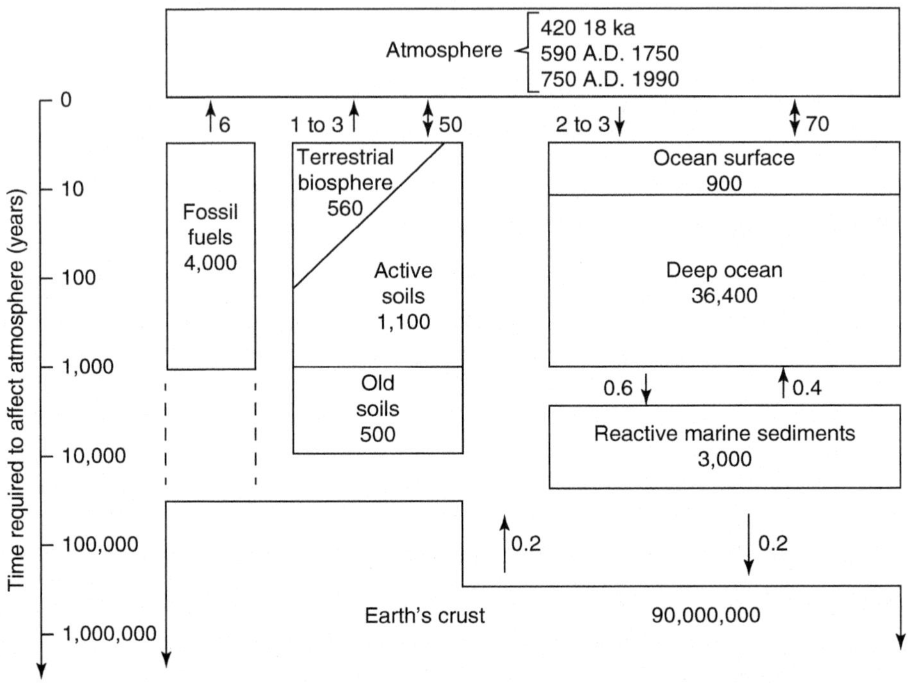

Fig. 3.1 Principal reservoirs and fluxes (*arrows*) in the global carbon cycle. Vertical placements relative to scale on the *left* show approximate time scales required for reservoirs and fluxes that affects atmospheric CO_2. *Double arrows* represent bidirectional exchange. *Single arrows* to and from the atmosphere are approximate estimates of anthropogenic fluxes for 1990. Terrestrial uptake of anthropogenic CO_2 is likely but not shown because of large uncertainties (From Sundquist 1993)

Carbon and nitrogen are two basic elements that constitute the components of proteins and are thus called "elements essential for life." In living organisms, the role of these elements is marked in the supra- and subsurface soil zones. Carbon and nitrogen, in particular, that are found in soil originate from external sources, and the conversion of these elements in the soil environment determines many physical, chemical, and biological soil properties (Mazur 1991; Freibauer et al. 2004; Gonet and Markiewicz 2007; Tan 2011; Kalembasa and Kalembasa 2015).

In relation to the title of the monograph, this section will present the scientific reports and the results of studies on labile forms since, in the context of biodynamics, these compounds are the most important.

Organic matter is a component of each soil type, and its quantity and proportion in the total soil mass depend on a variety of factors such as climatic zone, soil, temperature and precipitation, method of cultivation, and plant species (Lal 2004; Liaudanskiene et al. 2013).

The circulation of carbon and nitrogen is presented in Figs. 3.1 and 3.2.

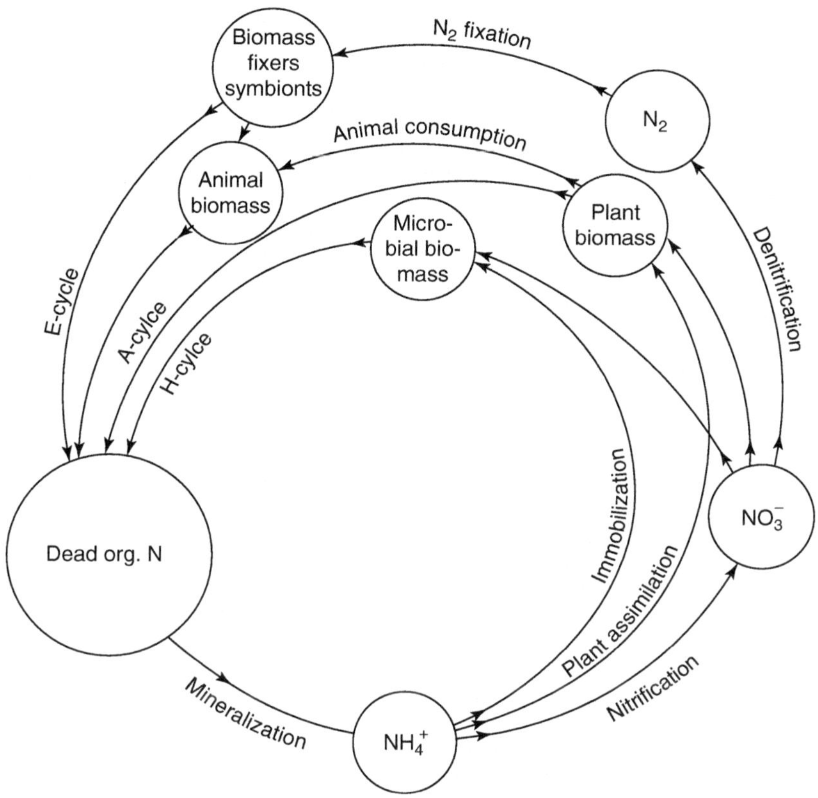

Fig. 3.2 The universal N cycle divided into its three subcycles: the elemental (*E*), the autotrophic (*A*), and the heterotrophic (*H*) (Jansson and Persson 1982)

3.1 Distribution and Sequencing of Organic Carbon Complexes

From an agricultural point of view, it is not only the total content of organic matter expressed as the quantity of carbon in organic complexes that is important but also the proportion of carbon in the individual fractions that are, to a different degree, prone to mineralization at relatively further stages of the synthesis. In soil, carbon is incorporated into inorganic and organic complexes. The quantitative ratio between these two groups of compounds depends on a variety of factors (Rees et al. 2005; Tan 2005) and, in general, does not undergo any major changes on the globe; however, in recent years there have been a growing number of reports on the increase of CO_2 concentration in the atmosphere, which may cause the so-called greenhouse effect. According to IPCC (2007), 4.1 10^9 tonnes are released annually into the atmosphere.

It should be strongly emphasized that organic matter in soil plays an important role in the total circulation cycle of carbon that is found in organic complexes as the

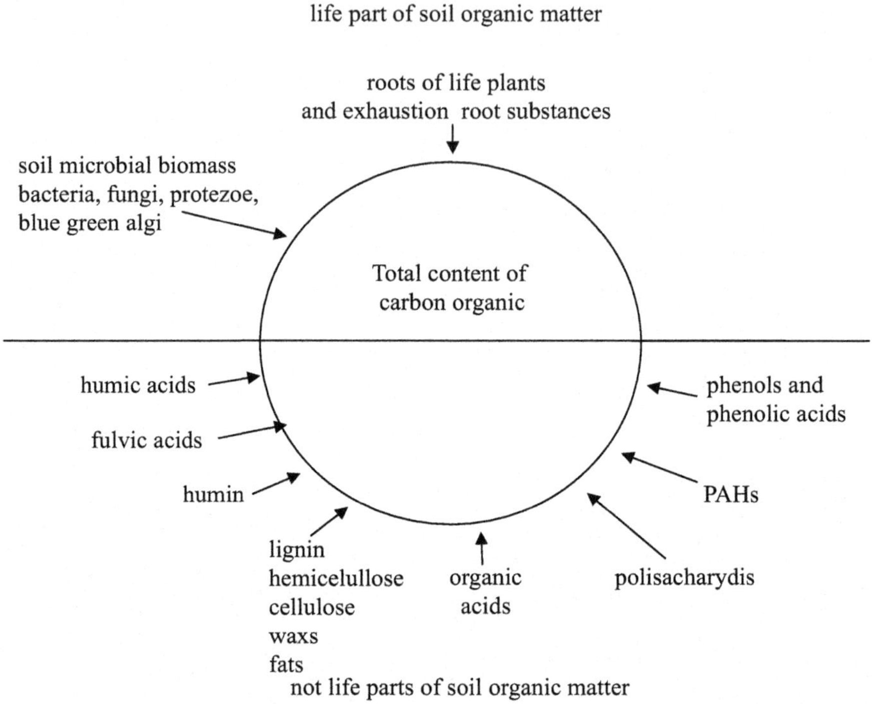

Fig. 3.3 Separated compounds of soil organic matter (abbreviation)

main source of CO_2 in the atmosphere. Bohn (1976) estimated that organic matter in soil contained 30.1×10^{14} kg of carbon in organic complexes and that this quantity was higher than in other compounds found on the globe ($\sim 20.8 \cdot 10^{14}$ kg) (Smith et al. 1993). The distribution of organic soil matter supplies the highest amount of CO_2 into the atmosphere. Organic carbon complexes hidden in deeper soil layers such as hard coal, brown coal, crude oil, deep-sea and deep-ocean sediments, and bottom deposits are obviously more abundant, yet they are less involved in the carbon circulation.

Of this abundant amount of carbon in organic complexes found in the profile of mineral soils, carbon in humic acids (HA), fulvic acids (FA), and humin constitutes 65–75 %. The rest of the carbon is incorporated in polysaccharides, proteins, and other organic compounds as presented in Fig. 3.3.

3.1.1 Humic and Fulvic Acids

Humic and fulvic acids are very important compounds of carbon in soil and often investigated in different conditions. Results of those investigations are published in monograph as well as papers (Flaig 1971; Flaig et al. 1975; Schnitzer 1982;

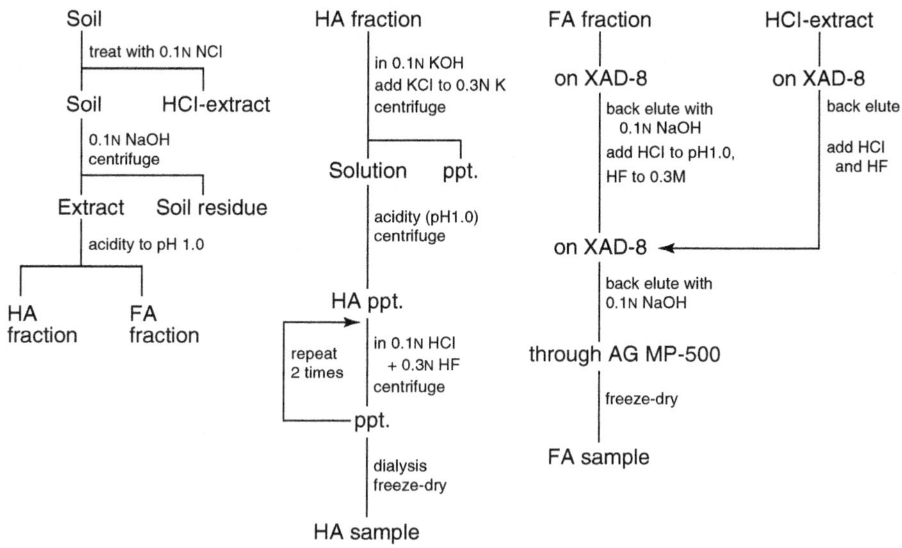

Fig. 3.4 Preparation of humic acid and fulvic acid samples (IHSS method) (Tan 2005)

Stevenson 1994; Sutton and Sposito 2005; Crompton 2012; Klavins et al. 2013). Extraction of humic and fulvic acids from soils was carried out by different reagents like 0.1 and 0.5 mol $NaOH\,dm^{-3}$ and $Na_2P_2O_7$, respectively; by a mixture of $Na_4P_2O_7 + NaOH$ and Na_2CO_3; or by pyridine (Schnitzer 1982; Kalembasa et al. 1999; Tatzber et al. 2007). For the unity of different determinations and investigations of humic and fulvic acids, the method of the International Humic Substances Society (IHSS) is recommended (Fig. 3.4).

The results of these investigations are analyzed in the aspects of physical, chemical, and biological properties. In the view of the turnover of organic carbon and nitrogen compounds in soils, the most important fractions of those elements are in labile and biologically active forms, which is soil participate in quick mineralization–immobilization processes and have stimulate effects on growth and development of plants (Trojanowski 1973) and they are sources of nitrogen for plants (Szajdak and Österberg 1996). Humic substances have also biologically active properties to plants and microorganisms (Trojanowski 1973).

3.1.2 *Soil Microbial Biomass (SMB)*

Microbiological soil biomass is very important in the dynamics of conversions of organic carbon complexes, and its quantity is determined by climatic and soil conditions, species of cultivated plants, and system of cultivation. A variety of methods for quantitative evaluation of microbiological biomass have been developed. Brooks et al. (1985), Beck et al. (1997), Wu et al. (1999), and Joergensen et al. (2011)

reported that microbiological soil biomass might contain up to 5 % of the total carbon and nitrogen in organic complexes. The importance of microbiological biomass consists in rapid and quantitatively diversified processes of its renewal and mortification, which is associated with the creation of organic carbon compounds prone to mineralization and the process of synthesis of more complex organic carbon compounds such as humic acids or other substances with defined structure. The role of microbiological biomass in soil is also related to the fact that it impacts not only the conversions of organic carbon and nitrogen compounds but also, to a high degree, of phosphorus and sulfur (Xu et al. 2013).

3.1.3 Polysaccharides

Polysaccharides in soil are the main source of carbon in very different forms. They are well described in a few monographs (Noltmann 1972; Trojanowski 1973; Tan 1998; Crompton 2012). Waxes and fats are also found mainly in acidic forest soil but in limited, not in big, amount, and therefore they do not play an important role.

3.1.4 Organic Acids

Organic acids such as malate, citrate, and oxalate play an important role in the conversions of organic carbon compounds, especially in the rhizosphere and in acquisition nutricut and metal detoxification (Jones 1998; Jones et al. 2003). Organic acids are produced in greater amounts in the rhizosphere of legumes than in other plants.

All the acidic organic compounds found in soil having a significant impact on the conversion of other compounds and biologically active substances, indole-3-acetic acid (Szajdak et al. 2003; Szajdak and Nowak 2013) and gibberellic acid (Mander 2003), are particularly important.

Szajdak and Marygonova (2007) reported the concentration of IAA in high-moor peats ranged from $124.4 \text{ to } 210\ \mu g \cdot kg^{-1}$ and in low-moor peats the content of IAA was much lower and reached 57.9 to $134.4 \mu g \cdot kg^{-1}$. The relationship between the concentration of IAA in peats and humus substances as well as dissolved organic carbon reached the values from 0.692 to 0.793 $\mu g \cdot kg^{-1}$.

3.1.5 Polycyclic Aromatic Hydrocarbons

Polycyclic aromatic hydrocarbons (PAH) are solid soil components and their content is determined by many factors, and therefore, the evaluation of their threshold content depends on the context of toxicity and the assessment criteria. Maliszewska-Kordybach (1996) reported three criteria for evaluating this toxicity: (1) sum of the contents, (2) number of exceeded reference value for individual PAH (according to Dutch regulations), and (3) initial proposals, as own method, for arable lands in Poland (the sum of PAH content normalized to "average soil" characteristics in the area under investigation). Organic matter content and the <0.02 mm size fraction were chosen as the soil properties most adequate for this purpose.

Maliszewska-Kordybach et al. (2008) have shown that the mean content of PAHs in Polish arable soils reached the value of 395 $\mu g \cdot kg^{-1}$ and with dominance of 4–6 hydrocarbon rings (74 % of total PAHs.) Soil properties affected PAH content to a limited extent. The organic matter content was the only parameter correlated significantly with the content of Σ16 PAHs. Fismes et al. (2002) reported the possibility of the taking up of PAHs by vegetable plants.

3.1.6 Phenols and Phenolic Acids

Lignin degradation in soil is the major source of these compounds. Phenolic acids that are found in soil are mainly composed of ferulic, p-coumaric, p-hydroxybenzoic, and vanillic acids. These acids exert a strong impact on the populations of bacteria, fungi, and actinomycetes and are rapidly degraded by soil microorganisms (Camberdella 2005). Phenolic acids have also an inhibitory effect on the growth and development of plants (Wang et al. 1967). Martens (2002a) reported that phenolic acids released from plants during their decomposition in soil are important elements in many different soil processes and may be entirely extracted from soil with 1 M NaOH. Phenolic acids are also a component of humic acids in soil, which results from one of the hypotheses on the process of humic acid formation (Kalembasa and Tengler 2004), and cause stabilization of the soil aggregates and, therefore, influence the structure of soil (Martens 2002b). The total content of phenol compounds in soil approximates 3–7 $kg \cdot ha^{-1}$ which, calculated per organic soil substance, equals 0.002–0.005 % (Flaig 1971; Whitehead 1964) and depends on many factors (Szajdak and Życzyńska-Bałoniak 1994).

In soil, phenol compounds may originate from three sources:

1. Microbiological degradation of lignin
2. Introduction with plants
3. Microbial synthesis

Higher plants contain phenol compounds that, having been introduced into soil (postharvest residues of plants collected at green, the stage of technological maturity), may be a source of these compounds in soil. The most important substances of this kind include flavonoids such as flavones and flavonols. Moreover, in higher plants, a variety of phenolic acids, e.g., p-hydroxybenzoic acid, vanillic acid, and protocatechuic acid, are found in glycoside complexes.

A comprehensive discussion of the conversions of phenyl substances generated by microbial synthesis until the 1970s was reported by Trojanowski (1973).

Phenol compounds exert a physiological effect on the growth and development of plants. Flaig (1968a) demonstrated, with the use of ^{14}C, the uptake of phenolic acids (p-hydroxybenzoic, vanillic, and syringic; generated by lignin degradation) by plants.

It should be, however, emphasized that physiological effects of biologically active substances are positively manifested if there is a lack of optimal conditions for the growth and development of plants (Michalak 2006).

Phenolic acids that are found in soil most often include ferulic acid, p-coumaric acid, p-hydroxybenzoic acid, and vanillic acid. These acids impact bacterial, fungal, and actinomycetal populations and are rapidly degraded by soil microorganisms. Moreover, phenolic acids act as plant growth inhibitors.

Apart from the stimulating effect, phenolic acids also have a toxic action. Toussoun et al. (1968) found that phytotoxic phenol acids such as benzoic, phenylacetic, and p-coumaric were present during the degradation of plant residues. The problem with toxic effects of phenol compounds is seen on farms where cereals are predominant in the structure of crops. Guenzi and McCalla (1966) detected that the quantity of ferulic acid and p-coumaric acid in an extract of pyrophosphate or 2 M NaOH (the solution used to extract humic substances from soil) was two times higher in the soil fertilized with straw than in the control object.

Ohno (2001) demonstrated that phenolic acids originating from legume green mass might inhibit weed growth by means of allelopathic mechanism. However, oxidation of phenolic acids may significantly limit their allelopathic action.

Batistic and Mayaudon (1970) showed with the use of ^{14}C that 15 % went to the fraction of fulvic acids, 41 % to humic acids, and 26 % to humins during degradation of plant material under aerobic conditions in soil.

In the context of carbon conversions in the environment, it is not only its total content that is important but also the quantity and quality of compounds that undergo oxygenation (mineralization) and synthesis of novel substances (polymerization). Many attempts are thus being undertaken to extract organic carbon compounds into different groups with varied lability and proneness to conversions as an important indicator of the changes related to the agricultural use of soils. While developing chemical trials, it was assumed that organic carbon compounds are found in nature and their identification consists in determining the proneness of organic matter to oxidation processes (mineralization) that occur under natural conditions. In addition, these biologically active compounds may impact the growth and development of plants and soil microorganisms. This issue is also important in relation to the role of carbon compounds (particularly CO_2 and CH_4) in the climatic changes that have been observed in recent years and the implications for organic soil matter. Further studies into organic conversions of carbon compounds and their vulnerability to oxidation in different ecosystems are warranted (Chan et al. 2001; Weil et al. 2003; Ryan et al. 2009).

It is associated with implementing more intensive methods of soil cultivation into the agricultural production and with a lack of balance between the amount of organic matter introduced into soil and its quantity that undergoes mineralization as a source of nutrients to plants (Salter and Green 1933). Since this source is supplemented with fertilizers that, to a major or minor degree, are mineralized and immobilized depending on the activity of soil microorganisms (soil microbiological biomass), some researchers claim that understanding the turnover rate of soil microbiological biomass (SMB) is crucial for comprehending the causes of changes in the content of organic matter in soil (Ocio et al. 1991), while others (Mazzarino et al. 1987) believe that SMB is an insufficient indicator of these changes since such factors as microbial species in SMB and soil temperature and humidity significantly influence the amount of SMB.

The continuous and increasing interest in research into the lability of organic carbon compounds (and nitrogen compounds) results from the fact that by determining the total content of these elements in soil, it is difficult to identify the changes in their concentration as their high and relatively stable content does not permit a complete interpretation of the changes in soil, especially when their natural variability is assumed. It is thus attempted to sequence different fractions that may be used as a good indicator of compound changes in soil (Jenkinson and Rayner 1977).

In soil, organic matter is an indicator of numerous properties in the context of soil usability in agricultural production and environmental protection. The structure and characteristics of organic matter are related to complex compounds which determine the physical, chemical, and biological properties of soil. Studies have been thus performed and are still being conducted in order to determine these structures and their impact on other soil properties (Myśków 1981; Myśków et al. 1996).

As carbon is the main component of soil organic matter, therefore the majority of papers have focused on this element, both quantitatively and qualitatively. In the context of soil bioactivity and its impact on other soil organisms and plants, carbon in soil is divided into several groups depending on the properties and conversion rates for the individual compounds, which determines their stability in soil. These facts have generated the following division into groups:

1. Carbon compounds that are easily degraded during the vegetation season
2. Stabilized carbon compounds (in their physical and chemical properties) that form complexes with the mineral soil fraction and undergo very slow conversions
3. Carbon complexes that constitute the non-labile fraction and are degraded over hundreds of years, which is confirmed by the data reported by Jenkinson (1966) and Brunn et al. (2010) from the studies with ^{13}C

Jenkinson and Rayner (1977) developed the model starting from decomposable pool with a radiocarbon age of less than 1 year through a biomass pod at 25.9 years to a chemically stabilized pool with a radiocarbon age of 25.65 years. Variations on those pools have been used in other organic matter cycling models (Hunt 1977; Smith 1979).

Studies on the biological activity of organic carbon compounds have been conducted in two directions:

- Biological–vegetation studies: extraction of different bindings of organic carbon complexes from soil and their use in model studies on plants as pot and field experiments.

- Chemical studies based on the possibility of oxidizing organic carbon compounds found in soil; they consist in using chemical reagents with different oxidizing capacity. Łoginow et al. (1987) used $KMnO_4$ at different concentrations (33–333 mM) assuming that in soil the conversion of organic carbon compounds is associated with the enzymatic nature of their oxidation. Similarly, Blair et al. (1995) used $KMnO_4$ at a constant concentration of 333 mM as an oxidizing reagent in order to determine the quantity of carbon in easily mineralized complexes. Tirol-Padre and Ladhe (2004) reported that carbon in organic compounds that was oxidized with a neutral $KMnO_4$ solution was correlated significantly higher ($P = 0.01$) with the total carbon content than with the amount of carbon in aqueous soil extract ($P < 0.05$) and no significant correlation with carbon in soil microbial mass. Carbon in organic complexes oxidized by $KMnO_4$ thus cannot be a reliable indicator of the quantity of labile carbon in soil and should not be recommended as an indicator that characterizes the fractions of labile carbon.

In addition, for the total carbon in organic soil matter, fractioning was performed by extracting humic acids (Schnitzer 1982) in dissolved organic carbon (Cook and Allan 1992; Zsolnay and Garlitz 1994) or in 0.01 M $CaCl_2$ (Zsolnay 2003) separated of particle size (Christensen 1986), natural ^{13}C abundance (Balestend et al. 1990; Lefroy et al. 1993), and microbiological biomass C (Sparling 1992). Lefroy et al. (1993) modified the method by Łoginow et al. (1987) by introducing one concentration of $KMnO_4$ and claiming that it was sufficient to characterize the lability of organic carbon compounds. The use of low $KMnO_4$ concentrations results in non-repeatable results, even though this compound is a strong oxidizer (although only in neutral environment) and becomes labile when introduced into soil. Moreover, an aqueous $KMnO_4$ solution is a rather labile reagent and its concentration changes rapidly. The abovementioned difficulties prompted us to oxidize the selected fractions of soil organic matter with an aqueous $K_2Cr_2O_7$ solution (Kalembasa and Kalembasa unpublished results).

The methods of fractioning and sequencing organic soil substances have been discussed in numerous publications (Schnitzer 1982; Stevenson 1994). Initially, the objective of these studies was to sequence these substances and characterize them in a cognitive sense, while the publications in this subject area that have been released in recent years have been directed at organic carbon and nitrogen conversions in the context of plant production. In this case, the majority of research into carbon and nitrogen conversions in soil is divided into two groups (McLauchlan and Hobbie 2004):

1. Carbon and nitrogen compounds that are easily mineralized
2. Non-labile carbon and nitrogen compounds

Ad. (1) This fraction is thought to be an important indicator of soil quality (Haynes 2005; Kolář et al. 2009; Strosser 2010). This is determined with different methods such as carbon compounds soluble in cold or hot water; compounds extracted in different salt solutions that extract soluble proteins, amides, and amino sugars; and

SMB, hemicelluloses, carbohydrates, and compounds that are mineralized with CO_2 release under aerobic conditions. Other strongly oxidizing chemical solutions introduced included $KMnO_4$ or $K_2Cr_2O_7$ (Blair et al. 1995; Chan et al. 2001; Rovira and Vallejo 2002, 2007).

Exposing soil samples to concentrated mineral acids, e.g., H_2SO_4 at a concentration of 6, 9, or 12 M $H_2SO_4\,dm^{-3}$, is a more drastic method of extracting carbon fractions (Walkey 1947; Plante et al. 2006).

Different ways and recommended methods of organic matter fractioning have prompted a division into two groups:

1. The degree of stability of plant residues introduced into soil and organic matter in soil which is susceptible to mineralization in the context of providing plants with nutrients. According to many authors (Rovira and Vallejo 2000, 2007; Shiroto and Yokozawa 2006), for this purpose, fractioning can be done in three groups using 1 M and 2.5 M H_2SO_4.
2. The degree at which organic matter is susceptible to oxidation. Some authors (Tirol-Padre and Ladhe 2004) recommend using a neutral solution of 33 mM $KMnO_4\,dm^{-3}$ with the division into two groups or $K_2Cr_2O_7$ in 6, 9, and 12 M H_2SO_4 by sequencing four fractions (Chan et al. 2001).

The organic substances in soil are divided into two fractions: the so-called "light" fraction, i.e., composed of plant and animal parts with different degrees of decomposition which is extracted from soil with a solution of specific density at 1.6–2.0 g cm^{-3}. This fraction contains up to 8 % of total carbon in organic complexes and app. 5 % of total nitrogen. The organic matter of this fraction also includes microbial biomass which constitutes 3–5 % of organic matter (Vance et al. 1987).

Dissolved organic matter contains organic carbon complexes the most mobile and most-rapidly mineralized fraction of organic carbon compounds is the source of energy for microorganisms involved in a variety of soil processes such as nutrient transport. They can also contain organic toxic compounds and are leached into soil profile. Dissolved organic matter is defined as organic compounds, found in the soil solution, which pass through 0.45 μm filters (Herbert and Bertsch 1995; Zsolnay 2003).

To sum up, it may be concluded that the studies on stability and fractioning of organic matter in soil are conducted in two directions:

- Its quantitative extraction and determination of organic compounds (proteins, amino acids, lipids, etc.)
- Extracting the individual fractions in relation to varied stability in acid hydrolyzates or susceptibility to oxidation

The results of analytical publications are valuable provided they can be used to develop models, e.g., the century model (Gijsman et al. 2002), and predict the conversions and construct indexes. In the case of papers on labile carbon forms, a variety of indicators have been developed that are used in practice. Blair et al. (1995) proposed several indices such as carbon management index (CMI), pool size index (CPI), and lability index (LI). In order to determine the values of these indices, it is necessary to identify the total carbon content in organic complexes and the quantity of carbon in the complexes of the labile fraction (LC) that are oxidized with $KMnO_4$ (333 mM dm^{-3}) in a neutral environment (Łoginow et al. 1987) and the fraction that is not oxidized, the so-called non-labile fraction of carbon complexes (NLC). The quotient of these fractions determines the lability of organic carbon compounds in the soil. The value of carbon lability in soil in relation to the lability of carbon complexes in the reference soil (i.e., having natural lability of carbon compounds) is described by the lability index (LI) for carbon and is expressed in percentage values. The pool size index (CPI) is the quotient of total content of carbon in organic complexes found in cultivated soil in relation to the reference soil. The value of carbon management index (CMI) is the product of pool size index (CPI) and lability index (LI) multiplied by 100.

The use of organic wastes and composts as well as postharvest residues has a substantial impact on the dynamics of the conversions of organic carbon and nitrogen complexes. The application of these waste products at doses containing nutrients (particularly nitrogen) above required levels has a profoundly negative influence in the context of soil biochemistry.

The ultimate objective of research into the fractioning of organic carbon compounds is to determine their impact on the growth and development of plants and microbiological activity of soil.

3.2 Distribution and Sequencing of Organic Nitrogen Complexes

The total content of nitrogen in soil ranges from 0.02 to 2.5 %, of which nitrogen in organic complexes constitutes 90 % (Table 3.1), as presented (Stevenson 1982; Schulten and Schnitzer 1998; Szajdak and Życzyńska-Bałoniak 2002).

There is thus a significant positive correlation between the content of carbon and nitrogen in organic complexes, which results in a similar distribution of these elements in the soil profile and the concentration decreases in its lower layers. Based on these relations, it is assumed that nitrogen constitutes 5 % of the organic matter in the soil surface layer. The factors influencing the accumulation of organic compounds in soil also facilitate the accumulation of nitrogen and vice-versa.

Based on these relations, a parameter has been created in soil science and chemical–agricultural studies: this is the C/N ratio determining two major processes in soil that significantly impact the content of these elements, i.e., mineralization (a decrease in the content) or sorption (which increases the content).

The advantage of organic compounds over mineral forms in the total content of this element is the reason behind a substantial number of studies that have aimed at explaining the conversion of organic carbon and nitrogen compounds as well as mineral nitrogen complexes in soil (Ladd and Jackson 1982); all the more so because the current studies have demonstrated that plants can uptake only NH_4^+ and NO_3^-.

Table 3.1 Geochemical distribution of nitrogen on the earth

Nitrogen pool	Nitrogen mass ($\times 10^{20}$ g)	Percentage of total nitrogen mass
Atmosphere	38.0	1.95
Lithosphere		
Igneous rocks	1930.0	97.86
Sedimentary rocks	4.0	0.20
Biosphere		
Living organisms	0.00038	
Terrestrial organic nitrogen	0.0082	0.02
Ocean-bottom organic nitrogen	0.0054	
Total nitrogen mass	1972.0	100.00

Table 3.2 Fractionation of soil N based on acid hydrolysis

Form	Definition and method	% of soil N (usual range)
Acid-insoluble N	Nitrogen remaining in soil residue following acid hydrolysis; usually obtained by difference (total soil N-hydrolyzable N)	20–35 %
NH_3-N	Ammonia recovered from hydrolyzate by steam distillation with MgO	20–35 %
Amino acid N	Usually determined by the ninhydrin-CO_2 or ninhydrin-NH_3 methods; recent workers have favored the latter	30–45 %
Amino sugar N	Steam distillation with phosphate–borate buffer at pH 11.2 and correction for NH_3-N; colorimetric methods are also available (see text); also referred to as hexosamine-N	5–10 %
Hydrolyzable unknown N (HUN fraction)	Hydrolyzable N not accounted for as NH_3, amino acids, or amino sugars; part of this N occurs as non-α-amino N in arginine, tryptophan, lysine, and proline	10–20 %

Bremner (1965, 1967)

Studies into diagnosing, characterizing, and describing the importance of organic complexes were undertaken after acid hydrolysis of the soil samples with 6 mol HCl dm^{-3} (Bremner 1965, 1967). An analysis of the acidic hydrolyzate of the soil samples demonstrated an approximate proportion of nitrogen in the different organic fractions: like amino acids and amino sugars (Table 3.2).

Nitrogen in organic compounds found in soil is a mixture of these compounds incorporated in the microbial biomass, plant and animal residues, and humic substances. These materials (the groups of organic matter) have different vulnerability to conversion in soil and to acid hydrolysis. Microbial biomass is most active and prone to these processes.

The current state of knowledge on organic nitrogen compounds found in soil was generated with the application of acid hydrolysis and with subsequent identification and quantitative analysis (Bremner 1965; Yonebayashi and Hattori 1980; Preston 1982).

The process of hydrolysis with 6 mol HCl dm^{-3} was adopted from the methodology of protein analysis and determination of amino acids in plant and animal materials. Therefore, it has some limitations when applied for soil due to its mineral fraction. It is also supposed that artifacts are generated during hydrolysis of soil in an acidic environment, which was proven by Asami and Hara (1970) by adding glucose to soil samples before hydrolysis. The hydrolyzate was found to contain less nitrogen, while its content increased in the non-hydrolyzing fraction. As demonstrated by Cheng and Kurtz (1975), an addition of HF acid into a soil sample before acid hydrolysis increases the proportion of nitrogen in a hydrolyzate, which probably results from a release of amino acids and other nitrogen compounds that were not deaminated during hydrolysis. Griffith et al. (1976) also found that the proportion of nitrogen in the soil fraction after hydrolysis may be released by treating it with mild acids and bases. This procedure causes an increase in the content of hydrolyzing nitrogen of up to 10 % of the total nitrogen content. During hydrolysis, a significant proportion of the nitrogen in ammonia form from the sorption complex and nitrogen from free amino acids and amides passes into a hydrolyzate (Bremner 1967).

According to Łoginow (1967), some of the nitrogen (as ammonia) may originate from humic acids in a buffered environment. It is also possible that ammonia is formed during the reaction of protein substances with phenols in the presence of phenyloxidase (Haider et al. 1965).

It is thus concluded that the correctness of hydrolysis of soil organic matter containing organic nitrogen complexes requires further additions and discussions.

The method of two-stage sequential acid hydrolysis is another approach toward extraction and identification of nitrogen compounds found in soil (Kalembasa 1995). This method is based on sequencing a fraction with a varied group of nitrogen compounds instead of an individual group of homogenous compounds by using acid hydrolysis at different concentrations of sulfur acid (VI). However, there is the potential to identify specific groups of nitrogen compounds and provide a quantitative evaluation of the total nitrogen and carbon content in the individual hydrolyzates (Fig. 3.5).

The possibility of determining the content of carbon in organic complexes in the individual hydrolyzates with the oxidative–reductive method and $K_2Cr_2O_7$ solution (for oxidation of organic carbon compounds) is the considerable advantage of using H_2SO_4 instead of HCl (Kalembasa and Kalembasa 1986, 1992). This results from the lack of Cl^- ions in a hydrolyzate, as their presence forms (with $K_2Cr_2O_7$) CrO_2Cl_2 which binds Cr^{+6} from $K_2Cr_2O_7$, causing erroneous and excessively high results.

The methods for investigating the dynamics of mineral and organic nitrogen compounds must be strictly defined by the environmental conditions and sampling time, which results from high variability of these compounds over time.

A similar procedure is based on this method with sulfuric acid used for the fractionation of nitrogen compounds in organic soils (Kalembasa and Becher 2009) as well as nitrogen and carbon in mineral soils (Becher and Kalembasa 2011).

The inability of plants to uptake nitrogen in organic compounds and the substantial time and spatial variability of mineral nitrogen forms in the superficial layers

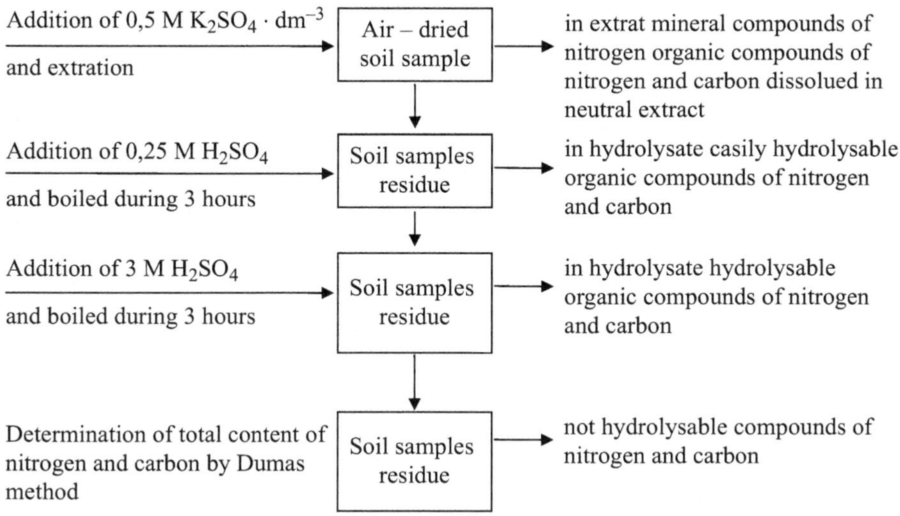

Fig. 3.5 Scheme of sequential acidic hydrolysis method

and levels of the soil profile have prompted studies on the development of methods that would allow for detecting the amount of nitrogen in the forms absorbable by plants in cultivated and forest soil types. The N_{min} test (Fotyma and Fotyma 2004) is used to determine the quantity of nitrogen in absorbable forms in a soil profile of up to 90 cm deep. Despite being recommended for agriculture, it does not determine variability in the quantities of absorbable forms throughout the vegetation season due to using only a single sampling in autumn; furthermore, it does not include the changes in the content of this element that are the result of mineralization–immobilization during the vegetation season. It is very difficult to determine the quantity of nitrogen in sown soil within the vegetation season due to the changes in chemical (mainly the oxidative–reductive processes), physical (water and temperature), and biological (organic substances in soil and activity of enzymatic reactions) processes.

These factors have prompted development of a variety of methods that stimulate the natural processes occurring under field conditions.

These methods can be divided into two groups:

- Biological
- Chemical

Mineralization and immobilization of nitrogen compounds in soil occur parallel to each other and determine the quantitative and qualitative composition of organic nitrogen and carbon compounds (Jansson and Persson 1982). In order to entirely comprehend the direction and rate of changes, it is necessary to use not only nitrogen compounds with elevated ^{15}N quantity but also the ^{14}C isotope. It should be emphasized that, at the current state of knowledge and analytical laboratory practice, in studies on nitrogen compound conversions in soil and plants, regardless of their objectives, it is necessary to use nitrogen compounds enriched with the ^{15}N isotope. The technique of these studies consists in isotopic dilution (Kalembasa 1995).

The studies on mineralization–immobilization are methodologically complex and (moreover) lengthy due to the impact of numerous factors that may be grouped in the following areas:

- The chemical structure of organic nitrogen and carbon compounds in the soil organic matter
- The activity of microorganisms involved in the conversion of organic matter
- The rate and range of nitrogen conversions under the influence of nitrogen added in a mineral or organic form as well as salts of other elements
- The impact of inhibitors of the processes occurring in soil as mineralization–immobilization
- The impact of temperature and aqueous–aerial conditions in soil
- The type of soil and methods of its use

The complexity of conversions and the main stages of mineralization–immobilization may be schematically presented as:

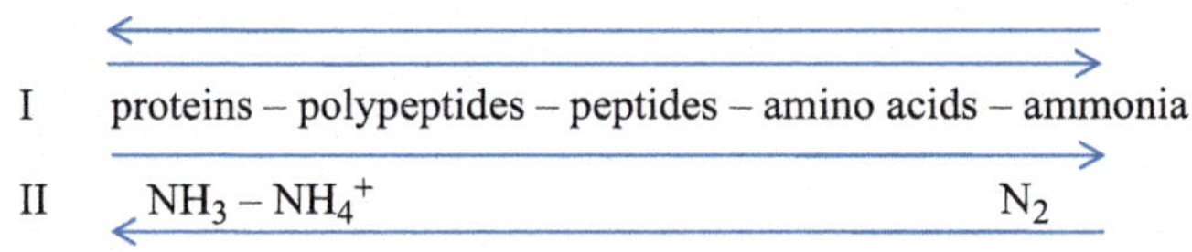

The above diagram presents two brief stages:

- Conversions of organic nitrogen compounds
- Conversions of mineral nitrogen compounds

The balance in stage I is usually slightly moved to the benefit of immobilization process, which favors the accumulation of a substantial part of the nitrogen in organic compounds that are relatively resistant to humus mineralization. Obviously, determining the content of humus in soil is impacted by a variety of other factors, such as physical, chemical, and biological factors.

The basic factor is, however, the carbon-to-nitrogen (C/N) ratio which determines the direction and nature of microbiological conversions. During conversions occurring in the soil, microorganisms use organic carbon complexes as energy and building materials. The ratio of the quantity of carbon extracted as CO_2 during decomposition to the total used carbon (E) depends on the species' microbial structure and environmental conditions in which the conversions take place, which makes this value variable. Assuming the carbon-to-nitrogen ratio in microbial biomass and decomposing organic matter as C_m: N_m and C_r: N_r, the transformation coefficient W_t can be calculated that describes the balance between nitrogen mineralization and immobilization:

$$W_t = \frac{C_m : N_m}{(1-E) \cdot C_r : N_r}$$

The transformation coefficient has different values: when $W_t = 1$, nitrogen appears in the environment in an amount that is necessary for the correct C/N ratio for microbial biomass (i.e., it is necessary for its growth). If the value is higher than 1, nitrogen in a mineral form is at an excessive amount in relation to the needs of microorganisms, which means that nitrogen is released from decomposing mass and can be uptaken by plants. A value of this coefficient lower than 1 indicates a lack of nitrogen in the mineral form in the environment, which is manifested as an advantage of immobilization and a deficit of nitrogen for plants.

Apart from the impact of C/N ratio on the quantity of nitrogen in the mineral form available to plants, the aqueous properties, redox, and temperature also have a big influence. Wide-spectrum studies on the impact of temperature and humidity in soil (Stanford et al. 1973,1974; Fotyma et al. 1987) have demonstrated that mineralization of organic nitrogen compounds in soil occurs according to the van't Hoff's rule specifying that an increase in environmental temperature by 10 °C increases the process of mineralization by 100 %.

The conversions of proteins in soil are, in the first stage, associated with their gradual degradation starting from two-stage hydrolysis that finally produces amino acids. The total content of amino acids, released from soil protein during acid hydrolysis depends upon several parameters and ranged from 10 up to 150 mg kg^{-1}. The second stage involves degradation of amino acids to ammonia and other inorganic nitrogen compounds.

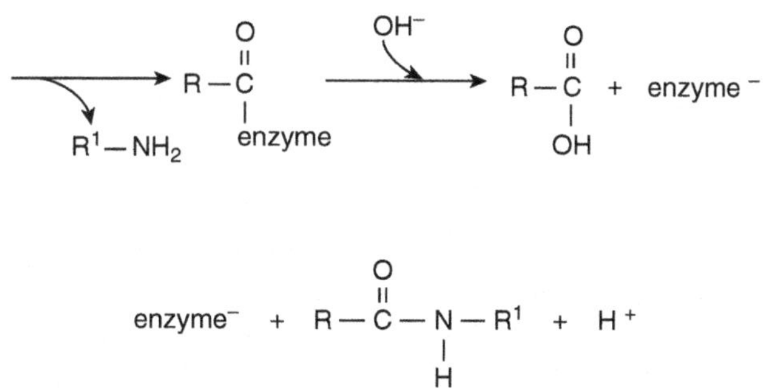

Hydrolysis of proteins that follows the above diagram is, under natural conditions, an enzymatic process, and the necessary set of enzymes (proteinases, polypeptidases, and peptidases) is of microbiological origin. Proteinases and peptidases [EC 3.4], according to their catalytic mechanism, are divided into four groups: (I) serine proteinases EC 3.4.21, (II) SH sulfhydryl proteinases EC 3.4.22, (III) acid proteinases EC 3.4.23, and (IV) metalloproteinases EC 3.4.24. The activity of proteinases decreases together with the depth of soil profile. Measuring the activity of proteinases in soil consists in determining the rate of protein hydrolysis into soil, incubation, and then cessation of hydrolysis (with toluene) with a subsequent quantitative measurement of the products of hydrolysis. Many methods for the evaluation of proteinase activity have been developed, and albumins (Ambroz 1965), hemoglobins (Ladd and Butler 1972), or casein and gelatin (Ambroz 1971) are used as substrates depending on the specific method. Amino acids that are generated with hydrolysis as well as other products of degradation and conversions of different nitrogen compounds constitute the so-called "free amino acid" pool present in soil, and they play an important role in chemical processes. Their quantity and durability depend on numerous physical, chemical, and biological processes in soil (Kalembasa and Niewiński 1990; Szajdak and Österberg 1996; Maysner et al. 2006; Szajdak 2011). Free amino acids may undergo different conversions, e.g., they may be synthesized into protein structures or be degraded (with NH_4^+ release) during deamination reaction.

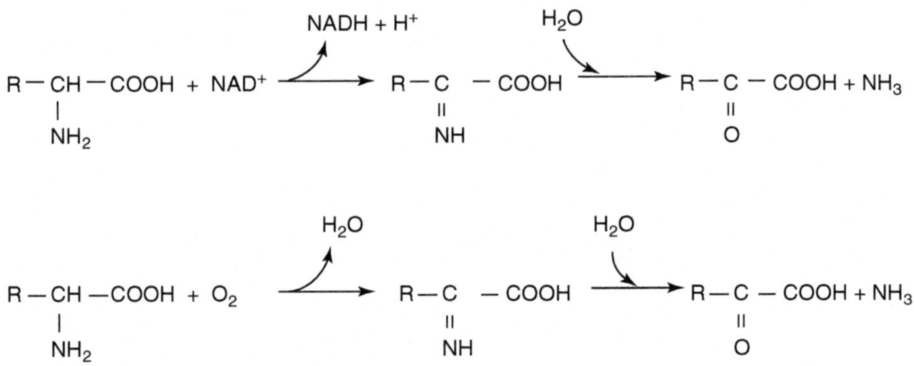

This is the main pathway of degradation of proteins into amino acids. However, amino acids may also be formed in soil as a result of hydrolysis of C-N bindings. This reaction is catalyzed by aminohydrolases [EC 3.51] and amidohydrolases [EC 3.5.3].

Peptidases [EC 3.4.11–15] include the enzymes that hydrolyze a dipeptide with a substrate and the ones that release single amino acids and dipeptide bindings from C-terminus and N-terminus of long peptide chains.

Asparaginase [EC 3.5.1.1] and glutaminase [EC 3.5.1.2] are enzymes that belong to the group of amidohydrolases, and they decompose asparagine and glutamine with the release of NH_4^+ NH_4^+ and aspartate and glutamate according to the scheme:

$$\mathrm{L-asparagine + H_2O + enzymes \rightarrow L-aspartate + NH_4^+ + e}$$

$$\mathrm{L-glutamine + H_2O + enzymes \rightarrow L-glutamine + NH_4^+ + e}$$

This deamination proceeds according to the schemes:

1. Hydrolytic

$$\mathrm{R-CHNH_2COOH + H_2O \rightarrow R-CHOHCOOH + NH_3}$$

2. Oxidative

$$\mathrm{R-CHNH_2COOH + O_2 \rightarrow R-COCOOH + NH_3}$$

3. Reductive

$$R-CHNH_2COOH+H_2 \rightarrow R-CH_2COOH+NH_3$$

In the majority of cases, deamination of amino acids in soil involves biological factors. The reactions presented above demonstrate that the products of amino acid deamination include hydroxyl acids, keto acids, and aliphatic acids. These products are rapidly mineralized to CO_2, which prevents their accumulation in the soil environment. Apart from enzymatic deamination, there are also other pathways of releasing NH_4^+ from amino acids, such as the so-called abiotic deamination represented by, for instance, oxidative deamination catalyzed by polyphenols (Łoginow 1967) according to the scheme:

$$\begin{array}{ccc} H_2O & \text{quinone} & \text{amino acid} \\ \uparrow & \downarrow\uparrow & \downarrow \\ O_2 & \text{polyphenol} & \text{keto acid}+NH_3 \end{array}$$

A simple oxidative and reductive reaction with nitrates (III) that generates particle nitrogen is another example of amino acid deamination:

$$R-CHNH_2COOH+HNO_2 \rightarrow R-CHOHCOOH+H_2O+N_2$$

Obviously, not all amino acids are deaminated in soil, and regardless of their origin, they may be used to build the soil microbial population, which constitutes the reverse of proteolysis, or eventually, the reactions of amino acids with polyphenols that, according to many authors (Flaig 1968b; Trojanowski 1973), are the basis for humification during which specific soil substances with a complex structure and a colloidal, polydispersive nature are formed. This process protects amino acids against deamination and, by incorporating them into humic substances, limits their rapid decomposition which can be separated (Fig. 3.6).

3.2.1 Free Amino Acids

The content of free amino acids is very low and is expressed in $\mu g \cdot kg^{-1}$ of soil. Jones et al. (2002) reported that the concept of free amino acid in the soil solution was low and independent of soil type. The average value of dissolved nitrogen compounds in amino acids is $24 \pm 8mM$, while in $NH_4^+ - 39.4mM$ and for $NO_3^- - 67mM$. Free amino acids contain on average 10–40 % of the total dissolved nitrogen compounds in the soil solution, and therefore, they constitute a considerable source in the total volume of absorbable nitrogen compounds in soil.

Gilbert and Altman (1966) demonstrated a high applicability of 20 % ethanol for extraction of free acidic and neutral amino acids from soil during 18–24 h, yet that method was insufficiently good for base amino acids. The recovery coefficient for

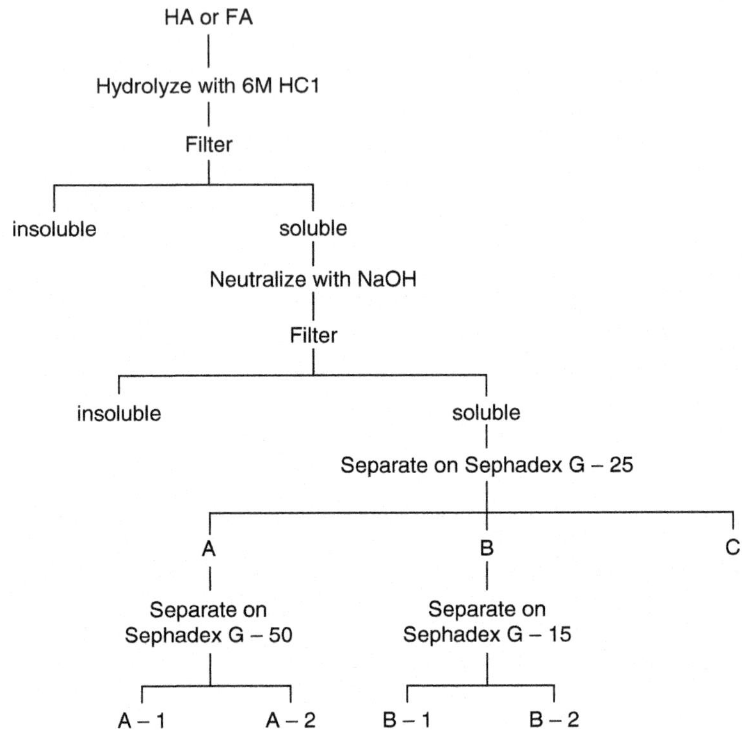

Fig. 3.6 Scheme for separating fractions rich in unidentified nitrogen (Schnitzer 1982)

amino acids added to soil and extracted with 20 % ethanol was 90–95 % for acidic amino acids, 80–85 % for neutral amino acids, and 1–5 % for base amino acids. Werdin–Pfisterer et al. (2009) detected mainly glutamic acid, glutamine, aspartic acid, asparagine, alanine, and histidine in aqueous soil extract, and they constituted app. 80 % of the total amino acid content in soil.

Studies of the presence of amino acids in soil depend on the methods of their extraction (Paul and Schmith 1960); their creation is soil, hydrolysis, or root exudates (Paul and Schmith 1961) and their eventual uptake by plants (Miettinen 1950).

3.2.2 Amino Polysaccharides and Amino Sugars

As shown in Table 3.2, amino polysaccharides and amino sugars are the second-largest group present in an acid hydrolyzate of soil organic matter.

These compounds originate mainly from microorganisms and soil fauna. In nature, amino sugars are found as polymers such as chitin, peptidoglycans, teichoic acids, and other compounds in the cell membranes (Parsons 1981). In soil, there is a large variety of polysaccharides, yet three are predominant: glucosamine, galactosamine, and muranic acid (Fig. 3.7).

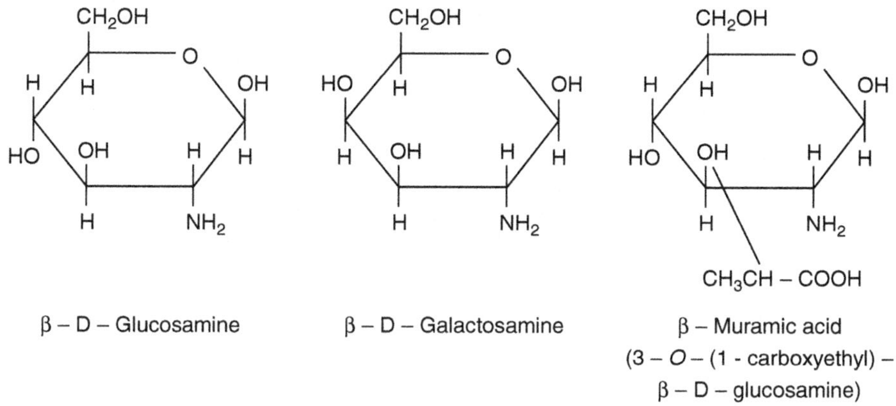

Fig. 3.7 Amino sugars which are predominant in soil

Hydrolysis of amino polysaccharides in soil is run by aminoglycan hydrolases. The adsorption of these enzymes by colloidal mineral and organic particles probably protects them against degradation by proteases and increases their stability. Adsorption may be, however, reduced by catalytic activity in relation to a substrate, especially with a high molecular mass and low solubility as chitin and peptidoglycans.

The first stage of glucosamine and N-acetylglucosamine decomposition to NH_4^+ is the formation of 6-phosphates by transferring phosphate groups from ATP, which is catalyzed by glucosamine kinase [EC 2.7.1.8.] and N-acetylglucosamine kinase [EC 2.7.1.59] as presented in Fig. 3.8 (the reactions 1 and 2).

Deamination of glucosamine 6-phosphate presented as the reaction 5 in Fig. 3.8 is a result of the catalytic activity of glucosamine 6-phosphate deaminase, but the first stage of this reaction is aldose–ketose isomerization, which the Enzyme Commission has named "glucosamine 6-phosphate isomerase" [EC 5.3.1.10]. The mechanism of deamination is presented below (Noltmann 1972).

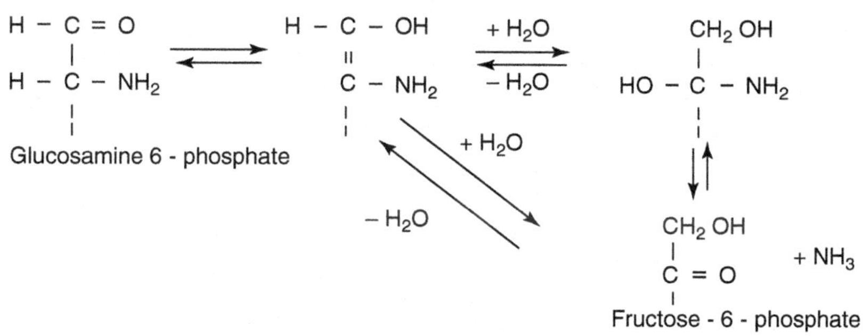

The balance of this reaction indicates that the final effect is NH_4^+, which can be uptaken by plants and fructose 6-phosphate.

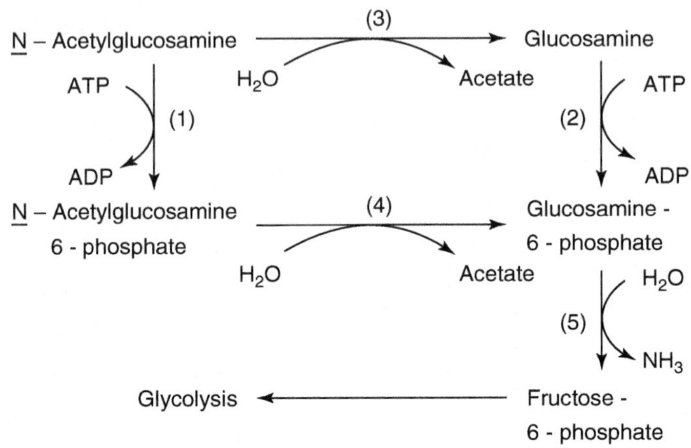

Fig. 3.8 Formation of NH_4^+ from glucosamine and N-acetylamine in microorganism. Enzymes involved in this reaction are (*1*) N-acetylglucosamine kinase, (*2*) glucosamine kinase, (*3*) N-acetylglucosamine deacetylase, (*4*) acetylglucosamine 6-phosphate deacetylase, (*5*) glucosamine 6-phosphate isomerase

As it is shown in Table 3.2, unknown nitrogen compounds in hydrolyzate as well as in the residue after hydrolysis equal up to 40–50 % of total nitrogen content in analyzed soil. However, these compounds do not quickly take part in mineralization process, but the new analytical techniques have been used for the identification of those nitrogen compounds (Leinweber and Schulten 1998).

3.2.3 Nucleic Acids

In soil, nitrogen is also found in such compounds as nucleic acids, nucleotides, nucleosides, purines, and pyrimidines. The conversion of nitrogen in nucleic acids requires a variety of enzymes. First, nucleic acids must be depolymerized to mononucleotides by nucleases [EC 3.1.4]. The nucleotides are dephosphorylated by nucleotidases [EC 3.1.3], yielding nucleosides which are N-glucosides of purines, pyrimidines, and pentoses.

Nucleases catalyze the hydrolysis of ester-bonding between phosphate groups and pentose in nucleic acids. Each phosphate group is linked to two pentose units and the nucleases are closed as phosphodiesterases [EC 3.1.4]. The group of nucleases is divided into two large subgroups according to the types of nucleic acids which they hydrolyze. Ribonucleases (RNases) hydrolyze ribonucleic acids (RNAs), and deoxyribonucleases (DNAses) hydrolyze deoxyribonucleic acids (DNAs).

Nucleic acids are transferred into soil with living or dead plants and animals and are found in all soil organisms. RNA and DNA added to soils are rapidly and extensively degraded (Greaves and Wilson 1970), and pure cultures of many soil microorganisms degrade nucleic acids (Antheunisse 1972). In most cases, measurement of nuclease activity is based on the production of inorganic PO_4^{3-}, so the assay embraces depolymerization of nucleic acids and dephosphorylation of the resulting mononucleotides.

Adsorption of nucleic acids on clay minerals in soils may protect them from degradation (Goring and Bartholomew 1952); on the other hand, nucleic acids are rapidly degraded in soils and inorganic PO_4^{3-} is released. Greaves and Wilson (1970) found that complexes of RNA or DNA with montmorillonite were degraded in soil, but X-ray diffraction studies indicated that RNA adsorbed in the central zones of the crystallites was partially protected from enzyme attack. The observed X-ray diffraction patterns, however, could have been caused by adsorption of degradation products such as adenine in the interlayer space of montmorillonite (Greaves and Wilson 1973).

Nucleotides that have been found in soil are thymidine 3′,5′-diphosphate and deoxyuridine 3′,5′-diphosphate, presumably derived from DNA (Anderson 1970). The lack of other nucleotides indicates that following their formation from nucleic acids, most nucleotides are rapidly degraded or assimilated by microorganisms. Nucleotidases may play only a minor role in the hydrolysis of nucleotides in soil because of the abundance of phosphatases with low specificity. Soil phosphatases have been reviewed in detail by Speir and Ross (1978). Hydrolysis of nucleosides to the constituent pentoses and purines, or pyrimidines, by nucleosidases has not been studied in soil.

The investigation of the turnover of RNA and DNA in soils has been accelerated due to new investigation method in the natural soil condition (Ceccherni et al. 2009).

3.2.4 Purines and Pyrimidines

Degradation of purines and pyrimidines in soil with the release of NH_4^+ has been investigated to a limited extent. Greaves and Wilson (1973) demonstrated that complexed adenine with montmorillonite is rapidly degraded in soils, but the products have not been identified. Many bacterial species of the *Enterobacteriaceae* family can use a variety of purines as the only source of N, C, and energy. In general, purines are degraded along pathways similar to those found in aerobic microorganisms. Degradation of pyrimidines can be carried out by a variety of soil microorganisms with the production of NH_4^+ and other end products, such as urea β-alanine and β-aminoisobutyrate, from each of which NH_4^+ can be subsequently produced. This has been stated in two ways, one involving oxidation at C-6 of the pyrimidine ring and the second reduction at this position. Prior to oxidation or reduction, cytosine is deaminated to produce uracil and NH_4^+, by the action of cytosine deaminase (cytosine aminohydrolase EC 3.5.4.1); this enzyme also deaminates 5-methyl cytosine, producing thymine and NH_4^+. The scheme of this reaction is presented in Figs. 3.9, 3.10, 3.11, 3.12, and 3.13.

The degradation of purines and pyrimidines by macroorganisms has been reviewed by Vogels and Van der Drift (1976).

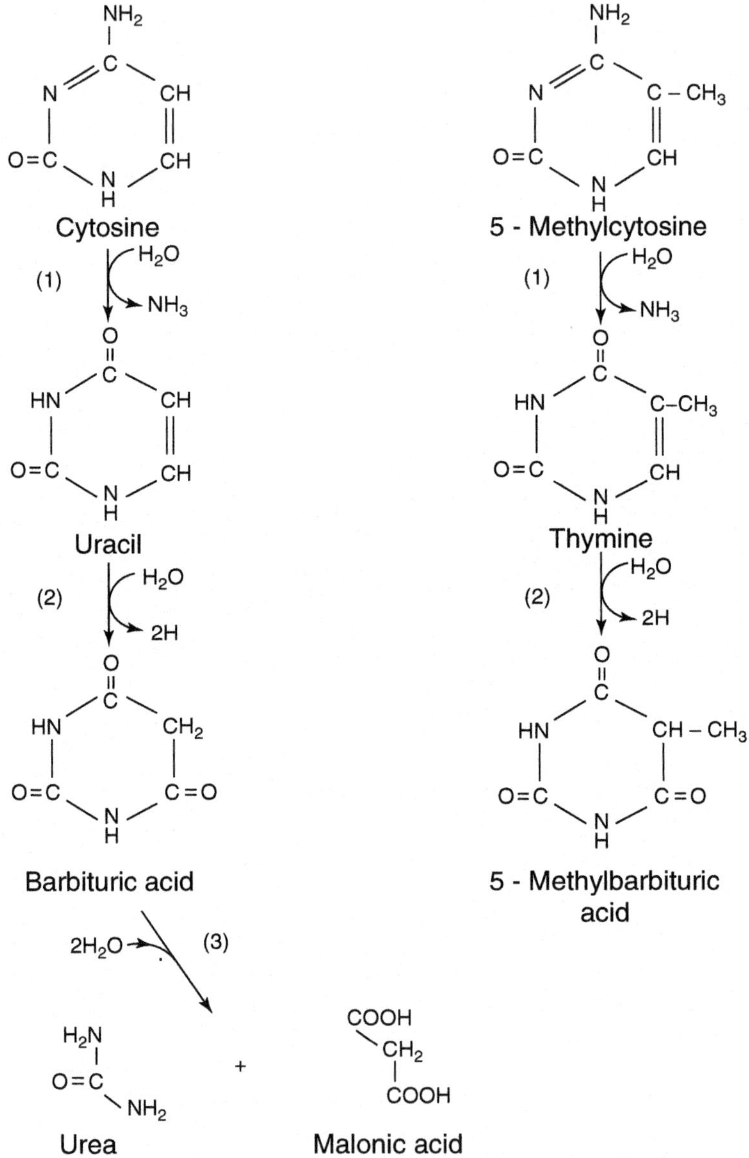

Fig. 3.9 Oxidative degradation of pyrimidines. Enzymes involved: (*1*) cytosine deaminase, (*2*) uracil dehydrogenase, (*3*) barbiturase

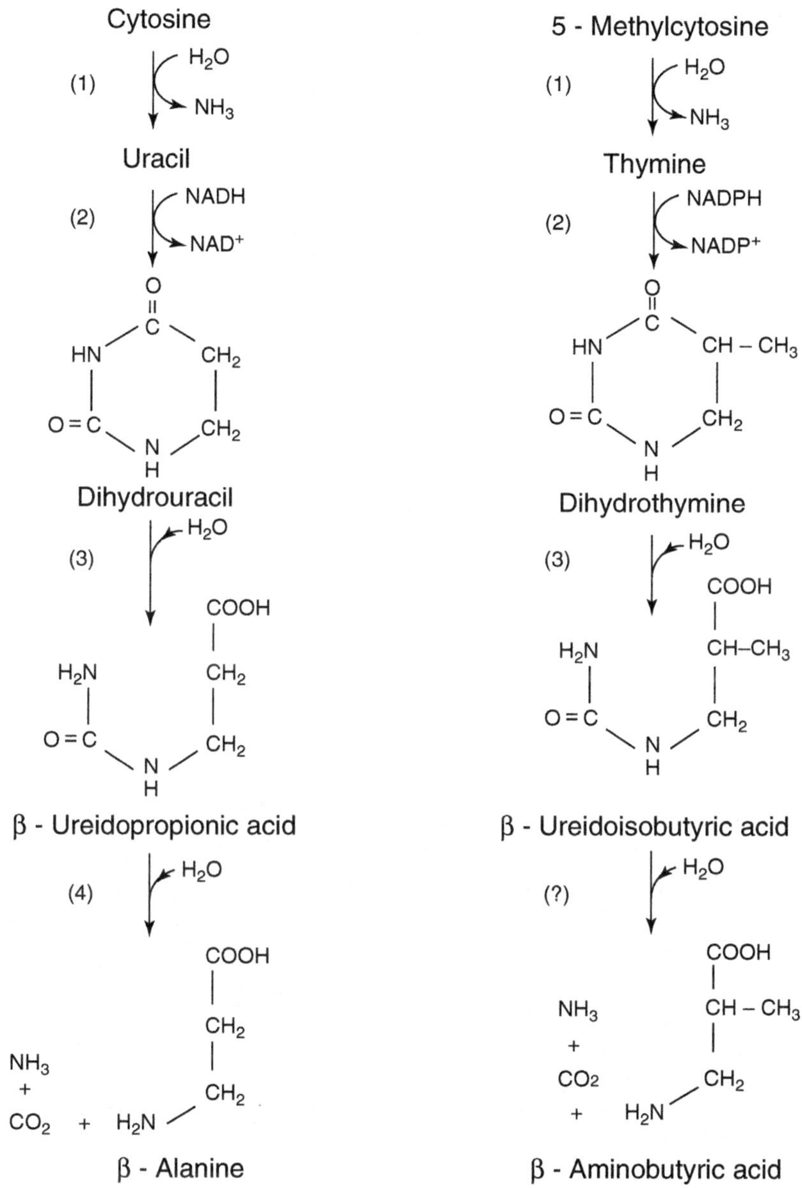

Fig. 3.10 Reductive pathways for degradation of pyrimidines. Enzymes involved: (*1*) cytosine deaminase, (*2*) dihydrouracil dehydrogenase, (*3*) dihydropyrimidinase, (*4*) β-ureidopropionase

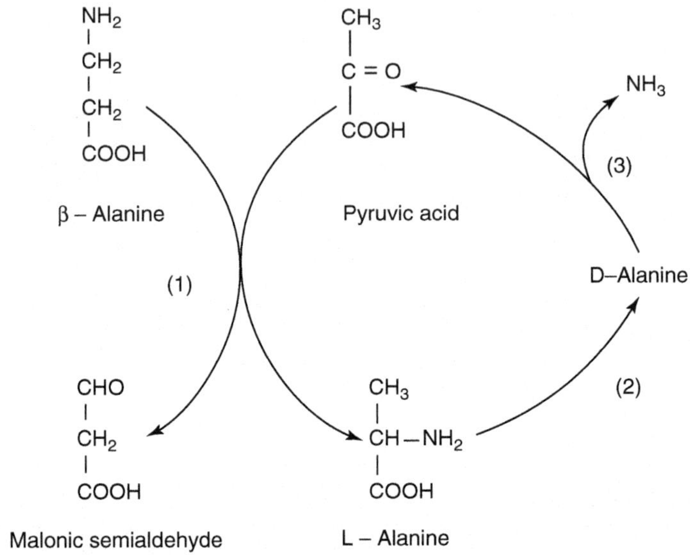

Fig. 3.11 Formation of NH_4^+ from β-alanine. Enzymes involved: (*1*) β-alanine/pyruvate aminotransferase, (*2*) alanine racemase, (*3*) D-alanine dehydrogenase

3.2.5 Amides

Amides are a group of compounds that have not been detected in acid hydrolyzates, although they are present in soils of different ecosystems in a free form similar to amino acids or amino sugars. The presence of amides in soils and their decomposition is well investigated and known, because urea is widely used as a nitrogen fertilizer (its proportion in the total volume of nitrogen fertilizers may reach as high as 60 %). Large amounts of urea are also found in natural, organic, and organic-mineral fertilizers and urea is introduced into arable soils and used on pastures. In soil, urea is enzymatically decomposed into CO_2 and NH_4^+ by urease.

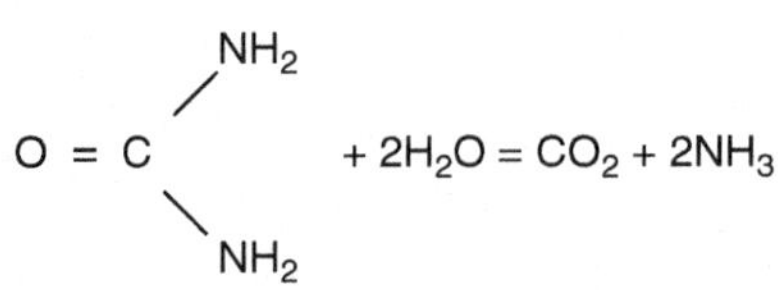

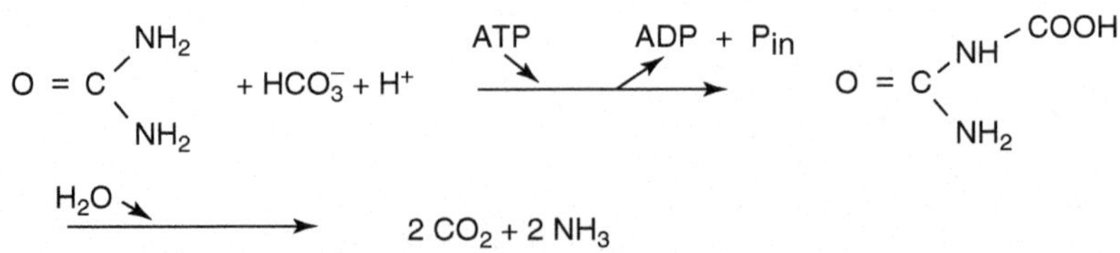

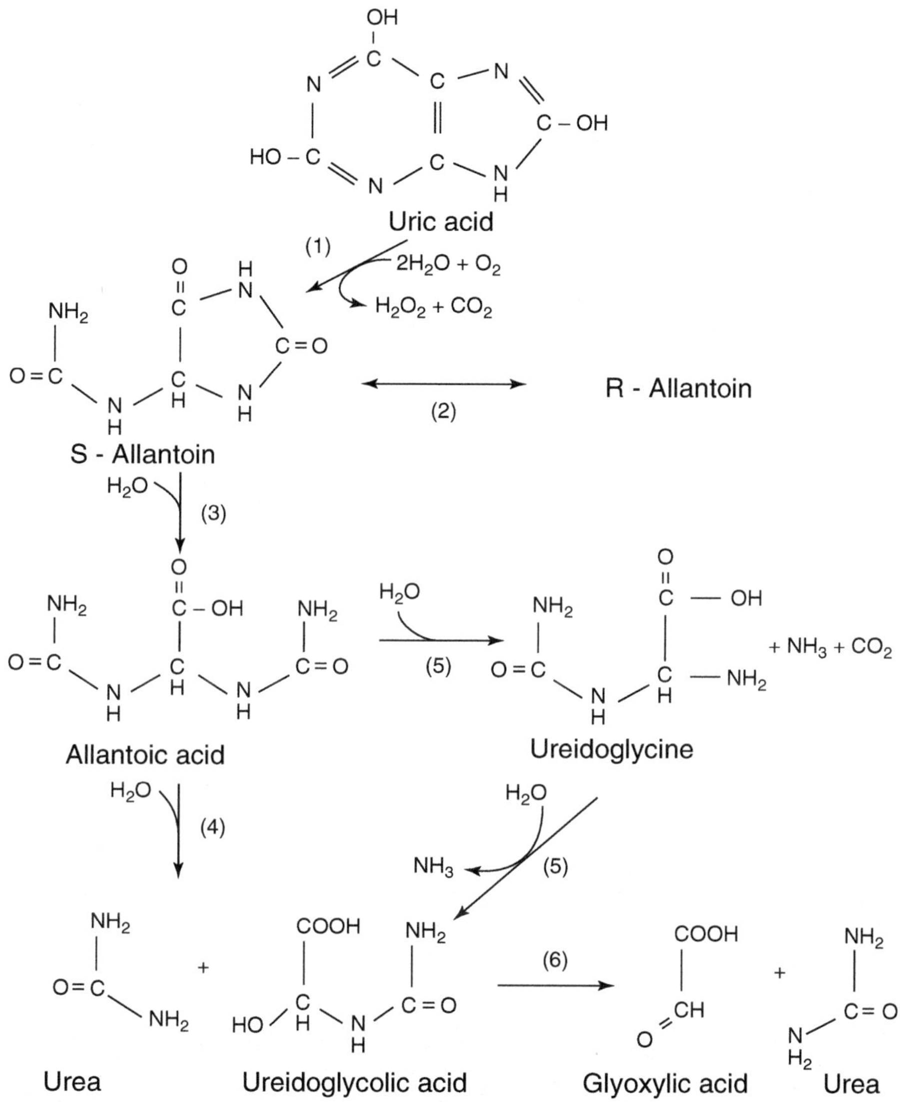

Fig. 3.12 Pathways of uric acid degradation under aerobic conditions. Enzymes involved: (*1*) urate oxidase, (*2*) allantoin racemase, (*3*) S-allantoinase, (*4*) allantoicase, (*5*) allantoate deiminase, (*6*) ureidoglycolate lyase

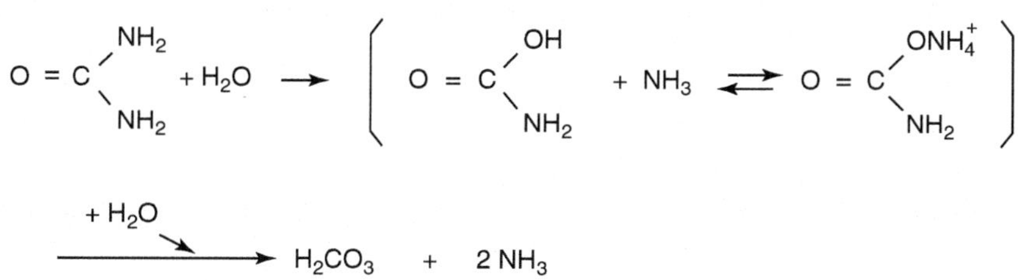

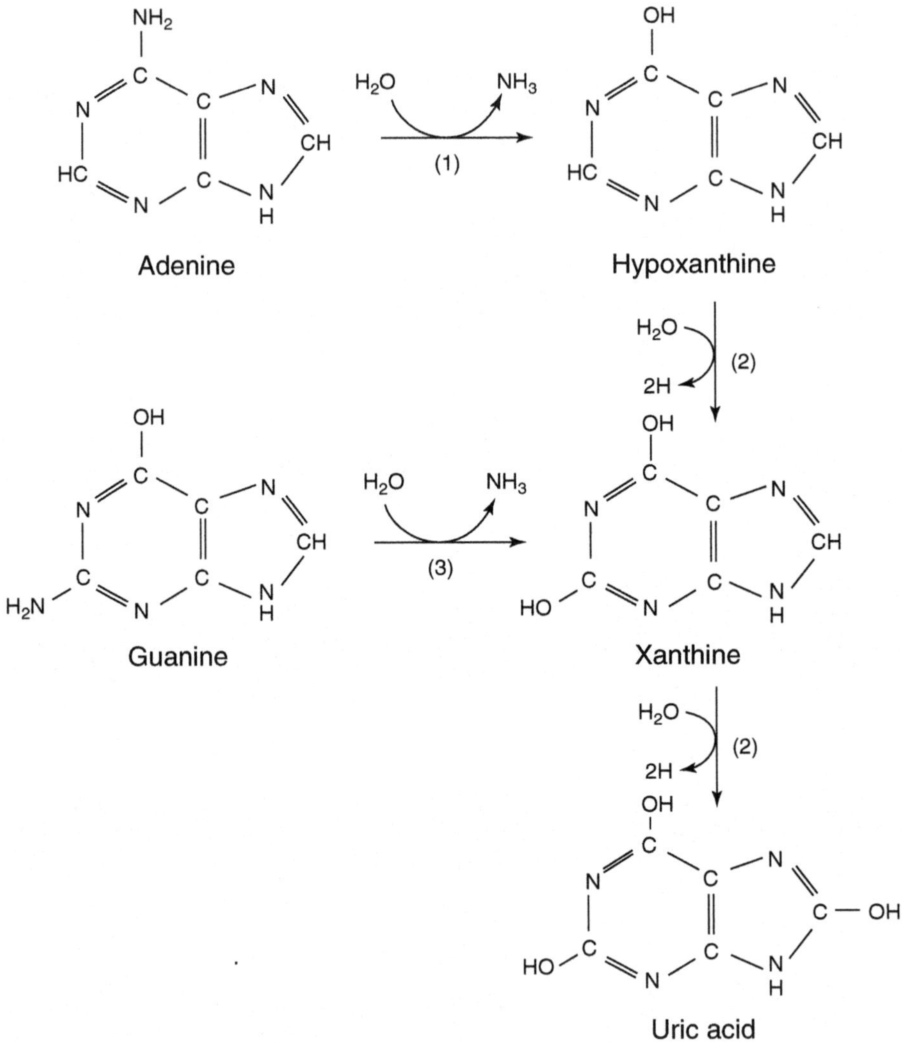

Fig. 3.13 Conversion of purines to uric acid under aerobic conditions. Enzymes involved: (*1*) adenine deaminase, (*2*) xanthine dehydrogenase, (*3*) guanine deaminase

Amidohydrolases (EC 3.5.1.5) (ureases) hydrolyze non-peptide C-N bonds in linear amides. Based on kinetic data, carbamate has been implicated as an obligatory intermediate in a two-step reaction catalyzed by the crystalline enzyme from jack bean (*Canavalia ensiformis*).

The reaction is presumed to involve a carbamoyl group transfer from a carbamoyl–enzyme complex to water. Urease produced by jack bean contains two essential atoms of bound Ni^{2+} per enzyme molecule (97,000 daltons), but the specific role of the metal ion in the catalysis has yet to be defined. It is possible that Ni^{2+} is related to amino acid residues and is positioned on the enzyme's active site (i.e., for the formation of the enzyme–substrate complex) and the metal enhances the electrophilic nature of the C atom of the $C=O$ group of urea, thus promoting nucleophilic displacement of the N atom. Among the enzyme ionization constants, SH, NH_4^+, and histidine groups are all thought to be involved at the urease catalytic site.

It is assumed that there is also an alternative pathway of urea hydrolysis which does not involve urease and is carried out by yeasts, alga, and some fungi. In this case, a two-stage sequential reaction takes place. In the first stage, urea is converted to allophanate by the action of urea carboxylase (urea: CO_2 ligase EC 6.3.4.6). Allophanate hydrolase (allophanate amidohydrolase EC 3.5.1.13) then converts allophanate to CO_2 and NH_4^+.

In the organic nitrogen compounds in soil, it should be taken under consideration the amount of nitrogen which is introduced into the soil as the results of biological reduction process by the diazotrophy. It is estimated then in this process yearly about 175.10^6 Mg of nitrogen is reduced and from this account nearly 30 % is reduced by non-symbiotic bacteria. The influence of herbicides on the growth and development as well as activity of microorganisms depends upon not only their activity but also from soil, climatic conditions and usually causes decreases of the amount of biologically reduced nitrogen. Roundup herbicides commonly used for the damage of about 70 plant species contain active substance glyphosate N-(phosphonomethyl)glycine which is decomposed (degraded) by enzyme produced by microorganisms to aminomethylphosphonic acid (AMPA), NH_3, CO_2, C_2H_5OH, H_2O, and phosphate. AMPA acid can be also cumulated in soil that caused increases in the amount of fungi and decreases of the diazotrophs' activity and disturbance in the biological nitrogen reduction process.

3.2.6 *Other Organic Nitrogen Compounds in Soils*

Bremner (1967) reported that minor amounts of nitrogen-containing compounds might be found in soil such as:

- Choline $-(CH_3)_3N-CH_2-CH_2OH$
- Ethanolamine $H_2N-CH_2-CH_2OH$
- Trimethylamine $-(CH_3)_3N$

Plants, soil microorganisms, and animals contain different amounts of phosphatidylcholine and phosphatidylethanolamine, of which the last bacteria can contain up to 40 % of the total lipid. Trimethylamine occurs in some fungi and higher plants.

Choline is degraded to glycine in a reaction sequence involving two successive oxidation steps followed by three successive demethylation reactions. The formation of NH_4^+ from glycine is catalyzed by amino acid oxidases and dehydrogenases or by a specific enzyme, glycine dehydrogenase [glycine: NAD^+ oxidoreductase (deaminating), EC 1.4.1.10].

Three pathways for NH_4^+ formation from ethanolamine have been found in microorganisms:

1. ethanolamine ammonia-lyase (EC 4.3.1.7),
2. oxidation of ethanolamine to NH_4^+ and glycolaldehyde is catalyzed by ethanolamine oxidase [ethanolamine: oxygen oxidoreductase (deaminating), EC 1.4.3.8], and
3. ethanolamine is phosphorylated by the transfer of PO_4^{3-} from ATP, catalyzed by an amino alcohol-ATP phosphotransferase, and the ethanolamine *O*-phosphate formed is hydrolyzed by a phospholyase (EC 4.2.99.7) to produce NH_4^+, acetaldehyde, and PO_4^{3-} .

Trimethylamine is oxidized to methylamine and formaldehyde, and primary amine dehydrogenase converts methylamine to NH_4^+ and formaldehyde.

Nitrosoamines in soils are the product of reaction between nitrates (III) which come as results of nitrification and denitrification processes with ammonium ions (NH_4^+) and amide group ($-NH_2$) liberated during mineralization process of organic nitrogen compounds in slightly acid and soils with good oxidation–reduction properties. Besides that, the secondary amines are in big amounts in plants which after harvesting parts are introduced into soil (Smith 1971). Investigation on the separation, determination, and metabolism of nitrosoamines in soil was carried out by several authors (Whitehead 1964; Pancholy 1976, 1978; Mallik and Tesfai 1981; Rostkowska et al. 1998). The application of mineral nitrogen fertilizers among others parameters influencing on the content of nitrosoamines in soil influenced in the biggest degree similarly like on the changes of microbiocenoses (Barabasz et al. 2002). Nitrosoamines did not show big stability in soils and on their decomposition in soil are carry out by microorganism (Tate and Alexander 1975).

Complex nitrogen compounds such as antibiotics and vitamins are found in soil as a result of microbial activity. In addition, alkaloids and chlorophyll are also introduced to soil with plant residues.

In the context of nitrogen dynamics and conversions in soil, the important compounds are those that are rapidly mineralized (amides, amino acids, amino sugars, soluble simple proteins) and undergo further conversions under redox conditions. This has prompted many scientists to develop chemical methods for extraction and quantitative analysis of nitrogen in the so-called easily mineralized nitrogen fraction. This fraction has a considerable theoretical and practical importance. After a complete analysis of an extract, conclusions may be drawn on the conversion of mineral and organic nitrogen compounds and on the quantitative measurement of nitrogen available to plants (which should be included in developing a scheme of nitrogen fertilization).

3.2.7 Models of Organic Nitrogen Processes

Studies on the conversion of nitrogen in soil during mineralization–immobilization can be summarized by attempts to develop mathematical models for their practical use under field conditions (Russel 1980; Broadbent 1986). These models have different forms and include from two to several factors that have an impact on these conversions. Jenkinson (1966) presented a mathematical model that determined the rate of conversion of organic nitrogen compounds in soil:

$$\frac{\mathrm{d}N}{\mathrm{d}t} = A - y \cdot N$$

where

$\frac{\mathrm{d}N}{\mathrm{d}t}$ – content of nitrogen (dN, a change of nitrogen quantity in dt time)

A – quantity of nitrogen in organic compounds introduced to soil

N – current quantity of organic nitrogen compounds at the moment of adding external nitrogen sources

y – nitrogen mineralization coefficient

The equation presented below is another example of a mathematical model that describes the conversions of nitrogen during mineralization–immobilization (Russel 1980):

$$\frac{\mathrm{d}N}{\mathrm{d}t} = K_1(t)N + K_2 + K_3(t)Y(t)$$

where

$\frac{\mathrm{d}N}{\mathrm{d}t}$ – content of nitrogen (dN, a change of nitrogen quantity in dt time)

K_1 – degradation coefficient for nitrogen in organic compounds

K_2 – addition of nitrogen as organic fertilizers

K_3 – addition of nitrogen as plant residue

Y – plant harvests

K_1, K_3, and Y relate to specific t time.

The model of mineralization–biological immobilization of nitrogen and carbon in soil was developed by Paul and Juma (1981). Under natural conditions, both of these processes take place parallel to each other with a tendency toward equilibrium. The net effect of nitrogen content in the mineral forms for these opposing processes is zero, which may change under the influence of agricultural engineering factors.

The mathematical model presented by Stanford et al. (1973) and Stanford et al. (1974) is an interesting example of a description of nitrogen mineralization in soil:

$$\log\left(N_0 - N_t\right) = \log N_0 - \frac{K \cdot t}{2303}$$

where

N_0 – mineralizing soil capacity
N_t – quantity of nitrogen in organic compounds mineralized within t time
K – mineralization coefficient

K coefficient shows an almost stable value for different soil categories and is app. 0.054 per week. This value largely depends on temperature and humidity in soil. The equation presented above calculates the amount of nitrogen that is potentially susceptible to mineralization, and for most of soils, it ranges from 5 to 40 % of its total content. Thus, it may be concluded that this amount includes not only nitrogen in microorganisms but also a substantial fraction of the nitrogen found in the humus.

References

Ambroz Z (1965) The proteolytic complex decomposing proteins in soil. Rostlinna Vyroba 11:161–170

Ambroz Z (1971) The determination of the activities of gelatinase and caseinase in the soil. Biology Sol 13:28–29

Anderson G (1970) The isolation of nucleoside diphosphates from alkaline extracts of soil. J Soil Sci 21:96–104

Antheunisse J (1972) Decomposition of nucleic acids and some of their degradation products by microorganisms. Antonie van Leeuwenhoek J Microbiol Serol 38:311–327

Asami T, Hara M (1970) On the fractionation of soil organic matter after hydrolysis with hydrochloric acid. J Soil Sci Manure 41 (From Soil Sci Plant Nutr Tokyo 17) p. 222

Balestend J, Mariotti A, Boisgontier D (1990) Effect of tillage on soil organic carbon mineralization estimated from ^{13}C abundance in maize fields. J Soil Sci 41:587–596

Barabasz W, Albińska D, Jaśkowska M, Lipiec J (2002) Biological effects of mineral nitrogen fertilization on soil microorganisms. Pol J Environ Stud 11:193–198

Batistic L, Mayaudon J (1970) Biological stabilisation in the soil of C^{14} ferulic acid, C^{14} vanillic acid and C^{14} p-coumaric acid. Ann Inst Pasteur 118:199–206

Becher M, Kalembasa D (2011) Fractions of nitrogen and carbon in humus horizons of arable Luvisols and Cambisols located on Siedlce upland. Acta Agrophysica 18:7–16 (In Polish)

Beck T, Joergensen RG, Kandeler E, Makeschin F, Nuss E, Oberholzer HR, Scheu S (1997) An interlaboratory comparison of ten different ways for measuring soil microbial biomass C. Soil Biol Biochem 29:1023–1032

Blair GJ, Lefroy RDB, Lisle L (1995) Soil carbon fractions based on their degree of oxidation and the development of a carbon management index for agricultural system. Aust J Agric Res 46:1459–1466

Bohn HL (1976) Estimate of organic carbon in world soils. Soil Sci Soc Am J 40:468–469

Bremner JM (1965) Organic forms of nitrogen. In: Black CA (ed) Methods of soil analysis Part 2 Chemical and microbiological properties, vol 9, Agronomy. American Society of Agronomy, Madison, pp 1238–1255

Bremner JM (1967) Nitrogenous compounds. In: McLoren D, Peterson GH (eds) Soil biochemistry, vol 1. Marcel Dekker Inc, New York, pp 19–66
Broadbent FE (1986) Empirical modeling of soil nitrogen mineralization processes. Soil Sci 141:208–213
Brooks PC, Landman A, Pruden G, Jenkinson DS (1985) Chloroform fumigation and the release of soil nitrogen: a rapid direct extraction method to measure microbial biomass nitrogen in soil. Soil Biol Biochem 17:837–842
Brunn S, Agren GI, Christensen BT, Jensen LS (2010) Measuring and modeling continuous quality distributions of soil organic matter. Biogeosciences 7:27–41
Camberdella CA (2005) Carbon cycle in soils formation and decomposition. In: Encyclopedia of soil in the environment. Elsevier Ltd, Amsterdam, pp 170–175
Ceccherni MD, Ascher J, Guervi G, Pietramellara G (2009) In field-detection and quantification of extracellular DNA. J Plant Nutr Soil Sci 172:626–629
Chan KY, Bowman A, Oates A (2001) Oxidizible organic carbon fraction and soil quality changes in an oxic Paleustalf under different pasture lays. Soil Sci 166(1):61–67
Cheng CN, Kurtz LT (1975) Extraction and desalting amino acids from soil and sediments, evaluation of methods. Soil Biol Biochem 7:319–322
Christensen BT (1986) Straw incorporation and soil organic matter in macro-aggregates and particle size separates. J Soil Sci 37:125–135
Cook BD, Allan DL (1992) Dissolved organic carbon in old field soils; compositional changes during the biodegradation of soil organic matter. Soil Biol Biochem 24:595–600
Crompton TR (2012) Organic compounds in soil sediments and sludges. Analysis and determination. Taylor & Francis Ltd/CRC Press, London, pp 3–456
Fismes J, Perrin-Ganier C, Empereur-Bissonnet P, Morel JL (2002) Soil-to-root transfer and translocation of polycyclic aromatic hydrocarbons by vegetables grown on industrial contaminated soils. J Environ Qual 31:1649–1656
Flaig W (1968a) Einwirkung von organischen Bodenbestandteilen auf das Pflanzenwachstum. Landwirtsch Forsch 21:103–127
Flaig W (1968b) Uptake of organic substances from soil organic matter by plants. Pontofiwae Academiae Scientiarum Scripta Varia 32:1–48
Flaig W (1971) Organic compounds in soil. Soil Sci 111(1):19–33
Flaig W, Beutelspacher H, Rietz E (1975) Chemical composition and physical properties of humic substances. In: Gieseking JE (ed) Soil components, vol 1. Springer, New York, pp 1–211
Fotyma E, Fotyma M (2004) Optimalisation of winter wheat nitrogen fertilization with the use of N_{min} soil test. Fertil Fertil 3(20):55–73 (in Polish)
Fotyma M, Fotyma E, Kęsik E, Krzyrzkowska T (1987) Mineralisation ability of soil in relations to nitrogen. Annu Soil Sci 38(3):79–81
Freibauer A, Rounsevell MDA, Smith P, Verhagen J (2004) Carbon sequestration in the agricultural soils in Europe. Geoderma 122:1–23
Gijsman AJ, Hoogenboom G, Paryon WJ, Kerridge C (2002) Modifying DSSAT crop models for low-input agricultural systems using a soil organic matter-residue module from CENTURY. Agron J 94:462–474
Gilbert RG, Altman J (1966) Ethanol extraction of free amino acids from soil. Plant Soil 24:229–238
Gonet SS, Markiewicz M (2007) The role of organic matter in environment. Polish Society of Humic Substances, Wrocław, pp 7–149 (In Polish)
Goring CAI, Bartholomew WV (1952) Adsorption of mononucleotides nucleic acids and nucleoproteins by clay. Soil Sci 74:149–164
Greaves MP, Wilson MJ (1970) The degradation of nucleic acids and montmorillonite – nucleic acid complexes by soil microorganisms. Soil Biol Biochem 2:257–268
Greaves MP, Wilson MJ (1973) Effects of soil microorganisms on montmorillonite-adenine complexes. Soil Biol Biochem 5:275–276
Griffith SM, Sowden FJ, Schnitzer M (1976) The alkaline hydrolysis of acid resistant soil and humic acid residue. Soil Biol Biochem 8:511–529

Guenzi WD, McCalla TM (1966) Phytotoxic substances extracted from soil. Soil Sci Soc Am Proc 30:214–216
Haider K, Fredrick LR, Flaig W (1965) Reaction between amino acid compounds and phenols during oxidation. Plant Soil 22:49–64
Haynes RJ (2005) Labile organic matter fractions as central components of the quality of agricultural soils. Adv Agron 85:221–268
Herbert BE, Bertsch PM (1995) Characterization of dissolved and colloidal organic matter in soil solution: a review. In: McFee WW, Kelly JM (eds) Carbon forms and functions in forest soil. Soil Science Society of America Inc, Madison, pp 63–88
Hunt HW (1977) A model simulation for decomposition in grasslands. Ecology 58:469–484
IPCC Climate Change (2007) The physical science bases. Cambridge University Press, Cambridge, p 996
Jansson SL, Persson J (1982) Mineralisation and immobilisation of soil nitrogen. In: *Nitrogen* in agricultural soils, Agronomy monograph no 22. American Society of Agronomy, Madison, pp 229–252
Jenkinson DS (1966) The turnover of organic matter in soil. In: The use of isotopes in soil organic matter studies, Brunswick Volkenrade 1963. Pergamon Press Inc, Oxford, pp 187–197
Jenkinson DS, Rayner JH (1977) The turnover of soil organic matter in some of the Rothamsted classical experiments. Soil Sci 123:298–305
Joergensen RG, Wu J, Brooks PC (2011) Measuring soil microbial biomass using an automated procedures. Soil Biol Biochem 43:873–876
Jones DL (1998) Organic acids in the rhizosphere – a critical review. Plant Soil 205:25–44
Jones DS, Owen AG, Fortar JM (2002) Simple method to enable the high resolution determination of total free amino acids in soil solution and extracts. Soil Biol Biochem 34(12):1893–1902
Jones DL, Dennis PG, Owen AG, van Hees PAW (2003) Organic acid behavior in soil misconceptions and knowledge gaps. Plant Soil 248:31–41
Kalembasa S (1995) The application of ^{15}N and ^{13}N isotopes in soil science and agrochemical research. W.N.T, Warszawa, pp 5–251 (in Polish)
Kalembasa D, Becher M (2009) Fractions of nitrogen in drained peat-muck soils located in the upper Liwiec River valley. Water Environ Rural Areas 9:73–82 (in Polish)
Kalembasa S, Kalembasa D (1986) Comparative investigations of the methods of organic carbon content determination in soil extracts. Annu Soil Sci 37(1):109–118
Kalembasa S, Kalembasa D (1992) The quick method for the determination of C/N ratio in mineral soils. Pol J Soil Sci 25(1):41–46
Kalembasa S, Kalembasa D (2015) Chemistry and biochemistry of soil – selected issues. University of Siedlce, Siedlce, pp 1–312
Kalembasa S, Niewiński S (1990) The occurrence of amino acids in humic and fulvic acids extracted from arable and forest grey – brown podzolic soil. Pol J Soil Sci 33(1):31–36
Kalembasa S, Tengler Sz (2004) The role of brown coal in fertilization and environment protection. Monograph no 52. University of Podlasie, pp 3–175 (in Polish)
Kalembasa S, Kalembasa D, Buchanan D (1999) Spectometric characteristic of humic acids extracted from soil by sodium hydroxide ultravibration and pyridine. Pol J Soil Sci 32(1):33–41
Klavins M, Purmalis O, Rodinov V (2013) Peat humic acid properties and factors influencing their variability in temperate bog ecosystem. Estonian J Ecol 62(1):35–52
Kolář L, Kužel S, Horáček J, Čechová V, Borová-Batt J, Peterka J (2009) Labile fractions of soil organic matter their quantity and quality. Plant Soil Environ 55(6):245–251
Ladd JN, Butler JHA (1972) Short-term assays of soil proteolytic enzyme activities using proteins and dipeptide derivatives as substrates. Soil Biol Biochem 38:19–30
Ladd JN, Jackson RB (1982) Biochemistry of ammonification. In: Nitrogen in agricultural soils, Agronomy monograph no. 22. American Society of Agronomy, Madison, pp 173–228
Lal R (2004) Soil carbon sequestration to mitigate climate change. Geoderma 123:1–22

Lefroy RDB, Blair GJ, Strong WM (1993) Changes in soil organic matter with cropping as measured by organic carbon fractions and ^{13}C natural isotope abundance. Plant Soil 155(156):399–402
Leinweber P, Schulten HR (1998) Nonhydrolyzable organic nitrogen in soil size separates from long-term agricultural experiments. Soil Sci Soc Am J 62:383–393
Liaudanskiene I, Slepeticne A, Velykis A, Satkus A (2013) Distribution of organic carbon in humic and granulodensimetric fractions of soil as influenced by tillage and crop rotation. Estonian J Ecol 62(1):53–69
Łoginow W (1967) Effect of humic acids on the deamination of amino acids. Pam Pulawski 29:3–43 (in Polish)
Łoginow W, Wiśniewski W, Gonet SS, Cieścińska B (1987) Fractionation of organic carbon based on susceptibility to oxidation. Pol J Soil Sci 20:47–52
Maliszewska-Kordybach B (1996) Polycyclic aromatic hydrocarbons in agricultural soils in Poland: preliminary proposals for criteria to evaluate the level of soil contamination. Appl Geochem 11:121–127
Maliszewska-Kordybach B, Smreczak B, Klimkowicz-Pawlas A, Terelak H (2008) Monitoring of the soil content of polycyclic hydrocarbons (PAHs) in arable soils in Poland. Chemosphere 73:1284–1291
Mallik MAB, Tesfai K (1981) Transformation of nitrosamines in soil and in vitro by soil microorganisms. Bull Environ Contam Toxicol 27:115–121
Mander LN (2003) Twenty years of gibberellin research. Nat Prod Rep 20:49–69
Martens DA (2002a) Identification of phenolic acid composition of alkali-extracted plant and soils. Soil Sci Soc Am J 66(4):1241–1248
Martens DA (2002b) Relationship between plant phenolic acids released during soil mineralization and aggregate stabilization. Soil Sci Soc Am J 66(6):1857–1867
Maysner T, Szajdak L, Kuś J (2006) Impact of the farming systems on the content of biologically active substances and the form of nitrogen in the soils. Agron Res 4:531–542
Mazur T (1991) Nitrogen in agricultural soil. PWN, Warszawa, p 239 (in Polish)
Mazzarino MJ, Oliva L, Abil A, Acoste M (1987) Factors affecting dynamics in a semiarid woodland (Dry Chaco Argentina). Plant Soil 138:85–98
McLauchlan KK, Hobbie SE (2004) Comparison of labile soil organic matter fractionation techniques. Soil Sci Soc Am J 68:1616–1625
Michalak A (2006) Phenolic compounds and their antioxidant activity in plants growing under heavy metal stress. Pol J Environ Stud 15(4):523–530
Miettinen K (1950) Assimilation of amino acids in higher plants. Symp Soc Exp Biol 13:210–229
Myśków W (1981) Microbiological activity indices as applied for an evaluation of soil fertility. Post Mikrobiol 20:173–192 (in Polish)
Myśków W, Stachyra A, Zięba S, Masiak D (1996) The biological activity of soil as an indicator of fertility. Soil Sci Annu 47:89–99 (in Polish)
Noltmann EA (1972) Altdose – Ketose isomerases. In: Boyer PD (ed) The enzymes vol. VI. Carboxylation and decarboxylation (nonoxidative) isomeration, 3rd edn. Academic, New York, pp 271–354
Ocio JA, Brooks PC, Jenkinson DC (1991) Field incorporation of straw and its effect on microbial biomass and soil organic N. Soil Biol Biochem 23:171–176
Ohno T (2001) Oxidation of phenolic acid derivatives by soil and its relevance to allelopathic activity. J Environ Qual 30:1631–1635
Pancholy SK (1976) Gas chromatographic analysis of carcinogenic nitrosamines in soil. Soil Biol Biochem 8:75–76
Pancholy SK (1978) Formation of carcinogenic nitrosamines in soils. Soil Biol Biochem 10:27–32
Parsons JW (1981) Chemistry and distribution of amino sugars in soil and soil organisms. In: Paul EA, Ladd JN (eds) Soil biochemistry. Marcel Dekker Inc, New York, pp 197–227

Paul EA, Juma NG (1981) Mineralization and immobilization of soil nitrogen by microorganisms. In: Clark FF, Roswall T (eds) Terrestrial nitrogen cycles. Stockholm. Ecol Bull 33:179–195
Paul EA, Schmith EL (1960) Extraction of free amino acids from soil. Soil Sci Soc Am Proc 24:195–198
Paul EA, Schmith EL (1961) Formation of free amino acids in rhizosphere and nonrhizosphere soil. Soil Sci Soc Am Proc 25:359–362
Plante AE, Conant RT, Paul EA, Paustian K, Six J (2006) Acid hydrolysis of easily dispersed and microaggregate – derived silt – and clay size fractions to isolate soil organic matter. Euro J Soil Sci 57:456–467
Preston LM (1982) The availability of residual fertilizer nitrogen immobilized as clay fixed ammonium and organic N. Can J Soil Sci 62:479–486
Rees RM, Bingham IJ, Baddeley JA, Watson CA (2005) The role of plants and land management in sequestering soil carbon in temperate arable and grassland ecosystems. Geoderma 128:130–154
Rostkowska K, Zwierz K, Różański A, Moniuszko-Jakoniuk J, Roszczenko A (1998) Formation and metabolism of N-nitrosamines. Pol J Environ Stud 7:321–325
Rovira P, Vallejo UR (2000) Examination of thermal and acid hydrolysis procedures in characterization of soil organic matter. Commun Soil Sci Plant Anal 31:81–100
Rovira P, Vallejo UR (2002) Labile and recalcitrant pools of carbon and nitrogen in organic matter decomposing at different depths in soil an acid hydrolysis approach. Geoderma 107:109–141
Rovira P, Vallejo UR (2007) Labile recalcitrant and inert organic matter in Mediterranean forest soils. Soil Biol Biochem 39:202–215
Russel JS (1980) Simulation of nitrogen behavior of soil/plant systems. In: Frissel MJ, van Veen JA (eds). Institute for Agricultural & Forest Environment, Wageningen, pp 493–504
Ryan J, Masri S, Singh M (2009) Seasonal changes in soil organic matter and biomass and labile forms of carbon as influenced by crop rotations. Commun Soil Sci Plant Anal 40:188–199
Salter RM, Green TC (1933) Factors effecting the accumulation and loss of nitrogen and organic carbon in cropped soils. J Am Soc Agron 25:622–630
Schnitzer M (1982) Organic matter characterization. In: Page AL (ed) Methods of soil analysis part 2, 2nd edn, Agronomy no 9. American Society of Agronomy, Madison, pp 581–594
Schulten HR, Schnitzer M (1998) The chemistry of soil organic nitrogen; a review. Biol Fertil Soils 26:1–15
Shiroto Y, Yokozawa M (2006) Acid hydrolysis to partition plant material into decomposable and resistant fractions for use in the Rothamsted carbon model. Soil Biol Biochem 38:812–816
Smith ITA (1971) The occurrence, metabolism and functions of amines in plants. Biol Rev 46:201–241
Smith OL (1979) An analytical model of the decomposition of soil organic matter. Soil Biol Biochem 11:585–606
Smith TM, Cramer WP, Dixon RK, Leemans R, Neilson RP, Solomon AM (1993) The global terrestrial carbon cycle. Water Air Soil Pollut 70:19–37
Sparling GP (1992) Ratio of microbial biomass carbon to soil organic carbon as a sensitive indicator of changes in soil organic matter. Aust J Soil Res 30:195–207
Speir TW, Ross DJ (1978) Soil phosphatase and sulphatase. In: Burns RG (ed) Soil enzymes. Academic, London, pp 197–250
Stanford G, Frere MH, Schwaninger DH (1973) Temperature coefficient of soil nitrogen mineralization. Soil Sci 115:321–323
Stanford G, Carter JN, Smith SJ (1974) Estimates of potentially mineralizable soil organic nitrogen based on short-term incubations. Soil Sci Soc Am Proc 38:99–102
Stevenson FJ (1982) Organic forms of soil nitrogen. In: Nitrogen in agricultural soil, Agronomy monograph no 22. American Society of Agronomy, Madison, pp 67–122
Stevenson FJ (1994) Humus chemistry: genesis, composition, reactions, 2nd edn. Wiley, New York, pp 3–469

Strosser E (2010) Methods for determination of labile soil organic matter. An overview. J Agrobiol 27(2):49–60
Sundquist ET (1993) The global carbon dioxide budget. Science 259(5097):934–941
Sutton R, Sposito G (2005) Molecular structure in soil humic substances. The new view. Environ Sci Technol 39(23):9009–9015
Szajdak L (2011) Amino acids in soil. In: Narwal SS, Szajdak L, Sampietro DA (eds) Soil allelochemicals. Research methods in plant sciences, vol 1. Stadium Press LLC, Houston, pp 89–113
Szajdak L, Marygonova V (2007) Occurrence of IAA auxin in some organic soils. Agron Res 5(2):175–187
Szajdak LW, Nowak J (2013) Impact of peat substrates with different concentration of indole-3-acetic on ornamental plant cultivation. Peatlands Int 1:25–28
Szajdak L, Österberg R (1996) Amino acids present in humic acid from soils under different cultivation. Environ Int 22:563–569
Szajdak L, Życzyńska-Bałoniak I (1994) Phenolic acids in brown soils under continuous cropping of rye and crop rotation. Pol J Soil Sci 27(2):113–121
Szajdak L, Marygonova V, Wylegalski S, Strojny Z (2003) The content of indole-3-acetic acid in municipal wastes compost as compared with other commercial media. In: International scientific conference. Municipal soils waste composts production and influence on the environment. Humic substances in ecosystems, vol 5, pp 55–59
Szajdak L, Życzyńska-Bałoniak I (2002) Influence of mid-field afforestation on the change of organic nitrogen compounds in ground water and soil. Pol J Enviro Stud 11(1):91–95
Tan KH (1998) Principles of soil chemistry, 3rd edn. Marcel Dekker Inc, New York/Basel/Hong Kong, pp 1–362
Tan KH (2005) Soil sampling, preparation and analysis, 2nd edn. Taylor & Francis Group, New York, pp 3–623
Tan KH (2011) Principles of soil chemistry, 4th edn. CRC Press, Boca Raton, pp 4–390
Tate RL, Alexander M (1975) Stability of nitrosamines in samples of lake water, soil, and sewage. J Natl Cancer Inst 54(2):327–330
Tatzber M, Stemmer M, Spiegel H, Katzlberger C, Mentler A, Gerzabek MH (2007) FTIR-spectroscopic characterization of humic acids and humin fractions obtained by advanced NaOH, $Na_4P_2O_7$, and Na_2CO_3 extraction procedures. J Plant Nutr Soil Sci 170:522–529
Tirol-Padre A, Ladhe JK (2004) Assessing the reliability of permanganate – oxidizable carbon as an index of soil labile carbon. Soil Sci Soc Am J 68:969–978
Toussoun TA, Weinhold AR, Linderman RG, Patrick ZA (1968) Nature of phytotoxic substances produced during plant residue decomposition in soil. Phytopathology 58:41–45
Trojanowski J (1973) Turnover of organic substances in soil. WNT, Warszawa, pp 5–331 (in Polish)
Vance ED, Brooks PC, Jensinson DS (1987) An extraction method for measuring soil microbial biomass C. Soil Biol Biochem 19:703–707
Vogels GD, Van der Drift C (1976) Degradation of purines and pyrimidines by microorganisms. Bacteriol Rev 40:403–468
Walkey A (1947) A critical examination of a rapid method for determining organic carbon in soils. Soil Sci 63:251–263
Wang TSC, Yang T-K, Chuang T-T (1967) Soil phenolic acids as plant growth inhibitors. Soil Sci 103(4):239–246
Weil RW, Islam KR, Stine M (2003) Estimating active carbon for soil quality assessment: a simplified method for laboratory and field use. Am J Altern Agric 18(1):3–17
Werdin-Pfisterer NR, Kiellan K, Boone RD (2009) Soil amino acid composition across a boreal forest successional sequence. Soil Biol Biochem 41:1210–1220
Whitehead DC (1964) Identification of p-hydroxybenzoic, vanillic, p-coumaric and ferulic acids in soils. Nature 202:417–418

Wu J, Joergensen RG, Pommerening B, Chussod R, Brookes PC (1999) Measurement of soil microbial biomass C by fumigation-extraction-an automated procedure. Soil Biol Biochem 22(8):1167–1169

Xu X, Thornton PE, Post WM (2013) A global analysis of soil microbial biomass, carbon, nitrogen and phosphorus in terrestrial ecosystems. Glob Ecol Biogeogr 22:737–749

Yonebayashi K, Hattori T (1980) Improvements in the method for fractional determination of soil organic nitrogen. Soil Sci Plant Nutr 26(4):469–481

Zsolnay A (2003) Dissolved organic matter; artifacts, definitions and functions. Geoderma 113:187–209

Zsolnay A, Garlitz H (1994) Water extractable organic – matter in arable soils – effects of drought and long term fertilization. Soil Biol Biochem 26:1257–1261

Chapter 4
Free Sulfuric Amino Acids and Rhodanese in Soils Under Rye Cropping and Crop Rotation

Lech Wojciech Szajdak and Teodor Rusu

Abstract During the entire period of study, the activity of rhodanese and the concentration of free sulfuric acids were higher in the soil under continuous rye cropping than in the soil under crop rotation. The transformations of organically bound sulfur in soil occur with the contribution of rhodanese enzyme (thiosulfate–cyanide–sulfurtransferase) [E.C. 2.8.1.1]. The enzyme catalyzes the conversion generating a thiocyanate group from thiosulfate and cyanide.

The linear and positive correlation between the concentration of free sulfuric amino acids and rhodanese activity in the soil under continuous rye cropping and in the soil under crop rotation was proved. Despite the correlation these two lines are not parallel. The slope of the curve of the soil under continuous rye cropping was about 1.5 times lower than the one determined for the soil under crop rotation. This indicates that the transformation processes of free sulfur amino acids and rhodanese were about 1.5 times slower in the soil under continuous rye cropping than in the soil under crop rotation.

Keywords Free sulfuric amino acids • Rhodanese • Continuous cropping of rye • Crop rotation

L.W. Szajdak (✉)
Institute for Agricultural and Forest Environment, Polish Academy of Sciences, Str. Bukowska 19, 60-809 Poznań, Poland
e-mail: lech.szajdak@isrl.poznan.pl

T. Rusu
University of Agricultural Sciences and Veterinary Medicine Cluj-Napoca (USAMV Cluj), Manastur Street 3-5, 400372 Cluj-Napoca, Romania
e-mail: rusuteodor23@yahoo.com

L.W. Szajdak (ed.), *Bioactive Compounds in Agricultural Soils*,
DOI 10.1007/978-3-319-43107-9_4

4.1 Introduction

Soil sulfur exists in both organic and inorganic substances; however, their proportion and form vary depending on soil type, fertilization, acidity, redox potential, cultivated plants, depth of profile, climate, and cultural conditions (Howarth et al. 1992). Inorganic soil sulfur comprises a small proportion (<25 %) of total sulfur, (S) and it is chiefly represented by soluble, adsorbed, insoluble, and coprecipitated SO_4^{-2} and S^- (Tabatabai 1986). Elementary sulfur (S^o) and other compounds of a lower oxidation state than SO_4^{-2} are mostly present as intermediates in the conversion (Wainwright 1979; Stevenson 1986; Watkinson 1989; Watkinson and Lee 1994; Watkinson and Bolan 1998). The assimilation of SO_4^{-2} under aerobic conditions results in the formation of an organic intermediate, PAPS (3-phosphoadenosine-5′-phosphosulfate) created due to the action between ATP and the sulfate. This intermediate acts as the source of the formation of sulfur-containing esters or amino acids. Dissimilatory sulfate reduction is largely caused by species of the bacteria genus *Desulfovibrio*, the product of this process being H_2S. Sulfate is used as the electron acceptor for the oxidation of organic compounds such as pyruvic, lactic, malic acid and ethanol, all formed by the fermentative activity of organisms. Sulfate reduction may be significant in terms of the removal of the available form of sulfur from the decomposition subsystem, and H_2S formation may be more important in other environmental respects (Swift et al. 1979). However, the formation of H_2S is a reversible process even under anaerobic conditions. The photoautotrophic sulfur bacteria such as *Chlorobium* and *Chromatium* use H_2S as the source of reducing power in the fixation of CO_2 and deposit elemental sulfur in or outside their cells. These organisms are relatively infrequent in terrestrial habitats except at the surface of bare waterlogged soils. Sulfur oxidation in soil containing sulfide or H_2S is largely dependent on the resumption of aerobic conditions. The most active group of sulfur-oxidizing organisms are the members of the chemoautotrophic family *Thiobacillaceae*.

The organic S mainly originates as plant and animal residues, which are subsequently degraded and remetabolized by soil organisms and microorganisms (Stevenson 1972; Steyn 1980). It is worth noting that sulfur exists mostly in the organic form (Vancura 1967; Thomson et al. 1986; McCaskil and Blair 1988; Mitchel et al. 1992; Lambert and Turner 1998). The main forms of organic S deposited in soil include S-containing amino acids and sulfonates. Sulfur in these compounds is directly bonded to carbon (C–S). It is also included in organic esters of sulfuric acid (C–O–S) in which S is bonded to oxygen in the form of C–O–SO_3 linkages (Saggar et al. 1981).

In addition, there are sulfonates in which sulfur occurs in the form N–O–SO_3^- and N–SO_3^-. Two percent of the total amount of sulfur in soil is available for plant uptake (Saggar et al. 1981, 1990a, b; Noggle et al. 1986; Wu et al. 1994; Benerjee and Chapman 1996). S-containing fertilizers are the main source of S in soil (Tabatabai 1984; Nguyen and Goh 1994; Fox et al. 1994; Boswell and Greg 1998). The amount of S supplied via fertilizers ranges from 10 to 30 kg ha^{-1} $year^{-1}$. Soil also obtains sulfur through rainfall (1–30 kg ha^{-1} $year^{-1}$) (Syers et al. 1987).

Table 4.1 Functional groups in the structures of sulfuric amino acids

Acidic	Basic
R–COOH	$-NH_2$
R–SH	=NH
$R-SO_2H$	–N=
$R-SO_3H$	

Table 4.2 The structures of sulfuric amino acids

Amino acids			MW
Acidic			
Cysteic acid	$HO_3S-CH_2-CH(NH_2)-COOH$		169.16
Taurine	$NH_2-CH_2-CH_2-SO_3H$	Tau	125.15
Cystine	$S-CH_2-CH_2-CH(NH_2)-COOH$ / $S-CH_2-CH_2-CH(NH_2)-COOH$ (S–S bonded)	Cys–S–S–Cys	240.30
Cysteine	$HS-CH_2-CH_2-CH(NH_2)-COOH$	Cys	121.16
Neutral			
Cystathionine	$HOOC-CH(NH_2)-CH_2-CH_2-S-CH_2-CH(NH_2)-COOH$		222.26
Methionine	$CH_3-S-CH_2-CH_2-CH(NH_2)-COOH$	Met	149.21

MW - molecular weight

Plants and microorganisms may create and transform different forms of sulfur. These compounds, located in soil organic matter, can be divided in two groups:

1. Compounds in which sulfur is carbon bonded:
   (i) Amino acids: methionine, cysteine, cysteine, taurine, cystathionine (Tables 4.1 and 4.2)
   (ii) Cofactors: biotin, thiamine, coenzyme A, ferredoxin (protein with including iron–sulfur) as well as lipoic acid (Figs. 4.2, 4.3, 4.4, and 4.5)
2. Compounds in which sulfur is included in the organic structure via –O-bonds and/or –N-bonds (Figs. 4.6, 4.7, 4.8, and 4.9). These compounds exhibit a high biological activity and instability in soil. They are the derivative of sulfatophosphates: adenosine 5-sulfatophosphate and adenosine 3-phosphate-5-sulfatophosphate. They participate in many sulfur conversions, transformations, and pathways of soil.

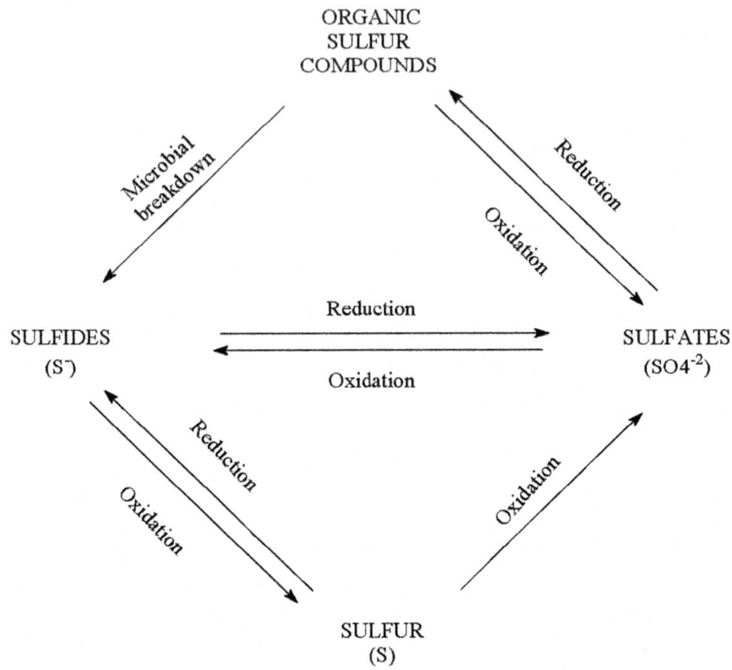

Fig. 4.1 Cycle of sulfur in the soils

Fig. 4.2 Biotin

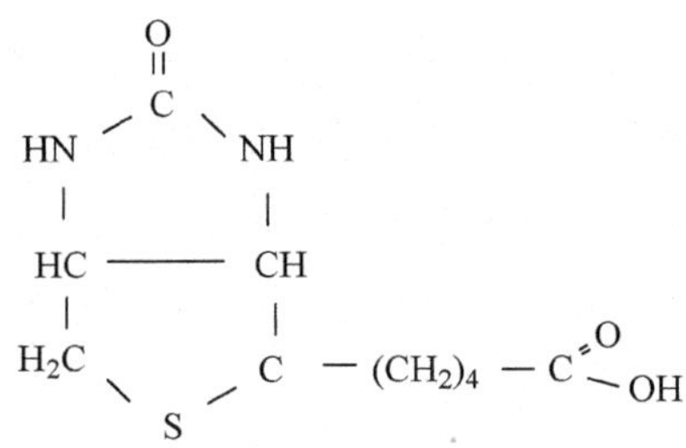

Fig. 4.3 Thiamine

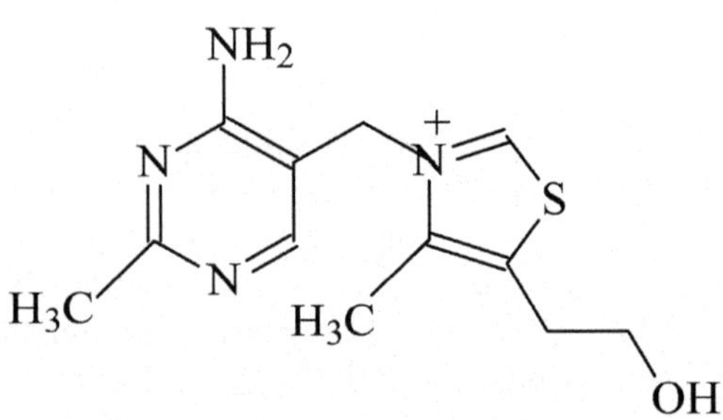

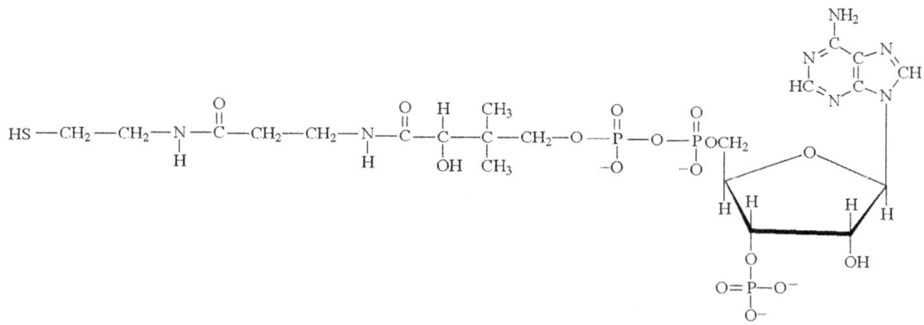

Fig. 4.4 Coenzyme A

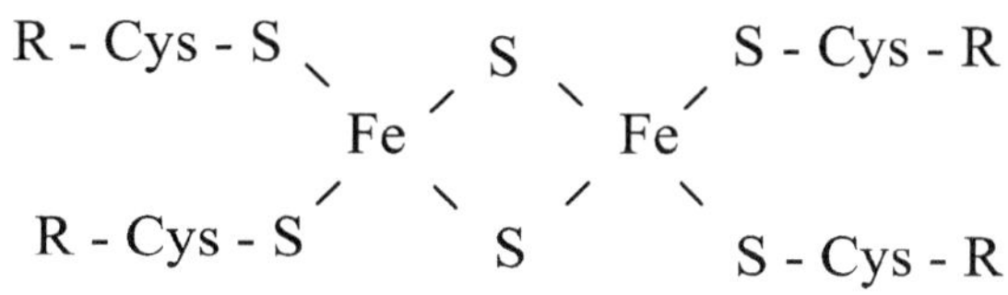

Fig. 4.5 Ferredoxin

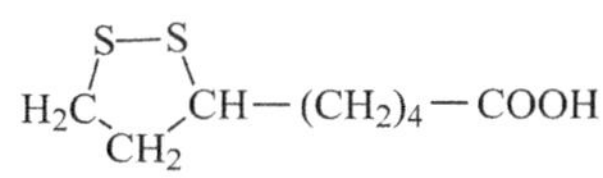

Fig. 4.6 Lipoic acid

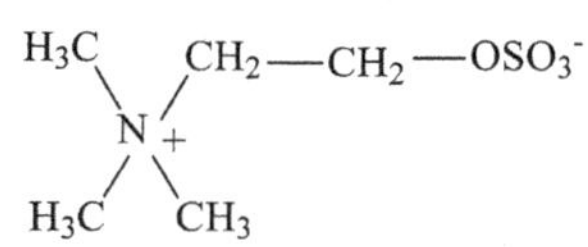

Fig. 4.7 Choline sulfate

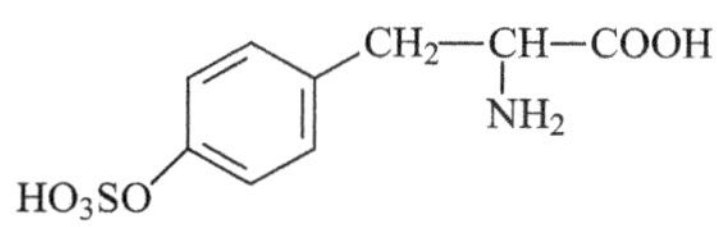

Fig. 4.8 Phenolic sulfate

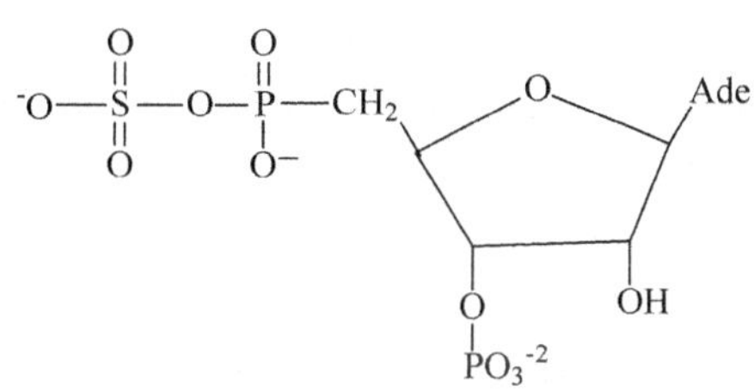

Fig. 4.9 3′ Phosphoadenosine-5′ phosphosulfate

Sulfur in the organic compounds of soil (R–C–S, C–O–S, and S–C–N), which originates from microorganisms, organisms, and plants, is transformed through aerobic and anaerobic conditions (Fitgerald 1976, 1978). The ester sulfates can hydrolyze to compounds with a hydroxylic group and sulfates. The hydrolysis leads to the cleavage of O–S bond. This conversion is catalyzed by sulfatase (Fig. 4.10).

The degradation of cysteine occurs in anaerobic conditions. The cysteine desulfhydrase and/or serine desulfhydrase catalyzes this conversion. In addition, methionine is transformed to mercaptan and ammonium (CH_3SH and NH_3).

Numerous studies on free amino acids reported that the concentration of sulfur amino acids was higher in the soils under continuous rye cropping than in the soils under Norfolk crop rotation (Szajdak and Ryszkowski 1997; Ryszkowski et al. 1994, 1998; Szajdak and Życzyńska-Bałoniak I 1995; Szajdak and Österberg 1996; Szajdak et al. 1996; Szajdak 1999). Sulfur amino acids convert to the derivatives of disulfide. These compounds activate the germination of spore fungus which leads to the infection of root system of cultivated plants.

The transformations of sulfur organically bound in soil occur with the contribution of rhodanese enzyme (thiosulfate–cyanide sulfurtransferase) [E.C. 2.8.1.1]. It catalyzes the conversion, generating a thiocyanate group from thiosulfate and cyanide (Starkey 1966; Mintel and Westley 1966; Nor and Tabatabai 1976, 1977; Tabatabai and Singh 1979; Page 1982; Stevenson 1986) (Figs. 4.11 and 4.12).

Fig. 4.10 Adenosine-5′ phosphosulfate

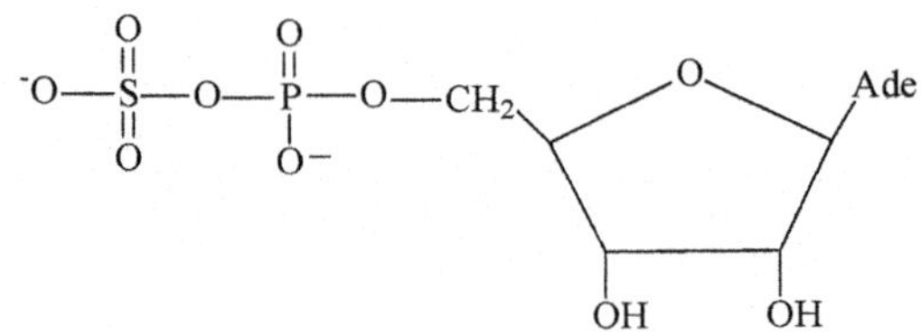

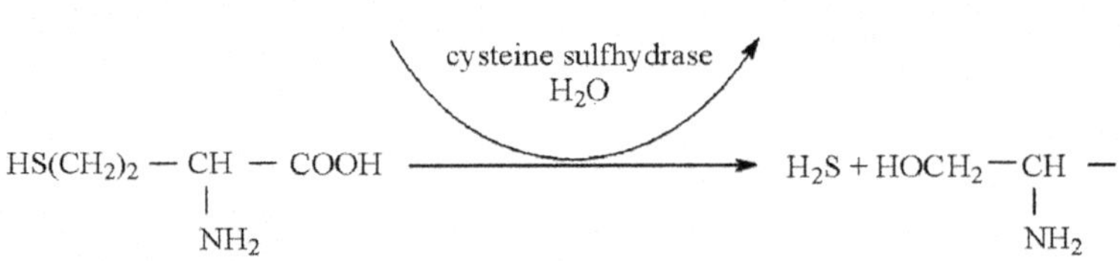

Fig. 4.11 Conversion of cysteine

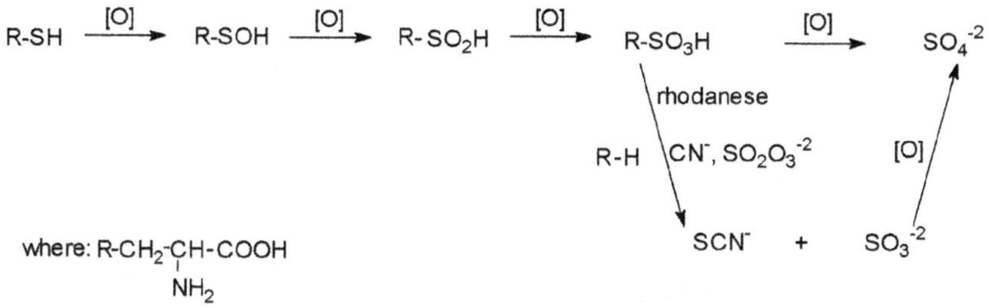

Fig. 4.12 Mechanism of the conversion of cysteine catalyzed by rhodanese

4.2 Methods for the Determination of Free Sulfuric Amino Acids in Soils

4.2.1 Materials and Equipment Required

Amino acids analyzer, pH meter, evaporator, lyophilizator, balance, freeze dryer, ultrasonic cleaner, autopipettes, and Pasteur pipettes

4.2.2 Reagents

(a) **Solution of standard amino acids** : The concentration of each amino acid equals 2.5 μM in 1 mL solution (Table 4.3) (Fig. 4.13), while the quantity of cystine is one-half of the amount of other amino acids. Standard commercial solutions should always be kept in the refrigerator. The manufacturer informs of the concentration of every amino acid in the solution.
(b) **Lithium citrate elution buffers per 1 L** :

   (i) *Lithium citrate buffer at pH = 2.9:* 9.4000 citric acid, 9.3880 g lithium chloride, 2.3000 g lithium citrate, 2 mL 30% Brij 35®, 0.1 mL caprylic acid, 2.5 mL thiodiglycol (Pierce Chemical Co., USA)

Table 4.3 Standard solution for the analyzing sulfuric amino acids. The amounts of amino acids are in 1 L

Amino acid	Amount per 1 L [g]
Cysteic acid	234.00
Taurine	156.40
Cystine	150.19
Methionine	186.50
Cystathionine	277.86

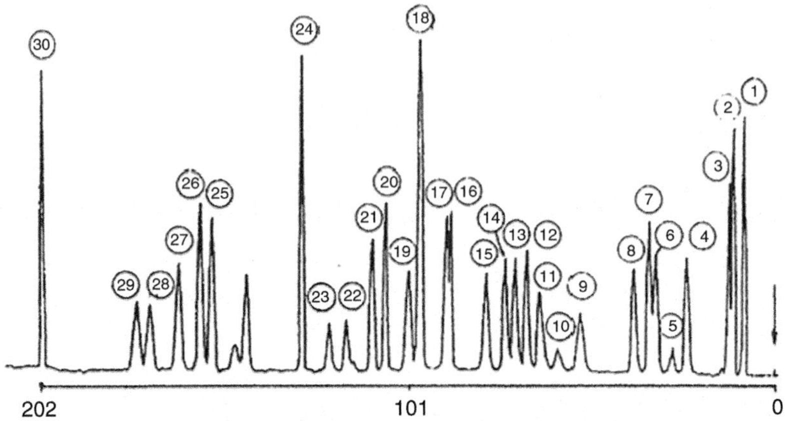

Fig. 4.13 Chromatograms of amino acids standards: (*1*) cysteic acid, (*2*) taurine, (*3*) phospoethanolamine, (*4*) urea, (*5*) hydroxyproline, (*6*) threonine, (*7*) serine, (*8*) glutamic acid, (*9*) aspartic acid, (*10*) proline, (*11*) glycine, (*12*) alanine, (*13*) citrulline, (*14*) α-aminobutyric acid, (*15*) valine, (*16*) cysteine, (*17*) methionine, (*18*) cystathionine, (*19*) isoleucine, (*20*) leucine, (*21*) tyrosine, (*22*) β-alanine, (*23*) β-aminobutyric acid, (*24*) γ-aminobutyric acid, (*25*) ornithine, (*26*) lysine, (*27*) histidine, (*28*) 1-methylhistidine, (*29*) 3-methylhistidine, (*30*) arginine (time of retention (min)

(ii) *Lithium citrate buffer at pH = 3.1:* 9.7000 g citric acid, 9.5700 g lithium chloride, 3.9000 g lithium citrate 2 mL 30 % Brij 35®, 0.1 mL caprylic acid, 2.5 mL thiodiglycol (Pierce Chemical Co., USA)

(iii) *Lithium citrate buffer at pH = 3.35:* 10.4980 g citric acid, 17.5000 g lithium chloride, 5.6500 g lithium citrate, 2 mL 30 % Brij 35®, 0.1 mL caprylic acid, 2.5 mL thiodiglycol (Pierce Chemical Co., USA)

(iv) *Lithium citrate buffer at pH = 4.05 :* 9.5000 g citric acid, 10.0000 g lithium chloride, 15.4500 g lithium citrate, 2 mL 30 % Brij 35®, 0.1 mL caprylic acid, 2.5 mL thiodiglycol (Pierce Chemical Co., USA)

(v) *Lithium citrate buffer at pH = 4.9 :* 8.5000 g citric acid, 39.9500 g lithium chloride, 50.6300 g lithium citrate, 2 mL 30 % Brij 35®, 0.1 mL caprylic acid

(c) **4 N acetate buffer at pH = 5.5** . Dissolve 1.0880 g sodium acetate in 800 mL of boiling deionized water. Decrease the solution to laboratory temperature and add 200 mL of ice acetic acid.

Flow rate of each buffer equals 12 mL/h.

(d) **Ninhydrin solution**: It must be prepared in a black bottle. Mix 1500 mL of methyl cellosolve (ethylene glycol monomethyl ether) and 500 mL of 4 N acetate buffer. Add 40 g of ninhydrin (analytical grade) to this solution and stir it under argon to dissolve ninhydrin. When ninhydrin is dissolved, add 0.8 g $SnCl_2$ $2H_2O$ and stir. Keep the solution of ninhydrin in the refrigerator. Flow rate of ninhydrin is 12 mL/h.

(e) **The separation of amino acids**: It is done in a glass column (0.37×20 cm) which includes resin Ostion LGFA^{0804}. Fill in the precolumn (1×8 cm) with resin Ostion LGKS^{0804}. Both Ostion LGFA and Ostion LGKS^{0804} are strong cation exchange resins.

The resin should be converted into the H^+ form. The process is carried out by 15 % HCl (1:5 ratio) on the water bath at 80 °C for 1 h. The suspension has to be stirred and next filtered with filter paper Whatman GT/C and washed with deionized water until pH neutral.

Resin is converted from H^+ form to Li^+ form using 2 N LiOH. After washing, resin should be diluted by the first elution buffer (pH = 2.9), (ratio 1:5) and left overnight. Then the buffer should be sucked off, while the dilution remains 1:1 and the column is ready for filling.

Norleucine is used as an internal standard. Norleucine is eluted immediately after leucine and before arginine. The concentration of commercially used norleucine as an internal standard equals 2.5 μM in 1 mL. The instrument is calibrated using the internal standard. The method of internal standard is based on calculating the ratio of the area under peak of the norleucine peak and the peat of investigated amino acid.

4.2.3 The Extraction of Free Sulfuric Amino Acids with Alcohol

The extraction should be done under mild conditions to prevent the degradation of unstable material. Thus, ethyl alcohol is used as the extraction agent as it does not degrade amino acids and can be easily evaporated.

500 g of soil with a known water content is mixed with 1500 mL of boiling alcohol (concentration should be adjusted with hot water depending on the water content in the soil sample). The final concentration of alcohol after mixing the sample should be 80 % by volume. The mixture is boiled for 5 min under a reflux condenser and allowed to cool to the ambient temperature. The mixture is centrifuged and filtered through Whatman GT/C filter paper (rinsed with 80 % alcohol). The filtrate extract should be evaporated and dried on the rotary vacuum evaporator at 40 °C.

4.2.4 The Determination of Free Amino Acids

The dry mass should be dissolved in 5 mL of alcohol and this concentrate should be stored in a refrigerator. 1 or 2 mL of this solution is taken in 25 mL volumetric flask and 0.5 mL of norleucine (internal standard, 10 % in isopropanol, 10 μmol mL^{-1}) and 1 mL of 1.5 N lithium citrate buffer (pH = 2.9) is added. The volume of solution is completed with deionized water up to the mark and the solution is kept in a refrigerator for 1 h. If necessary, the solution is centrifuged and used for analysis.

Other extraction agent instead of boiling ethanol leads to a larger or smaller amount of proteins in the extract solution. Thus, in order to interpret the result correctly, pay attention to the separation of low and high molecular portions. Proteins are precipitated by a sevenfold excess of acetone and by keeping them at −2 °C and mixing with 10 % of trichloroacetic acid in 1:1 ratio. After thorough washing, the precipitates are dried and weighed for hydrolysis, but please bear in mind that the presence of sulfates, heavy metals, and residues of organic agents cause losses.

4.2.5 The Extraction of Free Sulfuric Amino Acids with Supercritical Gas

4.2.5.1 Materials and Equipment Required

500-mL Erlenmeyer flask containing solvent, Milton–Roy mini pump: flow rate 16/60 mL/h, maximum pressure 36.0 MPa, three-way valve, pressure gauge 43.0 MPa, stainless steel tube (1.6×0.7 mm), HPLC tube, (stainless steel) 50 mL. Gas chromatographic oven with temperature control (from 25 to 500 °C) two-way valve. 250 Erlenmeyer flask to collect effluent solution (Schnitzer et al. 1986)

4.2.5.2 Reagents

(i) *n*-Pentane, temperature 250 °C, pressure 11.0 MPa; time is equal to 2 h.
(ii) Ethanol (EtOH) (100 v %), EtOH (75:25 v/v %), acetone–H_2O (40/60 v/v %), temperature 250 °C, pressure 14.0 MPa; time is equal to 2 h.

4.2.6 The Determination of Free Amino Acids

Each extract should be dried on a rotary evaporator at 45 °C to remove most of the solvent and then in a vacuum at 45 °C for 96 h and weighed (Fig. 4.14) (Schnitzer et al. 1986). Ten milligrams of dried sample is refluxed in a 50-mL conical flask with 25 mL of 6 N HCl for 24 h. Following hydrolysis, the insoluble residue is repeated by filtration and the filtrate taken repeatedly to dry on a rotary evaporator until most HCl is removed. The resulting residue is dissolved in Li–citrate buffer at pH = 2.9 and separated on amino acid analyzer. Coefficients of variation for amino acids between replicate analyses are 2.0 %.

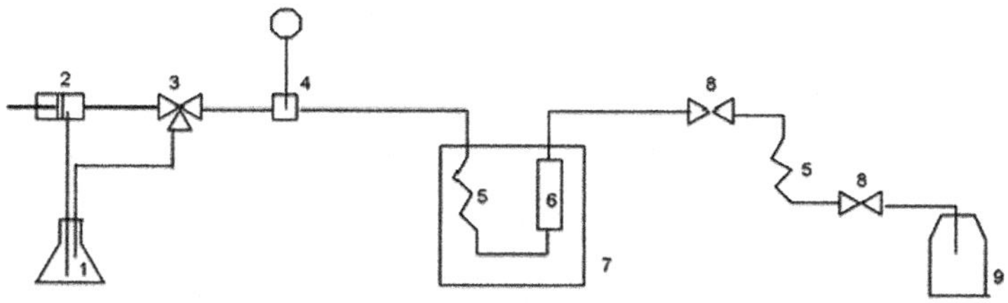

Fig. 4.14 Scheme of supercritical gas extractor. (*1*) 500-ml Erlenmeyer flask containing solvent; (*2*) mini pump, flow rate 16/60 mL/h, maximum pressure 36.0 MPa; (*3*) three-way valve; (*4*) pressure gauge 43.0 MPa; (*5*) stainless steel tube (1.6 by 0.7 mm); (*6*) HPLC tube, stainless steel 50 mL; (*7*) gas chromatographic oven with temperature control, 25–500 °C; (*8*) two-way valve; (*9*) 250-ml Erlenmeyer flask to collect effluent solution

4.2.7 The Determination of Rhodanese Activity in Soil

4.2.7.1 Reagents and Solutions

1. Toluene (Merck, Darmstadt, Germany)
2. Tris [Tris base tri(hydroxymethyl) aminomethane] (THAM, (Sigma Chemical Co.) in sulfuric acid solution (THAM-$H_2S)O_4$ buffer 0.05 M, pH = 6.0 ± 0.1)
3. 0.1 M $Na_2S_2O_3$ $5H_2O$
4. 0.1 KCN (Aldrich Chemical Co.)
5. Formaldehyde water solution of concentration 37 %
6. $CaSO_4$ $2H_2O$ (in formaldehyde solution)
7. 0.25 M $Fe(NO_3)_3$ 9 H_2O
8. Standard thiocyanide stock solution (Aldrich Chemical Co.) (0.02 M)
9. All the reagents used in the analyses are of analytical grade.

4.2.8 Method for the Determination of Rhodanese Activity in Soil

4 g of soil is placed in a Erlenmeyer flask (50 mL), and then 0.5 mL of toluene, 8 mL of THAM buffer, 1 mL 0.1 M 0.1 M $Na_2S_2O_3$ $5H_2O$, and 1 mL of KCN are added. The flask is swirled for a few seconds to mix the contents. The flask is stoppered and put in 37 °C water. After 1 h, the soil suspension is cooled to room temperature and 10 mL of $CaSO_4$ $2H_2O$ in formaldehyde solution is added.

The soil suspension is mixed for a few seconds and filtered through a Whatman No 2 folded filter paper. Then 5 mL of the filtrate is pipetted into a test tube, and 1 mL of the filtrate is pipetted into another test tube, and 1 mL of ferric nitrite water

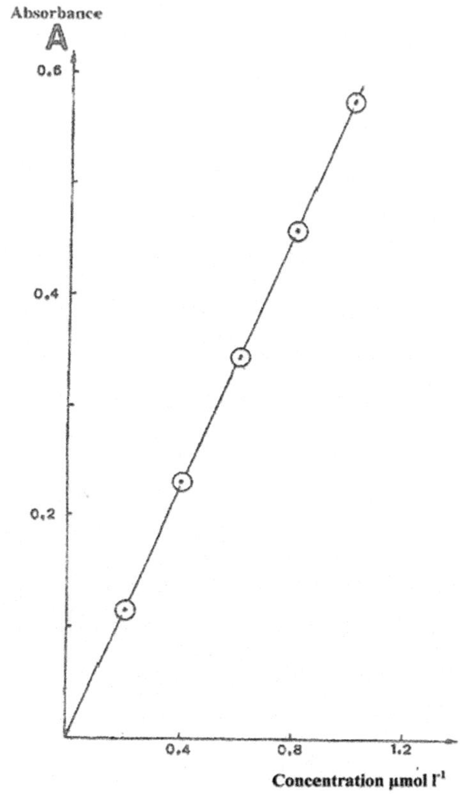

Fig. 4.15 Analytical curve of [A = f(c)] for iron thiocyanate at wavelength λ = 460 nm

solution is added. The absorbance of the solution is measured at the wavelength λ = 460 nm with 1-cm cell in relation to a reference sample.

To calculate the concentration of thiocyanate in filtrates, the analytical curve prepared earlier is used. Analytical curve is obtained with sodium thiocyanate content of 0.1, 0.4, 0.6, 0.8, and 1 µM of NaSCN. To obtain the following concentration range, 2, 4, 6, 8, 10 mL of basic thiocyanate solution is added to a 500-mL bulb. Afterward, 5 mL of each solution is placed in centrifugal test tubes of 10 mL volume, and 1 mL of ferric nitrite water solution is added to each of them. The subsequent procedure is the same as the one provided for the determination of rhodanese concentration in soil filtrates. The analytical curve of mean ferric thiocyanate absorbance values versus ferric thiocyanate concentration $[A = f(c_{Fe(SCN)3})]$ is constructed (Fig. 4.15).

The molar absorption coefficient of ferric thiocyanate is calculated as the slope of the analytical curve according to the Beer–Walter absorption law by means of the least squares formulas. The spectrophotometric measurements are carried out on Beckman Du-68 Spectrophotometer.

Rhodanese activity is determined by the method of Nor and Tabatabai (1976), which involves colorimetric determination of SCN^- produced when soil is incubated with buffered $S_2O_3^{-2}$ (pH 6.0) and CN^{-1} solutions and toluene at 37 °C for 1 h. The procedure used for the determination of SCN^- is based on the reaction of SCN^- with Fe^{+3} in acidic medium to form Fe–SCN colored complex. The intensity of the colored complex is measured spectrophotometrically at the wavelength $\lambda=460$ nm.

4.3 Conclusions

The quantity of S in soil organic matter varies due to a number of factors including vegetation, management practices, properties of soil, and climate. Organic S is the major form of S in most surface samples. The deficiency in crop production is becoming more common because of an increased use of S-free fertilizers, a reduction in the amount of S used as a pesticide, and higher crop yields, leading to higher demand for essential plants nutrients (Colemen 1966).

Winter rye has been grown continuously since 1957 on podzolic soils with a loamy sand texture (Table 4.4). The seven-course rotation comprised of potato, spring barley, alfalfa, oilseed rape, winter rye, and winter rye. Two types of fertilizers were NPK (N-nitrogen 90/kg/ha, P-phosphorus P_2O_5 60 kg/ha, K-potassium K_2O 120 kg/ha) and manure (30 t/ha). The mean temperature in winter was 0.1 °C, while in the growing season, it was 12.1 °C. Long-term mean rainfall amounted to 546 mm. Samples were analyzed immediately after collection from the soil layer at 20 cm depth.

The transformation of rhodanese activity in soils under crop rotation and continuous cropping of rye observed in the entire growing season was similar to the total transformations of sulfuric amino acids (Table 4.5).

During the entire period of study, the activity of rhodanese was higher in the soil under continuous rye cropping than in the soil under crop rotation. The mean annual values of rhodanese activity in the soil under continuous rye cropping were 483.1 nM g^{-1} h^{-1}, and they were 2.5 times higher than in the soil under crop rotation. The highest values of activity in both studied soils were noted at the beginning of March. In the soil under continuous rye cropping, they equaled 952.6 nM g^{-1} h^{-1}, whereas in the soil under crop rotation, they were 1.7 times lower. At the beginning of the

Table 4.4 Chemical properties of soils under crop rotation and continuous cropping of rye fertilized with NPK

		$C_{(organic)}$	$N_{(total)}$	$P_{(available)}$	$K_{(available)}$	$Mg_{(available)}$
Treatments	$pH_{(1\ M\ KCl)}$	g/kg				
CR control	5.8	6.80	0.727	0.073	0.059	0.020
CR NPK	5.9	6.48	0.696	0.109	0.680	0.019
CC control	5.7	7.23	0.727	0.060	0.036	0.021
CC NPK	5.8	6.96	0.696	0.099	0.090	0.025

Where *CR control* crop rotation – control, *CR NPK* rotation fertilized with NPK, *CC* control continuous cropping of rye – control, *CC NPK* continuous cropping of rye fertilized with NPK

Table 4.5 The activities of rhodanese in the soils under continuous cropping of rye and crop rotation

Type of cultivation	Activity of rhodanese ($\mu mol^{-1} \cdot g^{-1} \cdot h^{-1}$)									
	Time of sampling									
	5.III.	23.IV.	10.V.	18.V.	23.VI.	10.VII.	3.IX.	14.IX.	3.X.	25.X.
Continuous cropping of rye	0.9526	0.7268	0.5874	0.3134	0.2939	0.2928	0.2914	0.4058	0.4957	0.5013
	±0.036	±0.028	±0.022	±0.012	±0.011	±0.011	±0.011	±0.015	±0.019	±0.018
Crop rotation	0.5604	0.2845	0.1819	0.0979	0.0453	0.0837	0.1084	0.1103	0.1611	0.3042
	±0.021	±0.011	±0.007	±0.004	±0.002	±0.003	±0.004	±0.004	±0.006	±0.012

Szajdak (1996)

$x \pm \Delta x$—mean values with their confidence limit Δx at $\alpha = 0.05$

Table 4.6 Seasonal changes of the free sulfuric amino acids in the soil under continuous cropping of rye in μg·kg⁻¹ of soil

Sulfuric amino acids	The concentrations of the free sulfuric amino acids (μg·kg⁻¹) Time of sampling									
	5.III.	23.IV.	10.V.	18.V.	23.VI.	10.VII.	3.IX.	14.IX.	3.X.	25.X.
Cysteic acid	242.3	210.4	150.2	98.6	75.8	69.4	104.2	100.6	110.6	112.3
	±9.5	±8.2	± 5.9	±3.9	±3.0	±2.7	±4.1	±3.9	±4.3	±4.4
Taurine	48.2	40.3	20.1	12.9	16.3	20.9	31.7	28.7	30.7	39.2
	±1.9	±1.6	±0.8	±0.5	±0.6	±0.8	±1.2	±1.2	±1.2	±1.5
Cysteine	48.3	19.9	35.2	18.7	12.3	18.4	15.4	40.6	46.7	51.8
	±1.8	±0.8	±1.3	±0.7	±0.5	±0.7	±0.6	±1.6	±1.8	±2.0
Methionine	12.3	11.7±	10.8	7.9	–	15.2	10.2	35.6	33.8	31.2
	±0.5	0.5	±0.4	±0.3		±0.6	±0.4	±1.4	±1.3	±1.2
Cystathionine	27.8	10.3	26.1	13.9	15.2	12.4	11.7	10.8	9.8	12.4
	±1.1	±0.4	±1.0	±0.6	±0.5	±0.5	±0.5	±0.4	±0.4	±0.5
Total amount	378.9	292.6	242.4	152.0	119.6	136.3	173.2	216.3	231.6	246.9
	±14	±10	±8	±6	±6	±5	±6	±8	±8	±9

Szajdak (1996)
$x \pm \Delta x$—mean values with their confidence limit Δx at $\alpha = 0.05$

phase of plant intensive growth, a decrease in the values of rhodanese activity in both soils was observed. The activity remained low until the end of June—the beginning of flowering. In this period, both in the soil under continuous rye cropping and in the soil under crop rotation, the lowest values of rhodanese activity were noted. In the sample collected at the end of June from the soil under continuous rye cropping, the values of rhodanese activity amounted to 263.9 nM g^{-1} h^{-1} and were 5.8 times higher than the values of rhodanese activity determined in the soil under rotation. After the harvest, an increase in rhodanese activity was observed in both soils. The activity of rhodanese in soil could be taken as a measure of the abundance of fungi. A considerable density of microorganisms in soil participating in the transformation cycle of sulfur is closely associated with the total S^{-2}, $S_2O_3^{-2}$, $S_4O_6^{-2}$, SO_3^{-2}, and SO_4^{-2} concentration (Wainwright 1979). The increased content of these sulfur forms significantly contributes to a lowered soil pH. Therefore, in this study, the soil pH under continuous rye cropping was found to be about 0.25 pH units lower in comparison with the soil under crop rotation. A higher concentration of free sulfuric amino acid in the soil under continuous rye cropping and higher values of rhodanese activity may be interpreted to mean that fungi showing secondary metabolism increase participation in generating sulfuric amino acids in these conditions.

Total amount of free sulfuric amino acids in the soil under continuous rye cropping ranged from 119.6 to 378.9 μg kg^{-1} (Table 4.6).

In the soil under crop rotation, the concentration of these compounds was lower and reached from 80.7 to 328.0 μg kg^{-1}. The mean annual value of all determined sulfuric amino acids in the soil under continuous rye cropping equaled 218.9 μg kg^{-1} and was 27 % higher than in the soil under crop rotation. The highest concen-

Table 4.7 Seasonal changes of the free sulfuric amino acids in the soil under crop rotation in μg · kg⁻¹ of soil

Sulfuric amino acids	The concentrations of the free sulfuric amino acids (μg · kg⁻¹) Time of sampling									
	5.III.	23.IV.	10.V.	18.V.	23.VI.	10.VII.	3.IX.	14.IX.	3.X.	25.X
Cysteic acid	199.5	202.7	120.7	82.4	60.4	65.4	70.4	69.8	84.7	78.4
	±7.8	±7.6	±4.8	±0.1	±2.4	±2.6	±2.9	±2.8	±3.3	± 3.2
Taurine	42.7	28.6	18.5	9.5	12.8	10.6	15.2	8.4	9.7	15.8
	±1.7	±1.1	±0.7	±0.4	±0.5	±0.4	±0.6	±0.3	±0.4	±0.6
Cysteine	35.4	15.6	10.8	5.7	2.4	5.4	6.7	20.7	29.5	32.6
	±1.4	±0.6	±0.4	±0.2	±0.1	±0.2	±0.3	±0.8	±1.2	±1.3
Methionine	19.6	20.8	11.6	3.2	1.2	2.7	5.1	19.5	31.7	35.2
	±0.8	±0.8	±0.5	±0.1	±0.1	±0.1	±0.2	±0.9	±1.2	±1.4
Cystathionine	30.8	12.7	–	1.7	3.4	2.8	4.8	2.7	5.8	8.4
	±1.2	±0.53		±0.1	±0.1	±0.1	±0.2	±0.1	±0.2	±0.3
Total amount	328.0	280.4	162.6	102.5	80.7	86.9	102.2	121.1	161.4	170.4
	±12	±10	±6	±5	±3	±4	±4	±4	±6	±6

Szajdak (1996)
$x \pm \Delta x$—mean values with their confidence limit Δx at $\alpha = 0.05$

tration of free sulfuric amino acids was found in both soils at the beginning of March. At the time, the amount of free sulfuric amino acids in the soil under continuous rye cropping equaled to 243.3 μg kg⁻¹ and was 13.4 % higher than in the soil under crop rotation. However, in the flowering phase, which is the period of plant intensive growth (18.05–23.06), the amount of free sulfuric amino acids decreased rapidly. The decrease in free sulfuric amino acids in the soil under continuous rye cropping was fourfold. In contrast, this phenomenon was reversed in the period of decay after harvest when an increase in free sulfuric amino acids content was observed in the studied soils. The obtained results confirm the results of other studies showing that the changes in free sulfuric amino acids are more substantial in the soils under crop rotation, especially in the decay period after harvest (Życzyńska-Bałoniak and Szajdak 1991; 1992) (Table 4.7).

The analysis of the concentration of each amino acid in the samples collected in different periods of time showed that the dominating amino acid was cysteic acid. The highest concentration of cysteic acid was observed in March in both studied soils. At that time, the concentration of this amino acid was 18 % higher in the soil under continuous rye cropping than in the soil under crop rotation. The most remarkable difference between both concentrations was however observed in the middle of May when the concentration of cysteine acid in the soil under continuous rye cropping was 97 % higher than in the soil under crop rotation. Cysteine in the soils constitutes a substrate for transformation, the product of which is cysteic acid (Freney 1958; Freney et al. 1972; Freney and Stevenson 1972; Freney and Williams 1983; Freney 1986). Almost twice more cysteine was found in the soil under continuous rye cropping than in the soil under crop rotation. The transformation

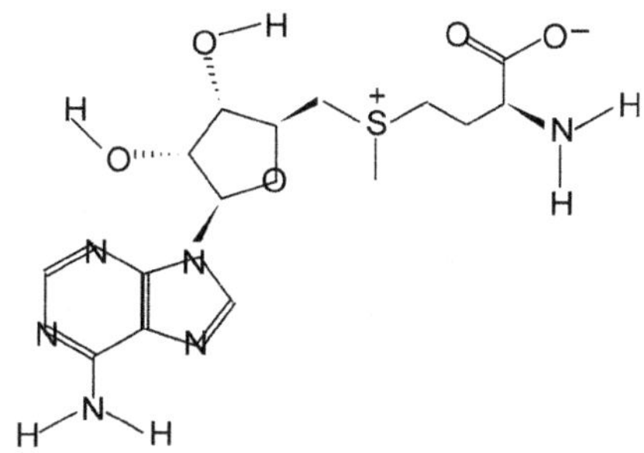

Fig. 4.16 S-Adenosylmethionine (also known as ademetionine), S-adenosyl-L-methionine, active methionine, methioninyl adenylate, adenosylmethionine. Molecular formula, $C_{15}H_{22}N_6O_5S$; molecular weight, 398.43738. Physiologic methyl radical donor involved in enzymatic transmethylation reactions and present in all living organisms

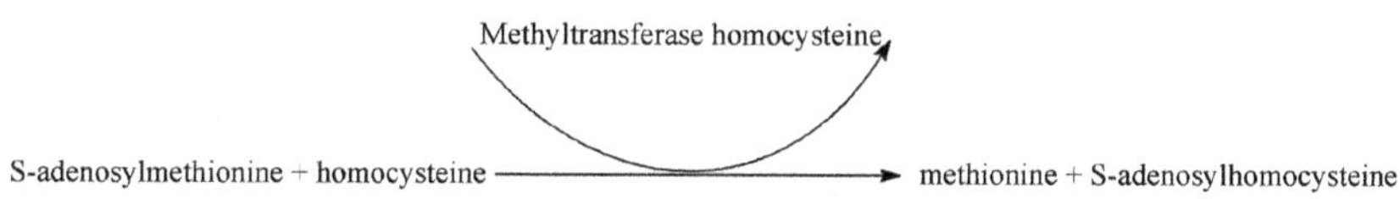

Fig. 4.17 Transformation of the homocysteine with formation of methionine

coefficient is defined as the ratio of the mean value of cysteic acid concentration to the mean value of cysteine concentration. In the soil under crop rotation, it equaled 5.8, whereas in the soil under continuous rye cropping, it equaled to 4.1. The higher transformation coefficient in the soil under crop rotation as compared to the soil under continuous rye cropping indicates a more efficient transformation of cysteic acid in the soil under crop rotation. It reflects higher oxidative conditions occurring in the soils under crop rotation than in the soils under continuous cropping of rye.

Methionine's content, among all studied sulfuric amino acids, was found to differ the least in both examined soils. This substance comes into soil from its active form, S-adenosylmethionine, and it is identified as a biologically active substance, a product of the secondary metabolism of fungi (Bu'Lock 1980; Trojanowski 1973). S-Adenosylmethionine is physiologic methyl radical donor involved in enzymatic transmethylation reactions and present in all living organisms (Fig. 4.16).

Methyltransferase homocysteine participates in the transfer of methyl group of S-adenosylmethionine to homocysteine with the creation of methionine (Fig. 4.17).

The higher concentration of methionine was found in the soil under continuous rye cropping which suggested that more fungi demonstrated the secondary metabolism in the soil under continuous rye cropping than in the soil under crop rotation. This conclusion is in line with the results of other studies carried out by microbiologists who discovered that in the soils under continuous rye cropping, there was a trend toward a rise in fungal biomass and a decrease in the number of bacteria (Durska et al. 1986).

Table 4.8 Statistical evaluation of the free sulfuric amino acids dependence in the soil under continuous rye cropping and crop rotation in function of rhodanese activity

Type of cultivation	Slopes with confidence limit	Movement	Correlation coefficients	Test values for comparison of slopes	
				Experimental value	Theoretical value
	$a \pm \Delta a$	b	r	t_d	t_t
Continuous cropping of rye	344.54 ± 12.7	52.53	0.975	2.2074	2.1199
Crop rotation	501.29 ± 19.1	62.49	0.926		

Szajdak (1996)
Where $a \pm \Delta a$—slope with confidence limit at $\alpha = 0.05$ and $(n-2)$ degrees of freedom, b-movement, r-correlation coefficient, t_D and t_r-test values for comparison of slopes

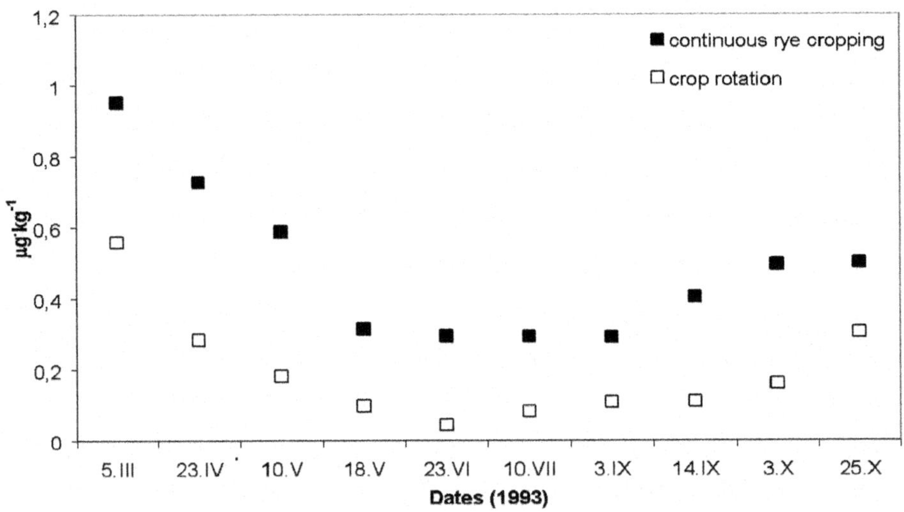

Fig. 4.18 Changes of the free sulfuric acids in the soils under continuous rye cropping (■) and crop rotation (□)

The linear and positive correlation between the concentration of free sulfuric amino acids and rhodanese activity in the soil under continuous rye cropping and in the soil under crop rotation was proved (Table 4.8) (Figs. 4.18 and 4.19).

Despite the correlation these two lines are not parallel. The slope of the curve of the soil under continuous rye cropping was about 1.5 times lower than the one determined for the soil under crop rotation. This indicates that the transformation processes of free sulfur amino acids and rhodanese were about 1.5 times slower in the soil under continuous rye cropping than in the soil under crop rotation. In the soil

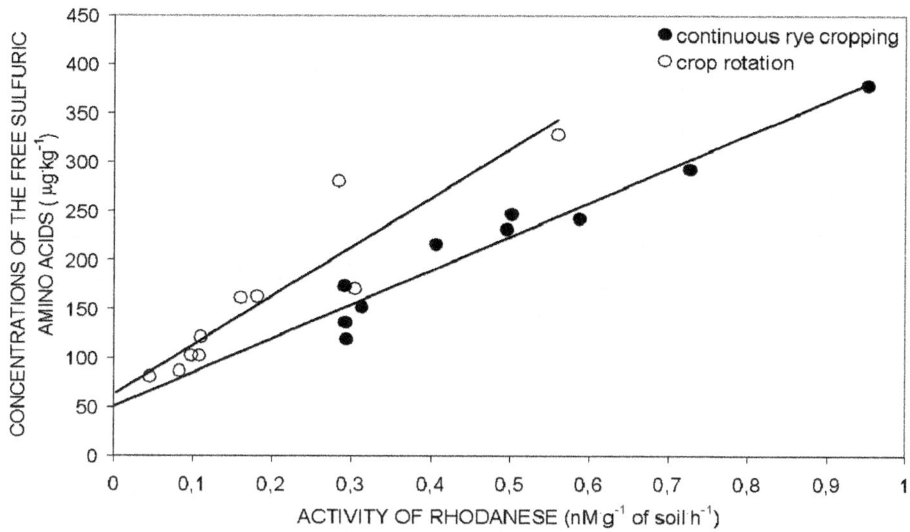

Fig. 4.19 Concentrations of the free sulfuric acids in the soils under continuous rye cropping (●) and crop rotation (○) in function of rhodanese activity

under continuous rye cropping, smaller microbe biodiversity and limits to biochemical transformations occurred in soil.

For many years, the role, the participation in the cycle, and the impact on crop yield of different types of S have been neglected. Data gathered in the studies indicated that in many soils the transformation involving organic form is of upmost importance for the S mineral nutrition of plants. The biotic and abiotic soil processes regulate S dynamic. Biotic conversions include the microbial and plant assimilation as well as the dissimilation of sulfate into organic S of plant and animal residues by soil microflora and microfauna. Abiotic processes include the complex physicochemical reactions, for instance, precipitation, reduction, and adsorption with soil mineral surfaces (Saggar and Bolan 2003). Thanks to new analytical methods allowing to determine S-containing compounds more accurately, the knowledge of their derivatives, degradation products, and metabolites is more profound, however still scarce.

References

Benerjee MR, Chapman SJ (1996) The significance of microbial biomass sulphur in soil. Biol Fert Soils 22:116–125

Boswell CC, Greg PEH (1998) Sulphur fertilizers for grazed pasture systems. In: Mauynard DG (ed) Sulphur in the environment. Marcel Dekker Inc, New York, pp 95–135

Bu'Lock JD (1980) Biosynthesis of ergot toxins. In: Steyn PS (ed) Biosynthesis of mycotoxins. A study in secondary metabolism. Academic, New York, pp 1–16

Colemen R (1966) The importance of sulphur as a plant nutrient in world crop production. Soil Sci 101:230–239

Durska G, Kaczmarek W, Kaszubiak H, Muszyńska M, Pędziwilk Z (1986) Występowanie bakterii i grzybów w glebie oraz ich aktywność oddechowa pod uprawą żyta w płodozmianach uproszczonych. Abundance of the bacteria and fungus and their respiratory activity under continuous cropping of rye. In: Myśków W, Kuś J, Kamińska M (eds) Ekologiczne skutki monokulturowej uprawy zbóż. Instytut Uprawy Nawożenia i Gleboznawstwa, Puławy, pp 95–106 (in Polish)

Fitgerald JW (1976) Sulphate esters formation and hydrolysis: A potentially important yet often ignored of the sulphur cycle of aerobic soils. Bact Rev 40:698–721

Fitgerald JW (1978) Naturally occurring organo suphur compounds in soils. In: Nriagu JO (ed) Sulphur in the environment. Part II Ecological impacts. Willey, New York, pp 391–443

Fox RL, Atesalp HM, Kampbel DH, Rhodes HF (1994) Factors influencing the availability of sulfur fertilizers to alfalfa and corn. Soil Sci Soc Am Proc 28:406–408

Freney JR (1958) Aerobic transformation of cysteine to sulphate in soil. Nature 8:1318–1319

Freney JR (1986) Form and reaction of organic sulphur compounds in soils. In: Tabatabai MA (ed) Sulphur in agriculture, Agronomy monograph 27. Society of Agronomy, Madison, pp 207–232

Freney J, Stevenson FJ (1972) Organic sulfur transformations in soils. Soil Sci 101:307–316

Freney JR, Williams CH (1983) The sulphur cycle in soil. In: Ivanov MV, Freney JR (eds) Global biogeochemical sulphur cycle. Willey, New York, pp 129–201

Freney JR, Stevenson FJ, Beavers AH (1972) Sulfur-containing amino acids in soil hydrolysates. Soil Sci 114(6):468–476

Howarth RW, Stewart JWB, Ivanov WV (1992) Sulphur cycling on the continents. Willey, New York, p 350

Lambert MJ, Turner J (1998) Sulphur nutrition and cycling in southern hemisphere temperate and subtropical forest ecosystems. In: Maynard DG (ed) Sulphur in the environment. Marcel Dekker Inc, New York, pp 253–294

McCaskil MR, Blair GJ (1988) Development of a simulation model of sulphur cycling in grazed pastures. Biogeochemistry 5:165–181

Mintel R, Westley J (1966) The rhodanese reaction, mechanism of thiosulfate binding. J Biol Chem 241:3386–3389

Mitchel MJ, David MB, Harrison RB (1992) Sulphur dynamics of forest ecosystems. In: Howarth RW, Stewart JWB, Ivanov WV (eds) Sulphur cycling on the continent. Willey, New York, pp 215–260

Nguyen ML, Goh KM (1994) Sulphur cycling and its implications on sulphur fertilizer requirements of grazed grassland ecosystems. Agric Eco Envir 49:173–206

Noggle JC, Meagher JF, Jones UC (1986) Sulphur in the atmosphere and its effects on plant growth. In: Tabatabai MA (ed) Sulfur in agriculture, Agronomy monograph 27. American Society of Agronomy, Madison, pp 251–278

Nor YM, Tabatabai MA (1976) Extraction and colorimetric determination of thiosulfate and tetrathionate in soils. Soil Sci 122:171–178

Nor YM, Tabatabai MA (1977) Oxidation of elemental sulfur in soils. Soil Sci Soc Am J 41:736–741

Page AL (1982) Methods of soil analysis. Part 2. Chemical and microbiological properties, 2nd edn, Agronomy No 9. American Society of Agronomy, Crop Science Society of America, Soil Science Society of America, Madison, pp 901–947

Ryszkowski L, Karg J, Szajdak L (1994) Effects of continuous cereal cropping on biota. In: Tallis JH, Norman HJ, Benton J (eds) VIth International Congress of Ecology, INTECOL, Book of Abstracts: Manchester 21–26 August 1994, pp 85–87

Ryszkowski L, Szajdak L, Karg J (1998) Effects of continuous cropping of rye on soil biota and biochemistry. Crit Rev Plant Sci 17(2):225–244

Saggar S, Bolan NS (2003) Secondary nutrients: sulphur, calcium, and magnesium. In: Benbi DK, Nieder R (eds) Handbook of processes and modeling in the soil-plant system. Food Products Press/The Haworth Reference Press, New York, pp 561–588

Saggar S, Bettany JR, Stewart JWB (1981) Sulphur transformation in relation to carbon and nitrogen in incubated soils. Soil Biol Bioch 13:499–511

Saggar S, Hedley MJ, Gillingham AG, Rowarth JS, Richardson S, Bolan NS, Gregg PEH (1990a) Predicting the fate of fertilizer sulphur in grazed hill country pasture by modeling the transfer and accumulation of soil phosphorus. N Z J Agric Res 33:129–138

Saggar S, Mackay AD, Hedley MJ, Lambert MG, Clark DA (1990b) A nutrient transfer model to explain the fate of phosphorus and sulphur in grazed hill-country pasture. Agric Eco Envir 30:295–315

Schnitzer A, Hindle AA, Meglic M (1986) Supercritical gas extraction of alkanes and alkanoic acids from soils and humic materials. Soil Sci Soc Am J 50:913–919

Starkey R (1966) Oxidation and reduction of sulfur compounds in soils. Soil Sci 101(4):297–306

Stevenson G (1972) Biologia grzybów, bakterii i wirusów. Biology of fungus, bacteria and viruses. PWRiL, Warszawa, p 280

Stevenson FJ (1986) Cycles of soils, A Wiley-Interscience publication. Wiley, New York, pp 285–320

Steyn PS (1980) The biosynthesis of mycotoxins. A study in secondary metabolism. Academic, London, p 355

Swift MJ, Heal OW, Anderson JM (1979) Decomposition in terrestrial ecosystems. Blackwell Scientific Publications, Oxford/London, pp 167–219

Syers JK, Skinner RJ, Curtin D (1987) Soil and fertilizer sulphur in UK agriculture. In: Proceedings of the Fertilizer Society, London, p 264

Szajdak L (1996) Impact of crop rotation and phenological periods on rhodanese activity and free sulfuric amino acids concentrations in soils under continuous rye cropping and crop rotation. Environ Int 22:563–569

Szajdak L (1999) Wpływ wieloletniej uprawy monokulturowej żyta na zawartość i właściwości fizykochemiczne związków biologicznie czynnych w glebach. Impact of long-term continuous cropping of rye on the content and physical properties of biologically active substance in soils. Zesz Nauk Akad Rol Wrocł. Akademia Rolnicza we Wrocławiu. pp 104 (in Polish)

Szajdak L, Österberg R (1996) Amino acids present in humic acids from soils under different cultivations. Environ Int 22(3):331–334

Szajdak L, Ryszkowski L (1997) Wpływ wieloletniej uprawy żyta w monokulturze na zawartość substancji biologicznie czynnych w glebie (Impact of long-term continuous cropping of rye on the concentrations of biologically substances in soils). Acta Acad Agric Tech Olst Agric 64:157–170 (in Polish)

Szajdak L, Życzyńska-Bałoniak I (1995) Biological active substances in soils under crop rotation and continuous cropping of rye. 5th Nordic symposium on humic substances and humex project seminar, Lund, 6–8 June, p 63

Szajdak L, Życzyńska-Bałoniak I, Blecharczyk A, Matuszewska T (1996) Amino acids in humic acids from soils under continuous cropping of rye and crop rotation. First world congress on allelopathy, Cadiz, Sept 16–20. p 86

Tabatabai MA (1984) Importance of sulphur in crop production. Biogeochemistry 1:45–62

Tabatabai MA (1986) Sulfur in agriculture. Agronomy No 27 in the series. Am Soc Agron, Crop Sci Soc Am, Soil Sci Soc Am, Madison, p 1–22

Tabatabai MA, Singh BB (1979) Rhodanese activity of soils. Soil Biol Bioch 11:9–12

Thomson JF, Smith IK, Madison JT (1986) Sulphur metabolism in plants. In: Tabatabai MA (ed) Sulphur in agriculture, Agronomy monograph 27. American Society of Agronomy, Madison, pp 59–122

Trojanowski J (1973) Przemiany substancji organicznych w glebie (Transformation of organic compounds in soils). PWRiL, Warszawa, p 331 (in Polish)

Vancura V (1967) Root exudates of plant. III. Effects of temperature and "cold shock' on the exudation of various compounds from seeds and seedlings of maize and cucumber. Plant Soil 27:319–328

Wainwright M (1979) Microbial S-oxidation in soils exposed to heavy atmospheric pollution. Soil Biol Bioch 11:95–98

Watkinson JH (1989) Measurement of the oxidation rate of elemental sulfur in soils. Aust J Soil Res 27:365–375

Watkinson JH, Bolan NS (1998) Modeling the rate of elemental sulphur oxidation in soils. In: Maynard DG (ed) Sulphur in the environment. Marcel Dekker Inc, New York, pp 135–172

Watkinson JH, Lee A (1994) Kinetics of field oxidation of elemental sulphur in New Zealand pastoral soils and the effects of soil temperature and moisture. Fert Res 37:59–68

Wu J, O'Donnel AG, He ZL, Syres JK (1994) Fumigation-extraction method for the measurement of microbial biomass-S. Soil Biol Bioch 26:117–125

Życzyńska-Bałoniak I, Szajdak L (1991) Sezonowe zmiany aminokwasów w glebie spod uprawy roślin w monokulturze i zmianowaniu (Seasonal changes of amino acids in soils under continuous cropping of rye). Badania nad bilansem substancji organicznej i składników pokarmowych w układzie gleba-roślina. Materiały sesji naukowej poświęconej pamięci Prof. dr. hab. Włodzimierza Łoginowa. Bydgoszcz 23 września, pp 85–87 (in Polish)

Życzyńska-Bałoniak I, Szajdak L (1992) Sezonowe zmiany wolnych aminokwasów w glebie spod monokultury żyta i zmianowania. In: Balazy S, Ryszkowski L (eds) Produkcja pierwotna, zasoby zwierząt i wymywanie materii organicznej w krajobrazie rolniczym (Primary production, resource of animals and leaching in agricultural landscape). ZBŚRiL PAN, Poznań, pp 7–15 (in Polish)

Chapter 5
Amino Acids, Indole-3-Acetic Acid, Stable and Transient Radicals, and Properties of Humic and Fulvic Acids as Affected by Tillage System

Lech Wojciech Szajdak, Miguel L. Cabrera, and Adam Jezierski

Abstract The concentration of bound amino acids and indole-3-acetic acid in humic acid (HA) and fulvic acid (FA) in samples collected from soils under no-tillage (NT) and conventional tillage (CT) management was measured. The samples were obtained from two long-term studies at the Horseshoe Bend (HSB) Experimental Area and at the Bledsoe Research Farm (BRF) in Georgia, USA. This study demonstrated the impact of NT and CT management on the content of amino acids in HA and FA. The total amount of bound amino acids in HA from NT was higher than in HA from CT. The contrary observation was noted for FA. Conventional tillage led to a decrease in the EPR signal intensity for HA, indicating a decrease in the dimension of the aromatic conjugation systems in these molecules. Transition from CT to NT is accompanied by a considerable rise in the signal intensity, which reflects more conjugation in the HA from NT compared to those from CT. Thus, HA from NT management from Horseshoe Bend Experimental Area and from Bledsoe Research Farm may be characterized by higher molecular weights and higher degree of condensation of aromatic constituents than HA from CT.

Keywords Tillage system • Humic and fulvic acids • Amino acids • Auxine • Transient radicals

---

L.W. Szajdak (✉)
Institute for Agricultural and Forest Environment, Polish Academy of Sciences, Str. Bukowska 19, 60-809 Poznań, Poland
e-mail: lech.szajdak@isrl.poznan.pl

M.L. Cabrera
College of Agricultural and Environmental Sciences, Department of Crop and Soil Sciences, The University of Georgia, Athens, GA 30602, USA

A. Jezierski
Faculty of Chemistry, Wrocław University, 50-383 Wrocław, Poland

L.W. Szajdak (ed.), *Bioactive Compounds in Agricultural Soils*,
DOI 10.1007/978-3-319-43107-9_5

5.1 Introduction

Long-term research on no-till systems in the 1970s showed positive impact on some physical, chemical, biochemical, and biological soil properties, which favored the development of no-till production systems (Wells and Touchton 1985). In no-tillage, residues from previous crops accumulate at the soil surface and lead to the development of a surface mulch layer. This mulch layer decreases evaporation and increases water infiltration, which results in soil moisture contents (during the growing season) that are 15–20 % greater in no-tillage than in conventional tillage. Furthermore, the mulch functions as a barrier between the soil surface and the atmosphere, thereby leading to slower changes in soil temperature when compared to conventional tillage. As a result, no-tillage has cooler soil temperatures in the spring and summer and warmer soil temperatures in late fall with overall lower temperature fluctuations (Herbek et al. 1984; NeSmith et al. 1985).

Under NT management, mechanical incorporation of fertilizers or crop residues within the upper soil layer is not possible. Consequently, nutrients taken up by plant roots from the subsoil and incorporated into the plant are subsequently deposited on the soil surface together with plant residues. Several studies have documented changes in the distribution of organic matter, nutrients, and acidity (pH) in the soil profile as a result of NT practices (Van Doren et al. 1976; Blevins et al. 1977; Juo and Lal 1979; Doran 1980; House et al. 1984; Hargrove et al. 1985; Worsham and Lewis 1985; Hendrix et al. 1986; Parmelee et al. 1989; Stearman et al. 1989a, b; Lipiec and Stępniewski 1995). Five- and ten-year tillage with corn indicated that soil organic matter increased significantly in NT soil compared to CT soil (Blevins et al. 1977, 1984). Tyler et al. (1983) showed significantly higher levels of organic matter in NT than in CT treatments after 2 years in soybean. Organic matter in the surface layer of NT soils is approximately twice that of tilled soils receiving the same inputs after 5 years (Phillips and Phillips 1984).

Most changes in soil organic matter due to no-tillage are observed in the upper few centimeters of soil (Franzluebbers 2004). Reduced tillage or no-tillage leads to accumulation of not only plant residues but also fine roots and microbial and microfaunal debris (Gregorich et al. 1994; Alvarez et al. 1998), which eventually lead to an increase in mineralized N from organic matter (Carter and Rennie 1984; Seiter and Horwath 2004). The accumulation of organic degradation products at the soil surface on no-tillage leads to greater microbial activity (aerobic and anaerobic bacteria) in no-tilled than in conventionally tilled soils. The larger number of aerobic and anaerobic bacteria together with greater organic matter and moisture contents provides a greater potential for the conversion of available nutrients by different processes such as ammonification, denitrification, and immobilization.

Degradation of the surface layer of organic plant residues leads to acidic compounds that significantly increase the acidity of the upper soil layer. These acidic compounds react with soil and accumulate in the surface in no-tillage systems

(Wells and Touchton 1985). Also, degradation of organic matter, decomposition of plant biomass, root exudates, transamination of respective keto acids, and autolysis of microorganisms in soils release amino acids, which typically account from 40 to 60 % of the total organic N present in soil, while inorganic forms of nitrogen account for only 1–10 % of the total N present in soils (Vancura 1967; Claudius and Merhotka 1973; Smith 1976; Stevenson 1982a, b, 1986; Szajdak et al. 1998). Amino acids created during these processes favored plant growth and served to explain, in part, how organic matter increased soil productivity. Thus, the study of amino acids in soil has won more interest (Lowe 1973; Schnitzer et al. 1974; Myhr et al. 1978; Goh and Edmeades 1979; Malhotra and Sarkar 1979; Holtzclaw et al. 1980; Liu et al. 1985; Dashman and Stotzky 1986; Tena et al. 1986; Życzyńska-Bałoniak and Szajdak 1993; Szajdak 1996; Szajdak and Österberg 1996; Szajdak and Sokolov 1997; Ryszkowski et al. 1998; Kögel-Knabner 2002; Ley and Schmidt 2002).

Despite the rather high input of amino acids into the soil, their actual concentrations of free forms are rather low. Amino acids in soil can undergo mineralization, migration down the soil profile, soil adsorption, and humification (Kuzyakov 1997). Free amino acids are rapidly degraded by microbes, so most of the amino acids in soils occur in bound form in the humino-peptides fraction. These amino acids are commonly bound to the central core of HA and FA (Bremner 1967; Sörensen 1967; Haworth 1971), which protects them from rapid degradation by microorganisms.

Natural processes of organic matter transformation lead to the formation of humic substances. It is known that under aerobic conditions (humification in terrestrial ecosystems) the main components of the formed humic substances are FA and HA. Stearman et al. (1989a, b) using ^{13}C-NMR suggested that differences in HA composition were more pronounced with depth than with tillage differences. The larger amounts of organic C in surface NT plots were correlated with greater amounts of aliphatic groups in HA extracted from these plots. These authors found that NT and CT treatments are similar in HA structural properties. Bayer et al. (2000) showed with the help of EPR and ^{13}C-NMR that under NT in weathered tropical and subtropical soils, stable organic matter originating from crop residues was less humidified than the original soil organic matter.

During the process of humification, the organic matter is known to undergo a variety of free radical reactions (Drozd et al. 1997; Jezierski et al. 1998, 2002; Jerzykiewicz et al. 2002). These free radicals are mainly of semiquinone type included in polyphenolic or Maillard-type polymeric matrices, which are sensitive to various physical and chemical agents, e.g., radiation, pH, redox reaction, acid–base reactions, etc. (Senesi 1990, 1992). These reactions can affect the nature and properties of the organic compounds present in humified organic matter.

With these antecedents in mind, the overall objectives of the study presented here were to evaluate the effects of CT and NT on amino acids, indole-3-acetic acid (IAA), stable and transient radicals, and properties of humic and fulvic acids in soil.

5.2 Materials and Methods

5.2.1 Soil Samples

Soil samples (20 cores per plot) were collected from the upper 20 cm of long-term CT and NT plots at the Horseshoe Bend Experimental Area, near Athens, Georgia (USA), and at the Bledsoe Research Farm, near Williamson, Georgia. CT (moldboard plowed, disked, and rotary-tilled) and NT (direct-drilled) plots at the Horseshoe Bend Experimental Area were established in 1978 on a Hiwassee sandy clay loam (clayey, kaolinitic, thermic Rhodic Kanhapludult). The organic C content of the soil was 2.34 %. For the 4 years prior to collection of the samples, all plots at the Horseshoe Bend Experimental Area were double cropped to grain sorghum (*Sorghum bicolor* L. Moench) and winter *rye* (*Secale cereale* L.). Additional information on the site and cultivation history can be found in Beare et al. (1992). CT (moldboard plowed, disked) and NT (direct-drilled) plots at the Bledsoe Research Farm were established in 1975 on a Cecil sandy clay loam (clayey, kaolinitic, thermic, Typic Kanhapludult). The content of organic C was 0.88 %. For the 4 years prior to collection of the samples, all plots at the Bledsoe Research Farm were on a 2-year rotation of winter rye, soybean (*Glycine maxima* L.), crimson clover (*Trifolium incarnatum* L.), and corn (*Zea mays* L.). A summary of selected soil characteristics is presented in Table 5.1.

5.2.2 Instruments

pH-meter, amino acids analyzer, centrifuge, BECKMAN DU®-68 spectrophotometer, Bruker NMR gaussmeter ER 035 M, Hewlett-Packard microwave frequency counter HP 5350B, Radiopan SE/X spectrometer, Bruker EPS300E spectrometer

Table 5.1 Selected soil properties and extraction yield (g/kg, +/− 95 % confidence interval) of humic acids (HA), fulvic acids (FA), and ash content for samples collected from long-term studies at the Horseshoe Bend Experimental Area (HSB) and at the Bledsoe Research Farm (BRF)

Sample	pH H_2O	pH 1 *M* KCl	C/N	Sand %	Silt %	Clay %	HA Yield [$g \cdot kg^{-1}$]	HA Ash content [%]	FA Yield [$g \cdot kg^{-1}$]	FA Ash content [%]
HSB NT	5.79	5.41	9.8	66.0	20.1	14.0	20.0±0.8	12.0	6.4±0.3	9.2
HSB CT	6.00	5.13	9.4	64.3	19.5	16.3	11.5±0.5	10.5	6.3±0.3	10.8
BRF NT	6.26	5.73	10.9	58.0	17.7	24.4	25.7±0.2	10.8	10.2±0.4	5.6
BRF CT	6.35	5.16	11.4	43.5	23.9	32.7	19.8±0.1	13.2	15.6±0.6	9.2

Szajdak et al. (2003)
Where *NT* no tillage, *CT* conventional tillage

5.2.3 The Isolation of HA and FA

Humic substances from air-dried soil samples were extracted with 0.1 NaOH using an extractant/soil ratio of 5:1 under N_2 atmosphere at room temperature (Swift 1996). The system was shaken for 4 h and subsequently stored for 20 h. The dark-colored supernatant solutions after separation from the residual soils by centrifugation (4000 g for 30 min) were adjusted to pH = 1.3 with 6 N HCl and allowed to stand for 24 h at room temperature for the coagulation of the HA fractions, which were separated by centrifugation. The purification of the HA was performed by repeat solution at pH = 7, centrifugation to remove insoluble materials, followed by reprecipitation at pH = 1.3, and the removal of the supernatants liquids using the following method. The HA fractions were dissolved in 400–500 ml of deionized water and adjusted to pH = 7.0. The solutions were centrifuged at 6.000 g at 2–4 °C for 1 h to separate a mixture of clay, adjusted to pH = 1.3, and centrifuged after 24 h of storage at room temperature. The procedure was repeated three times. The finally precipitated HA were freeze-dried and kept in a vacuum desiccator over P_2O_5. FA were recovered from the supernatants by sorbing on a column of XAD-8 resin, eluting with 0.1 NaOH, and passing the eluate through a column of Amberlite IR-120 in the H+-form. The FA solutions were evaporated under reduced pressure at 30 °C, freeze-dried, and dried in a vacuum desiccator over P_2O_5.

5.2.4 The Hydrolysis of HA and FA and the Determination of Amino Acids

To obtain bound amino acids, 20-ml 6 M HCl was added to 20 mg of HA or FA. The solution was left overnight at room temperature, after which it was saturated with gaseous argon to remove oxygen. Next, the sample was sealed, heated for 24 h at 105 °C, centrifuged at 5000 g and transferred to a measuring flask, and evaporated to dry mass under reduced pressure. Determination of 24 bound amino acids was done with a T 339 amino acids analyzer (Microtechan Praha). The separation of the bound amino acids was achieved by injecting 100 μl samples into an Ostion LGFA (0.37 × 20 cm) column. Lithium-citric buffers of the following pHs were used, 2.90, 3.10, 3.35, 4.05, and 4.90, and the absorbance of the eluent–ninhydrin complex was monitored at 520 nm (Szajdak 1996). The mobile phase was pumped at 12 $cm^3 \cdot h^{-1}$ and developed a pressure of 2.5 *M*Pa. A full-range recorder span of 100 mV was used to provide on-scale peaks (Szajdak 1996; Szajdak and Österberg 1996; Szajdak and Sokolov 1997; Szajdak et al. 1998).

Electron paramagnetic resonance (EPR) spectra were obtained with Radiopan SE/X and Bruker ESP300E spectrometers at X-band operating at room temperature. Solid samples were placed in quartz tubes of 5 mm outer diameter. The Bruker spectrometer with a 100-kHz magnetic field modulation was equipped with a Bruker NMR gaussmeter ER 035 M and Hewlett-Packard microwave frequency counter

HP 5350B. A Li/LiF standard was used for the g-value calibration; 4-hydroxy-TEMPO and Reckitt ultramarine were used as standards of spin concentration. The quantitative EPR technique was applied (microwave power 1 mW, modulation amplitude 1 G, 20.0 mg sample, standard quartz tubes, etc.). EPR parameters determined were: free radical concentrations and g-values.

EPR measurements were carried out for:

(i) HA and FA in air
(ii) HA and FA under gaseous ammonia flowing over the sample in quartz tube placed in the resonance cavity

According to Chen et al. (1977), 3 mg of HA or FA were dissolved in 10 ml of the following solutions: 0.05 M $NaHCO_3$, 0.025 M $NaHCO_3$, and 0.05 M $Na_4P_2O_7 \cdot 10\,H_2O$ at $pH = 7.00 \pm 0.01$. Absorbances at 464 nm (E_4) and 665 nm (E_6) of HA and FA in each solution were measured and $E_{4/6}$ ratios calculated from spectrums in visible region. BECKMAN DU®-68 spectrophotometer with 1 cm thickness of layer was used for spectrometric measurements. All the experiments were run in triplicate, and the results were averaged. All the chemicals used in this study were of analytical grade.

The method of the determination of IAA is described in the Chap. 10.

5.3 Conclusions

The yield of HA was from 42 to 59 % higher in NT than in CT soils (Table 5.1). In contrast, the concentration of FA was 54 % higher in CT than in NT soils.

Indole-3-acetic acid (IAA) belongs to phytohormones—auxins (for more information, see Chap. 9)—and activates germination and roots growth. In NT, the concentration of IAA in HA was from 37 to 80 % greater and in FA from 75 to 91 % greater than in HA from CT. This may indicate greater soil productivity under NT than CT (Table 5.2).

In general, the total concentration of bound amino acids in HA was higher in NT than in CT samples (Table 5.3).

The highest concentration (57 mg·kg^{-1} of soil) was found in HA extracted from NT samples collected at the Horseshoe Bend Experimental Area. This concentration was 20 % higher than that measured in NT samples collected at the Bledsoe Research Farm. The lowest concentrations of bound amino acids in HA were measured in CT soils from the Horseshoe Band Experimental Area; they were 27 % lower than those measured in CT at the Bledsoe Research Farm.

In contrast to bound amino acids in HA, bound amino acids in FA were present at higher concentrations in CT than in NT soils (Table 5.3).

The highest concentrations in FA were measured in CT soils from the Bledsoe Research Farm, which had a total concentration of 30 mg·kg^{-1} of soil, 43 % higher than the concentration found in FA from NT soils. In the FA extracted from samples collected at the Horseshoe Band Experimental Area, the total concentration of

Table 5.2 IAA in HA and FA (mg kg^{-1}) from long-term conventional-till (CT) and no-till (NT) studies at the Horseshoe Bend (HSB) Experimental Area and at the Bledsoe Research Farm (BRF) in Georgia (USA) (Szajdak unpublished data)

Sample		IAA
HA	HSB NT	1979.77
	HSB CT	739.12
	BRF NT	2718.89
	BRF CT	2441.72
FA	HSB NT	2032.57
	HSB CT	1847.79
	BRF NT	2217.35
	BRF CT	1663.01

bound amino acids was 19 % higher in CT than in NT soils. Stevenson (1982a) postulated an increase in the amount of amino acids in soil as an impact of long-term cultivation practices.

When considering amino acids as acidic, neutral, and basic, it was found that in all soil samples the highest concentrations were of neutral amino acids, ranging from 47 to 65 % of the total determined amino acids. This is consistent with results obtained by numerous other authors (Flaig 1971; Haworth 1971; Kalembasa and Niewiński 1990; Shu-Yen et al. 1985; Stevenson 1982a; Szajdak and Österberg 1996; Szajdak and Sokolov 1997; Szajdak et al. 1998). In all samples of HA and FA, basic amino acids had the lowest concentrations ranging from 11 to 16 % in HA and from 9 to 15 % in FA. The percentage of bound amino acids with a negative net charge in HA and in FA ranged from 22 to 40 %.

The results showed that glutamic acid, glycine, alanine, and valine dominated in all samples of HA and FA. The concentration of glutamic acid in FA from CT samples was highest at the Horseshoe Bend Experimental Area (4 mg·kg^{-1} of soil), 18 % higher than in NT soil. Among neutral amino acids, glycine, alanine, and valine dominated. The concentrations of these amino acids were higher in HA from NT soils and in FA from CT soils.

In addition, the results showed a much higher concentration of alanine and lysine in HA from NT than in those from CT. This indicated a higher microbial biomass in NT samples, since alanine and lysine are typical constituents of bacteria cell walls (Stevenson 1972; Durska and Kaszubiak 1980a, b, c). A similar phenomenon of accumulation of alanine and lysine was noticed in our previous study concerning free and bound amino acids in soils under continuous cropping of rye and crop rotation (Życzyńska-Bałoniak and Szajdak 1993; Ryszkowski et al. 1998).

In contrast, the study revealed a much higher concentration of proline in HA and FA from soils under CT than in HA and FA from NT soils. A similar phenomenon of accumulation of proline in continuous cropping of rye was observed earlier concerning free and bound amino acids (Życzyńska-Bałoniak and Szajdak 1993; Ryszkowski et al. 1998). Accumulation of proline in soil under CT may be due to its heterocyclic structure, which could make it more resistant to degradation. The accumulation of proline in soils could be considered as negative effect because proline is a secondary amine, which in the presence of nitrite ions may form N-nitrosamines (under acidic conditions), that are potent toxins, with carcinogenic

Table 5.3 Mean concentration (mg·kg^{-1} of soil +/– 95 % confidence interval) of bound amino acids in humic acids and fulvic acids and nitrogen in amino acids (italics) from soils under conventional tillage (CT) and no-tillage (NT)

	HA				FA			
	Horseshoe bend		Bledsoe farm		Horseshoe bend		Bledsoe farm	
Amino acids	NT	CT	NT	CT	NT	CT	NT	CT
Acidic								
Cysteic acid	1.30±0.1	0.83±0.03	0.73±0.03	0.65±0.02	1.69±0.1	2.14±0.1	1.65±0.1	2.76±0.1
	0.10±0.001	*0.05±0.002*	*0.06±0.001*	*0.06±0.001*	*0.13±0.01*	*0.16±0.01*	*0.12±0.01*	*0.20±0.01*
Taurine	0.22±0.01	0.10±0.004	0.58±0.02	0.43±0.02	0.12±0.01	0.10±0.004	0.07±0.003	0.18±0.01
	0.02±0.001	*0.01±0.001*	*0.06±0.001*	*0.06±0.001*	*0.01±0.001*	*0.01±0.001*	*0.01±0.002*	*0.02±0.001*
Phosphoethanolamine acid	0.44±0.01	0.18±0.01	0.34±0.01	0.26±0.01	0.17±0.01	0.22±0.01	0.15±0.01	0.32±0.01
	0.12±0.001	*0.12±0.001*	*0.59±0.02*	*0.57±0.01*	*0.02±0.001*	*0.02±0.001*	*0.01±0.001*	*0.05±0.001*
Aspartic acid	1.19±0.1	1.24±0.1	6.01±0.2	3.73±0.1	0.24±0.01	0.33±0.01	0.19±0.01	0.45±0.02
	0.13±0.01	*0.13±0.001*	*0.68±0.08*	*0.39±0.01*	*0.08±0.001*	*0.03±0.001*	*0.02±0.001*	*0.04±0.002*
Threonine	1.51±0.1	1.19±0.04	0.46±0.02	0.82±0.03	1.67±0.1	1.40±0.1	1.19±0.04	1.99±0.1
	0.17±0.01	*0.14±0.001*	*0.06±0.001*	*0.09±0.003*	*0.19±0.01*	*0.16±0.01*	*0.14±0.01*	*0.23±0.01*
Serine	2.42±0.1	1.84±0.1	2.42±0.1	0.60±0.02	1.56±0.1	1.74±0.1	1.67±0.06	2.85±0.1
	0.32±0.02	*0.23±0.001*	*0.32±0.01*	*0.08±0.002*	*0.21±0.01*	*0.23±0.01*	*0.22±0.01*	*0.38±0.02*
Glutamic acid	5.55±0.2	1.29±0.1	2.42±0.1	0.78±0.02	3.32±0.1	4.03±0.2	1.13±0.04	1.71±0.1
	0.52±0.02	*0.12±0.001*	*0.23±0.01*	*0.07±0.002*	*0.31±0.01*	*0.38±0.001*	*0.10±0.01*	*0.16±0.02*
α-Aminoadipic acid	0.29±0.01	0.11±0.01	1.24±0.05	0.39±0.01	0.19±0.01	0.26±0.01	0.14±0.01	0.64±0.001
	0.02±0.001	*0.01±0.001*	*0.11±0.01*	*0.08±0.002*	*0.01±0.001*	*0.02±0.0001*	*0.01±0.001*	0.06±0.002
Neutral								
Proline	3.11±0.1	0.85±0.1	0.73±0.03	2.02±0.1	0.69±0.03	0.88±0.03	0.58±0.02	1.40±0.1
	0.38±0.01	*0.10±0.01*	*0.08±0.003*	*0.24±0.01*	*0.08±0.003*	*0.11±0.001*	*0.07±0.008*	*0.17±0.01*
Glycine	8.41±0.3	3.73±0.1	6.86±0.3	3.96±0.2	2.95±0.1	3.58±0.1	2.69±0.1	3.07±0.1
	1.03±0.04	*0.45±0.02*	*0.84±0.01*	*0.48±0.02*	*0.36±0.001*	*0.43±0.002*	*0.32±0.01*	*0.37±0.002*

Alanine	4.58±0.2	2.04±0.1	4.04±0.2	2.69±0.1	1.31±0.01	1.94±0.1	1.34±0.05	2.43±0.1
	0.72±0.08	*0.32±0.01*	*0.68±0.01*	*0.42±0.02*	*0.21±0.001*	*0.31±0.002*	*0.21±0.01*	*0.38±0.02*
Valine	6.18±0.2	1.85±0.1	5.25±0.2	5.61±0.2	1.44±0.1	2.72±0.1	1.35±0.06	2.26±0.1
	0.73±0.08	*0.22±0.001*	*0.62±0.01*	*0.67±0.06*	*0.17±0.01*	*0.32±0.02*	*0.16±0.01*	*0.27±0.01*
Cysteine	0.64±0.1	0.27±0.01	0.74±0.3	0.37±0.02	1.26±0.1	1.19±0.1	0.28±0.01	0.38±0.01
	0.07±0.002	*0.03±0.001*	*0.08±0.003*	*0.04±0.002*	*0.14±0.01*	*0.37±0.02*	*0.03±0.001*	*0.04±0.002*
Methionine	1.18±0.1	0.34±0.01	0.52±0.02	0.66±0.03	1.05±0.1	0.25±0.01	0.76±0.02	0.89±0.03
	0.11±0.004	*0.03±0.001*	*0.06±0.0002*	*0.06±0.002*	*0.09±0.001*	*0.02±0.001*	*0.07±0.002*	*0.08±0.002*
Cystathionine	1.58±0.1	1.07±0.04	1.24±0.1	1.34±0.04	0.05±0.002	0.67±0.02	0.42±0.02	1.02±0.04
	0.21±0.001	*0.14±0.001*	*0.16±0.01*	*0.17±0.007*	*0.01±0.001*	*0.09±0.004*	*0.06±0.002*	*0.13±0.01*
Isoleucine	5.58±0.2	2.33±0.1	2.44±0.1	3.03±0.1	1.02±0.1	1.42±0.1	0.78±0.03	1.26±0.04
	0.59±0.02	*0.24±0.001*	*0.26±0.01*	*0.32±0.01*	*0.11±0.004*	*0.15±0.008*	*0.08±0.002*	*0.13±0.01*
Leucine	0.75±0.03	0.18±0.01	0.52±0.02	0.63±0.03	0.07±0.003	0.36±0.01	0.29±0.01	0.62±0.02
	0.08±0.008	*0.02±0.001*	*0.06±0.002*	*0.07±0.003*	*0.01±0.004*	*0.04±0.002*	*0.03±0.002*	*0.07±0.001*
Tyrosine	3.91±0.1	1.65±0.1	3.25±0.1	2.22±0.1	0.67±*0.02*	0.94±0.03	0.73±0.03	0.81±0.03
	0.30±0.01	*0.12±0.001*	*0.34±0.01*	*0.28±0.01*	*0.06±0.002*	*0.07±0.008*	*0.006±0.001*	*0.07±0.001*
β-Alanine	0.49±0.01	0.17±0.1	0.39±0.01	0.16±0.01	*0.21*±0.01	0.18±0.01	0.23±0.01	0.27±0.01
	0.07±0.006	*0.02±0.001*	*0.06±0.002*	*0.02±0.001*	*0.08±0.002*	*0.07±0.001*	*0.03±0.002*	*0.04±0.001*
γ-Aminobutyric acid	0.13±0.01	0.09±0.003	0.09±0.003	0.02±0.001	0.09±0.004	0.10±0.01	0.07±0.002	0.14±0.01
	0.002±0.002	*0.01±0.001*	*0.01±0.001*	*0.01±0.001*	*0.01±0.008*	*0.01±0.001*	*0.01±0.002*	*0.02±0.001*

(continued)

Table 5.3 (continued)

	HA				FA			
	Horseshoe bend		Bledsoe farm		Horseshoe bend		Bledsoe farm	
Amino acids	NT	CT	NT	CT	NT	CT	NT	CT
Basic								
Ornithine	0.85±0.03	0.27±0.01	0.56±0.02	0.53±0.02	0.17±0.01	0.22±0.01	0.15±0.01	1.02±0.04
	0.18±0.01	*0.06±0.0001*	*0.11±0.01*	*0.11±0.01*	*0.04±0.002*	*0.04±0.002*	*0.03±0.008*	*0.21±0.01*
Lysine	3.04±0.1	1.58±0.1	2.33±0.1	1.96±0.1	1.12±0.1	1.41±0.1	0.91±0.03	1.26±0.05
	0.58±0.02	*0.30±0.01*	*0.44±0.02*	*0.37±0.01*	*0.21±0.01*	*0.27±0.01*	*0.17±0.01*	*0.24±0.01*
Histidine	1.45±0.1	0.73±0.03	0.86±0.03	0.75±0.03	0.33±0.02	0.24±0.01	0.27±0.01	1.00±0.05
	0.26±0.01	*0.13±0.01*	*0.16±0.001*	*0.13±0.005*	*0.06±0.002*	*0.04±0.002*	*0.04±0.001*	*0.18±0.01*
Arginine	2.17±0.1	1.40±0.05	1.50±0.040	1.24±0.1	0.42±0.02	0.62±0.02	0.36±0.01	1.40±0.06
	0.34±0.01	*0.22±0.01*	*0.24±0.01*	*0.19±0.001*	*0.07±0.008*	*0.09±0.004*	*0.05±0.002*	*0.22±0.01*
Total amount	56.97±2.5	56.97±2.5	56.97±2.5	56.97±2.5	21.81±1.4	26.94±1.6	17.4±1.2	30.131±4
	7.07±0.4	*3.23±0.2*	*6.17±0.03*	*4.66±0.2*	*2.56±0.1*	*3.39±0.2*	*2.05±0.1*	*3.74±0.12*

Szajdak et al. (2003)

and mutagenic effects (Plaa 1980; Kofoed et al. 1981; Von Hofe et al. 1987; Larsson et al. 1990).

The occurrence of different forms of amino acids in the soils is expected, because these substances are created from protein, by heterotrophic organisms.

$$\text{Proteins} \rightarrow \text{peptides} \rightarrow \text{amino acids}$$

Furthermore, some amino acids are precursors or intermediate agents of other amino acids in soils (α, ε-diaminopimelic acid to lysine, tryptophan to indole-3-acetic acid, etc.) (Tena et al. 1986) and are also included in the bacteria cell walls (α, ε-diaminopimelic acid, β-alanine, and lysine) (Stevenson 1972). However, complex factors may affect the quantity and quality of amino acids in soils, including formation, conversion, and/or degradation by indigenous biota, adsorption by clay minerals, and connection with reducing sugars and quinones.

The ratio of absorbances at $\lambda = 465$ and 665 nm referred as $E_4/_6$ has been widely used for characterization of the properties of humic substances. The $E_{4/6}$ ration decreases with increasing molecular weight and condensation and is believed to serve as an index of humification. Therefore, low value of $E_{4/6}$ may be indicative of relatively high degree of condensation of aromatic constituents; and a high ratio reflects a low degree of aromatic condensation and infers the presence of relatively more aliphatic structures (Stevenson 1982b).

The $E_{4/6}$ ratios for HA and FA were calculated for different solutions (Table 5.4). However, the values of these factors carried out in 0.05 M $NaHCO_3$ are the most recommended (Chen et al. 1977). Significant differences were observed between $E_{4/6}$ ratios from CT and NT. $E_{4/6}$ ratios of HA from CT in 0.05 M $NaHCO_3$ from Horseshoe Bend Experimental Area and from Bledsoe Research Farm were from 1.02 to 1.19 larger than ratios of HA from NT (Table 5.4). This may indicate that the degree of condensation of aromatic constituents in HA from NT is higher than in HA from CT (Orlov 1983).

EPR is a very sensitive and well-established technique for measuring minute structural changes involving electron transfer with the formation of free radical

Table 5.4 $E_{4/6}$ rations of HA and FA from long-term conventional-till (CT) and no-till (NT) plots at the Horseshoe Bend (HSB) Experimental Area and at the Bledsoe Research Farm (BRF) in Georgia (USA)

Sample	HA			FA		
	0.05 M $Na_4P_2O_7 \cdot 10H_2O$ pH = 7.00 ± 0.01	0.025 M $NaHCO_3$	0.05 M $NaHCO_3$	0.05 M $Na_4P_2O_7 \cdot 10\ H_2O$ pH = 7.00 ± 0.01	0.025 M $NaHCO_3$	0.05 M $NaHCO_3$
HSB NT	8.16	7.53	6.82	17.4	17.0	12.5
HSB CT	8.88	7.88	7.72	18.6	18.3	14.1
BRF NT	9.17	6.57	6.52	18.3	14.6	13.3
BRF CT	8.18	6.64	6.65	20.2	14.2	12.7

Szajdak et al. (2003)
Where *NT* no-tillage, *CT* conventional tillage

Table 5.5 HA and FA from long-term conventional-till (CT) and no-till (NT) plots at the Horseshoe Bend (HSB) Experimental Area and at the Bledsoe Research Farm (BRF) in Georgia (USA)

Sample		*g*-value		Spin concentration per gram (10^{-17})		Multiplicity of spin concentration enhancement between gaseous ammonia and air
		Air	Gaseous ammonia	Air	Gaseous ammonia	
HA	HSB NT	2.0042	2.0044	1.06	1.30	1.22
	HSB CT	2.0038	2.0041	0.35	1.98	5.74
	BRF NT	2.0041	2.0041	3.29	3.79	1.15
	BRF CT	2.0039	2.0041	1.50	1.55	1.03
FA	HSB NT	2.0041	2.0047	0.16	1.66	10.4
	HSB CT	2.0041	2.0047	0.23	1.51	6.54
	BRF NT	2.0038	2.0047	0.08	1.51	18.4
	BRF CT	2.0037	2.0047	0.07	1.62	22.2

Szajdak et al. (2003)
Where *NT* no-tillage, *CT* conventional tillage

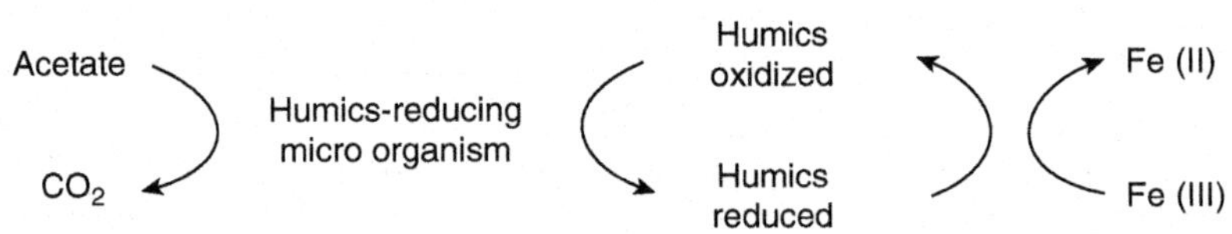

Fig. 5.1 Proposed electron mediator role of humic substances in the reduction of Fe(III) by Scott et al. (1998)

intermediates in carbonaceous materials (Senesi 1990, 1992; Czechowski and Jezierski 1997). The EPR spectra of HA and FA isolated from soils under the NT and CT exhibited a broad signal due to the presence of Fe(III) bound to organic and inorganic matrices and narrow lines at $g=2.0040$ (average value for HA samples) and $g=2.0039$ (average value for FA samples) (Table 5.5).

The *g*-value for HA and FA reflects to a considerable degree dimensions and structure of the aromatic conjugation systems in HA and FA (Strigutski et al. 1982; Bambalov et al. 2000). This parameter (g-value) increases from 2.0038 to 2.0042 for HA from CT to NT and from 2.0037 to 2.0041 for FA from CT to NT management. Also the free radicals newly generated upon conversion of CT to NT management are mainly of semiquinone type. The increase in the *g*-value during the process is attributed to the formation of oxygen-rich groups (e.g., semiquinone and other oxygen-rich functionalities) in organic matter under different tillage intensity. Organic radicals in humic substances, which are primarily quinone groups, are reduced when humic-reducing microorganisms transfer electrons to humic substances (Scott et al. 1998) (Fig. 5.1).

As mentioned earlier, greater concentrations of alanine in NT than in CT indicate a higher microbial biomass in NT samples than in CT ones. Hendrix et al. (1986)

postulated also that NT residues show a tendency toward greater nutrient immobilization (i.e., microbial uptake of nutrients from decomposing organic matter, which makes them unavailable to plants until the nutrients are released from microbial biomass). Supplementary information on the oxidation rate and the extent of bacterial activity in two different tillage systems is obtained from the characteristic maximum of the free radical concentration (Jerzykiewicz et al. 1999; Drozd et al. 1997; Jezierski et al. 1998). Additionally, it is known that further changes of the *g*-value of free radicals in organic substances may be used as an indicator of humification. With increasing humification, coals show a steady increase in overall spin concentration. This follows the generally accepted view on the coal metamorphism, where condensation and size of aromatic lamellas increase with maturation (Czechowski and Jezierski 1997).

The conditions of the oxidation in the soil influence the *g*-value and the radical spin concentration of HA and FA. In HA from sources of lower humification degree (CT), *g*-values are generally lower, and spin concentrations are from 2 to 3 times lower than those in HA from soil under long-term NT management (the Horseshoe Bend Experimental Area and the Bledsoe Research Farm). Higher spin concentrations and also higher *g*-value for HA extracted from both NT managements have a higher nitrogen content than HA from CT (Jezierski et al. 2002). The data agree with those of amino acids in HA and FA (Table 5.3). The content of nitrogen in HA is related to the total amount of amino acids in HA and also with the amount of nitrogen in amino acids (Table 5.3). From 20 to 40 % of nitrogen associated with HA may consist of amino acids or peptides bound to the central core of HA by hydrogen bonds (Haworth 1971). More N was lost through leaching from CT than from NT (Doran 1980; Blevins et al. 1984; Holland and Coleman 1987; House et al. 1984; Hendrix et al. 1986; Stinner et al. 1984).

It is well known that for aqueous solutions of HA and FA, the pH of the solutions strongly affects the free radical concentration. In alkaline solutions, a drastic increase of the free radical concentration occurs (Senesi 1990, 1992). The sterically small gaseous ammonia molecule possesses the ability to easily penetrate the matrix of HA and FA giving a similar effect to that of aliphatic amines. This effect also occurs for aqueous solutions, i.e., by the shift in quinone–semiquinone–hydroquinone equilibrium (involving anionic forms of the radicals). This effect exhibits also the opening up of the HA and FA structure because of interaction with the base, taking place in the solid phase. The above equilibrium for HA and FA is rapidly achieved (usually in less than 1 min). Furthermore, air oxidation at 150 °C results in generation of structural units sensitive to gaseous NH_3. This produces additional free radicals, higher than that for unutilized HA and FA. For our samples the ammonia effects caused from 1.03 to 22.19 times more increase in the spin concentration.

Examinations of the EPR spectra of HA and FA derived from the two tillage systems revealed two opposite effects on the signal characteristics achieved by chemical treatments. Molecules possessing electron donor capacity like gaseous ammonia and aliphatic amines, compared to the level of native (indigenous) radicals estimated for raw HA, exhibit strong effect on spin concentration enhancement.

Concentration of the former upon ammonia treatment "transient" radicals achieved saturation under gaseous ammonia. Lowering of ammonia molecular fraction in gas stream flowing over HA sample is manifested by lower enhancement of spin concentration. The formed "transient" radicals are associated with shift of *g*-value to the higher values, which indicates shift in:

$$\text{quinone} - \text{hydroquinone} - \text{semiquinone equilibrium}$$

Conventional tillage led to a decrease in the EPR signal intensity for HA, indicating a decrease in the dimension of the aromatic conjugation systems in these molecules. Transition from CT to NT is accompanied by a considerable rise in the signal intensity, which reflects more conjugation in the HA from NT compared to those from CT. These do not agree with those of absorption spectroscopy in the visible region (Tables 5.5).

This study demonstrated the impact of NT and CT management on the content of amino acids in HA and FA. The total amount of bound amino acids in HA from NT was higher than in HA from CT. The contrary observation was noted for FA. Conventional tillage led to a decrease in the EPR signal intensity for HA, indicating a decrease in the dimension of the aromatic conjugation systems in these molecules. Transition from CT to NT is accompanied by a considerable rise in the signal intensity, which reflects more conjugation in the HA from NT compared to those from CT. Thus, HA from NT management from Horseshoe Bend Experimental Area and from Bledsoe Research Farm may be characterized by higher molecular weights and higher degree of condensation of aromatic constituents than HA from CT.

References

Alvarez CR, Alvarez R, Grigera MS, Lavado RS (1998) Associations between organic matter and fractions of the active microbial biomass. Soil Biol Biochem 30:767–773

Bambalov NN, Smychnik T, Maryganova V, Strigutsky V (2000) Peculiarities of the chemical composition and the molecular structure of peat humic substances. Acta Agroph 26:149–177

Bayer C, Neto LM, Mielniczuk J, Ceretta CA (2000) Effect of no-till cropping systems on soil organic matter in sandy clay loam Acrisol from Southern Brazil monitored by electron spin resonance and nuclear magnetic resonance. Soil Till Res 53(2):95–104

Beare MH, Parmelee RW, Hendrix PF, Chang W, Coleman DC, Crossley DA Jr (1992) Microbial and faunal interactions and effects on litter nitrogen and decomposition agroecosystems. Ecol Mon 62:569–591

Blevins RL, Thomas GW, Cornelius PL (1977) Influence of no-tillage and nitrogen fertilization on certain soil properties after 5 year of continuous corn. Agron J 69:383–396

Blevins RL, Smith MS, Thomas GW (1984) Changes in soil properties under no-tillage. In: Phillips RE, Phillips SH (eds) No-tillage agriculture: principles and practices. Van Nostrad Reinhold, New York, pp 190–230

Bremner JM (1967) Nitrogenous compounds. In: McLaren AD, Peterson GH (eds) Soil biochemistry, vol 1. Marcel Dekker, New York, pp 19–66

Carter MR, Rennie DA (1984) Dynamics of soil microbial biomass N under zero and shallow tillage for spring wheat, using 15 urea. Plant Soil 76:157–164
Chen Y, Senesi N, Schnitzer M (1977) Information provided on humic substances by E_4/E_6 ratios. Soil Sci Soc Am J 41:352–358
Claudius G, Merhotka R (1973) Root exudates from lentil/*Lens culinaris Medik*/seedlings in relation to wilt disease. Plant Soil 38:315–320
Czechowski F, Jezierski A (1997) EPR studies on petrographic constituents of bituminous coals, chars of brown coals group components, and humic acids 600°C char upon oxygen and solvent action. Energy Fuels 11:951–958
Dashman T, Stotzky G (1986) Microbial utilization of amino acids and peptide bound on homoionic montmorillonite and kaolinite. Soil Biol Biochem 18(1):5–14
Doran JW (1980) Soil microbial and biochemical changes associated with reduced tillage. Soil Sci Soc Am J 44:765–771
Drozd J, Jezierski A, Chen Y (1997) Chemical and ESR properties of municipal solid waste compost. In: Rosen D (ed) Modern agriculture and environment. Kluwer Academic Publishers, Dordrecht, pp 395–400
Durska G, Kaszubiak H (1980a) Occurrence of α, ε-diaminopimelic acid in soil. I. The content of α, ε-diaminopimelic acid in different soils. Pol ecol Stud 6(2):189–193
Durska G, Kaszubiak H (1980b) Occurrence of α, ε-diaminopimelic acid in soil. II. Usefulness of α, ε-diaminopimelic acid determination for calculations of microbial biomass. Pol ecol Stud 6(2):195–199
Durska G, Kaszubiak H (1980c) Occurrence of α, ε-diaminopimelic acid in soil. III. α, ε-diaminopimelic acid as the nutritional component of soil microorganisms. Pol ecol Stud 6(2):201–206
Flaig W (1971) Organic compounds in soil. Soil Sci 111:19–33
Franzluebbers AJ (2004) Tillage and residue management effects on soil organic matter. In: Magdoff R, Weil RR (eds) Soil organic matter in sustainable agriculture. CRC Press, Boca Raton, pp 227–268
Goh KM, Edmeades DC (1979) Distribution and partial characterisation of acid hydrolysable organic nitrogen in six New Zealand soils. Soil Biol Bioch 11:127–132
Gregorich EG, Carter MR, Angers AD, Monreal CM, Ellert BH (1994) Towards a minimum data set to assess soil organic matter quality in agricultural soils. Can J Soil Sci 74:367–385
Hargrove WL, Boswell FC, Langdale GW (eds) (1985) The rising hope of our land. Proceedings, Southern Region No-Till conference. July 16–17, 1985. Griffin, University of Georgia, Athens, USA. p 147
Haworth RD (1971) The chemical nature of humic acid. Soil Sci 106(3):188–192
Hendrix PF, Parmelee RW, Crossley DA Jr, Coleman DC, Odum EP, Groffman PM (1986) Detritus food webs in conventional and no-tillage agroecosystems. Bioscience 36(6):374–380
Herbek J, Murdock JT, Blevins R (1984) Effect of planting dates of no-till and conventional corn on soils with restricted drainage. Agronomy Notes. University of Kentucky. Department of Agronomy. 17.3
Holland EA, Coleman DC (1987) Litter placement on microbial and organic matter dynamics in an agroecosystems. Ecology 68:425–433
Holtzclaw K, Schaumberg GD, LeVesque-Madore CS, Sposito G, Heick JA, Johnston CT (1980) Analytical properties of the soluble, metal-complexing fractions in sludge-soil mixtures: V. Amino acids, hexosamines, and other carbohydrates in fulvic acid. Soil Sci Soc Am J 44:736–740
House GJ, Stinner BR, Crossley DA Jr, Odum EP (1984) Nitrogen cycling in conventional and no-tillage agro-ecosystems: analysis of pathways and processes. J App Ecol 21:991–1012
Jerzykiewicz M, Drozd J, Jezierski A (1999) Organic radicals and paramagnetic metal complexes in municipal solid waste composts. An EPR and chemical study. Chemosphere 39:252–268

Jerzykiewicz M, Jezierski A, Czechowski F, Drozd J (2002) Influence of metal ions binding on free radical concentration in humic acids. A quantitative electron paramagnetic resonance study. Org Geochem 33:265–268

Jezierski A, Drozd J, Jerzykiewicz M, Chen Y, Kaye KJ (1998) EPR in the environmental control: copper complexes and free radicals in soil and municipal solid waste compost. Appl Magn Res 14:275–282

Jezierski A, Czechowski F, Jerzykiewicz M, Golonka I, Drozd J, Bylińska E, Chen Y, Seaward MRD (2002) Quantitative EPR study on free radicals in the natural polyphenols interacting with metal ions and other environmental pollutants. Spectrochim Acta 58:1293–1300

Juo ASR, Lal R (1979) Nutrient profile in a tropical Alfisol under conventional and no-till system. Soil Sci 127:168–173

Kalembasa S, Niewiński S (1990) The occurrence of amino acids in humic and fulvic acids extracted from arable and forest-grey-brown podzolic soils. Pol J Soil Sci 23:31–36

Kofoed D, Nemming O, Brunfeldt K, Nebelin E, Thomsen J (1981) Investigations on the occurrence of nitrosamines in some agricultural products. Acta Agric Scand 31:40–48

Kögel-Knabner I (2002) The macromolecular organic composition of plant and microbial residues as inputs to soil organic matter. Soil Biol Biochem 34:139–162

Kuzyakov YV (1997) The role of amino acids and nucleic bases in turnover of nitrogen and carbon in soils humid fractions. Eur J Soil Sci 48:121–130

Larsson BK, sterdahl BG, Regner S (1990) Polycyclic aromatic hydrocarbons and volatile N-nitrosamines in some dried agricultural products. Swed J Agric Res 20:49–56

Ley RE, Schmidt SK (2002) Fungal and bacterial responses to phenolic compounds and amino acids in high altitude barren soils. Soil Biol Biochem 34:989–995

Lipiec J, Stępniewski W (1995) Effects of soil compaction and tillage systems on uptake and losses of nutrients. Soil Till Res 35:37–52

Liu S-Y, Freyer AJ, Minard RD, Bollag J-M (1985) Enzyme-catalyzed complex-formation of amino acid esters and phenolic humus constituents. Soil Sci Soc Am J 49:337–342

Lowe LE (1973) Amino acids distribution in forest humus layers in British Columbia. Soil Sci Soc Am J 37:569–572

Malhotra SS, Sarkar SK (1979) Effects of sulphur dioxide on sugar and free amino acid content of pine seedlings. Plant Phys 47:223–227

Myhr K, Solberg Y, Selmer-Olsen AR (1978) The content of minerals, fibre, protein and amino acids in reed canary grass, timothy and meadow fescue. Acta Agric Scand 28(3):269–278

NeSmith DS, Hargrowe WL, Radcliffe DE, Tollner EW (1985) Tillage and residue management effects on soil physical properties. In: Hargrove WL, Boswell FC, Langdale GW (eds) The rising hope of our land. The rising hope of our land. Proceedings, Southern Region No-Till conference, July 16–17, 1985. Griffin, University of Georgia, Athens, USA. pp 87–92

Orlov DS (1983) Humus acids of soils. Balkema A.A, Rotterdam, p 378

Parmelee R, Beare MH, Blair JM (1989) Decomposition and nitrogen dynamics of surface weed residues in no-tillage agroecosystems under drought condition: influence of resources quality on the decomposer community. Soil Biol Biochem 21(1):97–103

Phillips RE, Phillips SH (1984) No-tillage agriculture: principles and practices. Van Nostrand Reinhold, New York, pp 171–189

Plaa GL (1980) Toxic responses of the liver. Toxicology. The basic sciences of poisons. Macmillan, New York, pp 216–231

Ryszkowski L, Szajdak L, Karg J (1998) Effects of continuous cropping of rye on soil biota and biochemistry. Crit Rev Plant Sci 17(2):225–244

Schnitzer M, Sowden FJ, Ivarson KC (1974) Humic acid reactions with amino acids. Soil Biol Bioch 6:401–407

Scott DT, McKnight DM, Blunt-Harris EL, Kolesar SE, Lovley DR (1998) Quinone moieties act as electron acceptors in the reduction of humic substances by humic-reducing microorganisms. Environ Sci Technol 32:2984–2989

Seiter S, Horwath WR (2004) Strategic for managing soil organic matter to supply plant nutrients. In: Magdoff R, Weil RR (eds) Soil organic matter in sustainable agriculture. CRC Press, Boca Raton, pp 269–293

Senesi N (1990) Molecular and quantitative aspects of the chemistry of fulvic acid and its interactions with metal ions and organic chemicals. Part I. The electron spin resonance approach. Anal Chim Acta 232:51–75

Senesi N (1992) Application of electron spin resonance and fluorescence spectroscopies to the study of humic substances. In: Kubat J (ed) Humus its structure and role in agriculture and environment. Elsevier Science B.V, Amsterdam, pp 11–26

Shu-Yen L, Freyer AJ, Minard RD, Bella JM (1985) Enzyme-catalyzed complex-formation of amino acids esters and phenolic humus constituents. Soil Sci Soc Am J 49:337–340

Smith WH (1976) Character and significance of forest tree root exudates. Ecology 57:324–331

Sörensen JH (1967) Duration of amino acid metabolites formed in soil during decomposition of carbohydrates. Soil Sci 104:204–241

Stearman GK, Lewis RJ, Tortorelli LJ, Tyler DD (1989a) Characterization of humic acids from no-tilled and tilled soils using carbon-13 nuclear magnetic resonance. Soil Sci Soc Am J 53:744–749

Stearman GK, Lewis RJ, Tortorelli LJ, Tyler DD (1989b) Herbicide reactivity of soil organic matter fractions in no-tilled and tilled cotton. Soil Sci Soc Am J 53:1690–1694

Stevenson G (1972) Biology of fungus, microbes and viruses. PWRiL, Warszawa, p 280 (in Polish)

Stevenson FJ (1982a) Amino acids. In: Humus chemistry, genesis, composition, reactions. Wiley, New York, pp 65–78

Stevenson FJ (1982b) Amino acids. In: Humus chemistry, genesis, composition, reactions. Wiley, New York, pp 264–284

Stevenson FJ (1986) Cycle of soil: carbon, nitrogen, phosphorus, sulfur and micronutrients. Wiley, New York, pp 150–230

Stinner BR, Crossley DA Jr, Odum EP, Todd RL (1984) Nutrient budgets and internal cycling of N, P, K, Ca, and Mg in conventional-tillage, no-tillage, and old-field ecosystems on the Georgia piedmont. Ecology 65:354–369

Strigutski VP, Novosha YuYu, Smychnik TP, Bambalov NN (1982) Investigation of the humic acids by non-linear EPR-spectroscopy. Sov Soil Sci (Pochvovedenie, in Russian) 1:14–151

Swift RS (1996) Organic matter characterization. In: Methods of soil analysis. Part 3. Chemical methods-SSSA Book series no. 5. Madison. pp 1011–1069

Szajdak L (1996) Impact of crop rotation and phenological periods on rhodanese activity and free sulfuric amino acids concentrations in soils under continuous rye cropping and crop rotation. Environ Int 22(5):563–569

Szajdak L, Österberg R (1996) Amino acids present in humic acids from soils under different cultivations. Environ Int 22(3):331–334

Szajdak L, Sokolov G (1997) Impact of different fertilizers on the bound amino acids content in soils. Int Peat J 7:29–32

Szajdak L, Matuszewska T, Gawlik J (1998) Effect of secondary transformation state of peat-muck soils on their amino acids content. Int Peat J 8:76–80

Szajdak L, Jezierski A, Cabrera ML (2003) Impact of conventional and no-tillage management on soils amino acids, stable and transient radicals and properties of humic and fulvic acids. Org Geochem 34:693–700

Tena M, Pinilla JA, Magallanes M (1986) L-phenylalanine deaminating activity in soil. Soil Biol Biochem 18(3):321–325

Tyler DD, Overton JR, Chambers AY (1983) Tillage effects on soil properties, diseases, cyst nematodes, and soybean yields. J Soil Water Conserv 38:374–376

Van Doren DM, Triplett GB, Henry JE (1976) Influence of long-term tillage, crop rotation, and soil type combinations on corn yield. Soil Sci Soc Am J 40:100–105

Vancura V (1967) Root exudates of plant. III. Effects of temperature and "cold shock" on the exudation of various compounds from seeds and seedlings of maize and cucumber. Plant Soil 27:319–328

Von Hofe E, Schmerold I, Lijiński W, Jeltsch W, Kleihus P (1987) DNA methylation in rat tissues by series of homologous aliphatic nitrosamines ranging from N-nitrosodimethylamine to N-nitrosomethyldodecylamine. Carcinogenesis 8(9):1337–1341

Wells KL, Touchton JT (1985) Soil management and fertility for no-till production. In: Hargrove WL, Boswell FC, Langdale GW (eds) The rising hope of our land. Proceedings, Southern Region No-Till conference, July 16–17, 1985. Griffin, University of Georgia, Athens, USA. pp 71–77

Worsham AD, Lewis WM (1985) Weed management: key to no-tillage crop production. Tillage and cropping sequence effects on yields and nitrogen use sufficiency. In: Hargrove WL, Boswell FC, Langdale GW (eds) The rising hope of our land. Proceedings, Southern Region No-Till conference, July 16–17, 1985. Griffin, University of Georgia, Athens, USA. pp 177–188

Życzyńska-Bałoniak I, Szajdak L (1993) The content of bound amino acids in soil under rye monoculture and Norfolk crop rotation in different periods of plants development. Pol J Soil Sci 26(2):111–117

Chapter 6
Rate of Leaching of Organic and Inorganic Compounds in Tilled and Orchard Soils

Lech Wojciech Szajdak, Jerzy Lipiec, Anna Siczek, Urszula Kotowska, and Artur Nosalewicz

Abstract The first-order kinetic reaction rate model is used in predicting the leaching of atrazine and inorganic compounds (K^{+1} , Fe^{+3} , Mg^{+2}, Mn^{+2} , NH_4^+, NO_3^-, and PO_4^{-3}) from tilled and orchard silty loam soils. This model provided an excellent fit to the experimental concentration changes of the compounds vs. time data during leaching. Calculated values of the first-order reaction rate constants for the changes of all chemicals ranged from 3.8 to 19.0 times higher in orchard than in tilled soil. Higher first-order reaction constants for orchard than tilled soil are in line with both higher total porosity and contribution of biological pores in the former. The first-order reaction constants for the leaching of chemical compounds enable the prediction of the actual compound concentration and the interactions between compound and soil as affected by management system. The study demonstrates the effectiveness of simultaneous chemical and physical analyses as a tool for the understanding of leaching in variously managed soils.

Keywords Tilled and orchard soils • Leaching kinetics • Organic and inorganic compounds

6.1 Introduction

Chemical compounds for many years have been an important means of controlling weeds afflicting food crops. Informed opinion is a consensus that with the growing world food crisis, chemicals for weed control will continue to be vital in the production of food. This does not minimize other nonchemical approaches or integrated

L.W. Szajdak (✉)
Institute for Agricultural and Forest Environment, Polish Academy of Sciences, Str. Bukowska 19, 60-809 Poznań, Poland
e-mail: lech.szajdak@isrl.poznan.pl

J. Lipiec • A. Siczek • U. Kotowska • A. Nosalewicz
Institute of Agrophysics, Polish Academy of Sciences, 20-290 Lublin, Poland

L.W. Szajdak (ed.), *Bioactive Compounds in Agricultural Soils*,
DOI 10.1007/978-3-319-43107-9_6

weed control approaches, but rather recognize that herbicides remain in many instances our sole weapon defense.

Eliminating tillage causes shifts in weed species present (Triplett and Lytle 1972). Weeds botanically related to the crop and others that escape control increase in number to become a dominant problem. At first atrazine controlled most annual weeds found in corn fields. Fall panicum, never a problem weed before atrazine was widely used, tolerates atrazine and increased dramatically in continuous corn. Coupled with reduced cultivation, fall panicum pressure rendered atrazine inadequate as a sole herbicide in corn. Within weed species, biotypes that tolerate herbicides have appeared. Biotypes of pigweed and lamb's-quarter resistant to atrazine have been identified and have become problem weeds in parts of the USA and Canada (Bandeen et al. 1982). However, these species are susceptible to several other herbicides and can be controlled. Furthermore, it was pointed out that many annual broadleaf weeds are suppressed if mulches (small grain cover crops) are left on the soil surface (Liebl and Worsham 1983; Shilling et al. 1985). However, this beneficial effect, due to allelopathic interactions, should help suppress difficult-to-control annual broad leaf weeds in many broadleaf crops and possibly reduce the need for postemergence herbicide applications.

Leaching of chemicals through soil is an environmental concern because of the possibility that they will reach the water table and contaminate the groundwater. However, whether a chemical will reach the groundwater will depend not only upon its movement through the soil but also upon its disappearance from the soil. If, for instance, the rate of degradation is sufficiently rapid compared to the rate of leaching, the chemical will disappear before it can reach the groundwater and, therefore, will not pose the environmental problem. Therefore, the determination of soil leaching rates is important because the rate of leaching of a chemical indicates how long a chemical is retained in the top soil where it is most subject to degradation or dissipation.

Knowledge of land use and management effects is important due to environmental and economic impacts. The chemical concentration in the soil solution and leaching can be significantly affected by the type of land use and associated aggregate and pore structure (Holland 2004; Lipiec et al. 2011). In last decades herbicides have been increasingly used in agriculture. Atrazine (2-chloro-4-ethylamino-6--isopropylamino-1,2,3-triazine) is a derivative of symmetrical triazine (Fig. 6.1).

Symmetrical triazines are planer six-membered heterocyclic compounds with three angular nitrogen atoms. Atrazine represents organic weak base with p*Ka* 1.66.

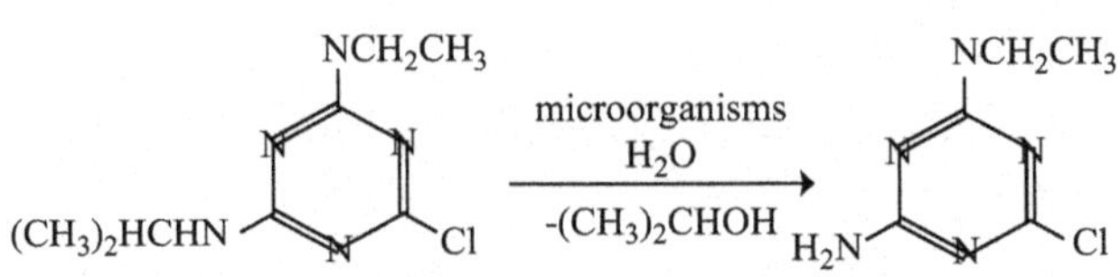

Fig. 6.1 Microbial conversion of atrazine in soil

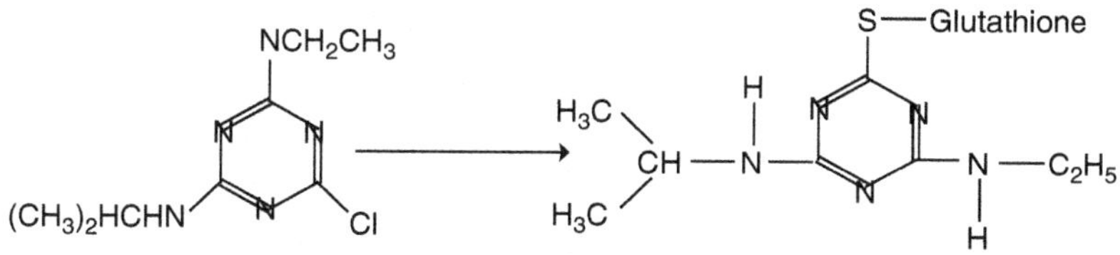

Fig. 6.2 Conversion of atrazine in corn and sorghum

This compound is dissociated in 50 % at pH 1.66; however, atrazine is dissociated in closely to pH 4. Atrazine is one of the most widely used herbicide for broadleaf weed and certain annual grass controls and its popular because of its effectiveness and low cost. It is relatively a nontoxic substance with LD_{50} for rats of 3 mg kg^{-1} or greater. Microorganisms are responsible for the environmental degradation of this compound (Fig. 6.2), but this is a slow process since its half-life in soil is 10–12 months. Recent investigations have pointed out that atrazine my enhance or suppress the toxicity of DDT and parathion depending on the soil type, content of pesticides, temperature, activity of enzymes, etc. (Bailey et al. 1978).

Therefore, it may persist in aquatic environment for many years. The first step is the cleavage of the N-alkyl C–Cl bond to form C–OH group which is another decomposition pathway (Bailey et al. 1978). The dynamics of atrazine in agricultural soils include several chemical, biochemical, physical, and biological processes:

(i) The application to soil
(ii) The leaching
(iii) The loss from surface runoff
(iv) The volatilization
(v) The sorption
(vi) The degradation
(vii) The root uptake

Atrazine, a derivative of S-triazine group of herbicides, is one of the most widely used pesticides in the world, due to its lowest cost per hectare associated with ease of application. The consumption of atrazine amounts ranged from 70 000 to 90 000 tons per year in the world.

Atrazine can be successfully applied using a broad range of application timings and tillage practices. It can be applied early preplant, preplant incorporated, preemergence, or postemergence. This substance is one of the most effective soil-applied herbicides for season-long weed control annual grasses and broadleaved weeds in many cereal crops, fruit orchards, vegetables maize, sugar cane, vines, citrus groves, grassland, and forestry (Graymore et al. 2001). However, recent studies have shown that atrazine in the field and in vitro inhibits the most famous auxin/phytohormone indole-3-acetic acids (IAA) formation (Grapelli and Rossi 1979; Rossi et al. 1984; Szajdak and Maryganova 2007). This negatively effect of atrazine has been observed in root development of vegetation cuttings, thinning in apples, plant height of several species of greenhouse-grown flowers and nursery-produced ornamental plants.

Due to its extensive use, long half-life in soil, and broad and serious toxic properties, atrazine has very high environmental significance. The high mobility of atrazine in soil (Tindall and Vencill 1995) and its potential contamination of groundwaters (Ritter et al. 1994) may represent a serious human health hazard because of the potential carcinogenic effects of S-triazines (Biradar and Rayburn 1995). This herbicide has shown acute and chronic toxicity effects in humans and animal studies. Acute toxicity has negative effects due to short-term exposure, while chronic toxicity occurs from long-term exposure to a harmful substance. Acute atrazine toxicity may cause abdominal pain, diarrhea, and vomiting. It is a mild irritant capable of producing skin rashes, mucous membrane irritation, and eye irritation. In vivo and in vitro atrazine converts to mono- and dialkylated atrazine.

According to high mobility between ecosystems, atrazine with its metabolites and decomposition products is transported to surface and subsurface water bodies. Therefore this herbicide had been found in groundwater, in rivers, in high mountain lakes, in drinking water supplies, in rainwater, and even in fog (Glotfelty et al. 1987). Thus, up to 22 $mg L^{-1}$ of atrazine was determined in groundwater (Long 1987; Habecker 1989).

High concentrations of this compound in drinking water and its significant negative effect on human health need consideration. Therefore some international organizations and countries defined acceptable limit of atrazine in drinking water. In Canada, this limit is equal to 0.06 mg L^{-1}; as per Indian Standards, no pesticide should be present in drinking water (Pranab and Ligy 2006). Furthermore, for the World Health Organization (WHO) and for the Environmental Protection Agency (EPA), the acceptable limit is restricted as two $\mu g L^{-1}$ and three $\mu g L^{-1}$, respectively (Trotter et al. 1990). In addition, Directive 98/83 European Community has set the maximum content of atrazine to 0.1 $\mu g L^{-1}$ and the total concentration for all pesticides to 0.5 $\mu g L^{-1}$ in regard to quality of water human consumption (Julali and Rowell 2003). Therefore taking into account of atrazine (i) the acute toxicity, (ii) high mobility between ecosystems in agricultural landscape, and (iii) long half-life of degradation, the rate of leaching in every soil system should be the great of interest.

Knowledge of land use and management effects is important due to environmental and economic impacts. Leaching and the chemical concentration in the soil solution can be significantly affected by the type of land use and associated aggregate and pore structure (Holland 2004; Lipiec et al. 2011). Research has shown that orchard grasses compared with tilled soils are characterized by a greater contribution of continuous biological pores made by soil fauna and plant roots (Słowińska-Jurkiewicz et al. 2001) that raise infiltration under ponded conditions. Quantification of pore-size distribution over a wide range of pore size revealed that the nature of the pore system is more heterogeneous in untilled orchard than tilled soil (Hajnos et al. 2006). Change of land use from tilled to grass or no-till system leads further to a higher soil organic carbon content and modified soil acidity (Amellal et al. 1998; Six et al. 2004; Gajda 2010; Lipiec et al. 2011). The differences in aggregate and pore structure and soil chemistry affect pore water residence time and sorption capacity and thereby leaching kinetics of agricultural chemicals.

The effect of land-use system on the chemical leaching is related to type and nature of the given chemical compound. It was shown that the presence of earthworm burrows can lessen the potential for leaching of pesticides due to adsorption in organic burrow linings and contribute to preferential flow and thereby permit solutes to bypass large parts of soil (Stehouwer et al. 1994). Moreover, eartworm feeding motion, due to investing and carrying herbicide residues away from the soil surface, can increase the amount of non-leachable herbicide residues in the soil (Farenhorst et al. 2000)

Jalali and Rowell (2003) postulated that the leaching of potassium increased with increasing calcium concentrations due to the capability of calcium ions to move potassium ions from exchange pools into solution. Manganese compared with iron is mobilized under higher pH and less reductive conditions (McDaniel and Buol 1991).

Less intense tillage compared with conventionally tilled soil declines nitrate concentration in drainage water and total N losses and can be more advantageous from the point of view of N conservation, recycling, and quality of the groundwater (Lipiec and Stępniewski 1995; Stout et al. 2000; Holland 2004). Studies of Shipitalo et al. (1990) revealed that no-till compared to tilled soil stimulated more losses of NH_4^+ than of NO_3^-. Brye and Norman (2004) reported that nitrate leaching losses were positively correlated with leaching losses of K^{+1}, Ca^{+2}, Mg^{+2}, and Na^{+1} in the maize agroecosystems, but not for the prairie, indicating that nitrate-N leaching affects the concomitant loss of cations. Intense leaching can increase nitrate concentration in drainage water above the level of 10-mg $NO_3^- L^{-1}$ being considered the safe limit for drinking water (Drury et al. 1993; Pervanchon et al. 2005).

Leaching of phosphate is rather small (< 1kg ha^{-1} $year^{-1}$) (Sharpley et al. 1994), but it can be environmentally important when joined with the surface runoff losses (Feiza et al. 2003; Puustinen et al. 2005) and may go beyond critical concentration levels for eutrophication (0.08–0.12 mg P L^{-1}) (Grant 1997).

Although general atrazine conversions are available, however, the specific pathways in every soil and water conditions are not yet clearly defined (rate of the formation of metabolites, interaction with enzymatic systems, etc.). Understanding the leaching kinetics of agricultural chemicals under different land uses is important in predicting the effect of management practices on environmental quality. Therefore, the aim of this study was to verify first-order reaction rate model in predicting the leaching of atrazine and inorganic compounds (K^{+1}, Fe^{+3}, Mg^{+2}, Mn^{+2}, NH_4^+, NO_3^-, and PO_4^{-3}) in tilled and orchard silty loam soils.

6.2 Materials and Methods

6.2.1 Soil and Treatment Descriptions

The experiment was performed on an Orthic Luvisol developed from loess, over limestone, at the experimental field of the University of Life Sciences in Lublin (51°15'N, 22°35'E), Poland. Long-term annual mean temperature and precipitations at the site are 7.4 °C and 572 mm, respectively. The experimental objects included (CT) conventionally tilled field (100 × 150 m) with main tillage operations including pre-plow (0.1 m depth) + harrowing, moldboard plowing (0.2 m depth), and crop rotation involving selected cereals, root crops, and legume crops and (OR) 35-year-old apple orchard (100 × 200 m) with permanent sward that was mown in the inter-rows during growing season (Siczek et al. 2008a, b; Lipiec et al. 2011).

The present management practices on CT were used for 30 years. Both field sites were close to each other. The research area characterizes rather uniform soils with respect to genesis and textural composition (Dobrzański and Zawadzki 1951).

Core samples of 100 mL volume and 0.05 m diameter from the top layer of 0–0.1 m and 0.1–0.2 m depths (four replicates) were used to determine bulk density on Bouwer (1986).

Total organic carbon was estimated by TOC 5050A with Solid Sample Module, SSM-5000A, Shimadzu, Japan. Some characteristics of the soils are given in Table 6.1.

The leaching of atrazine and inorganic compounds (K^{+1}, Fe^{+3}, Mg^{+2}, Mn^{+2}, NH_4^+, NO_3^-, and PO_4^{-3}) was determined in soil columns of undisturbed structures, taken with steel cylinders of 0.215 m diameter and 0.2 m height from the depth of 0–0.20 m (three replicates) (Klute and Bouwer 1986; Mallants et al. 1997). The sampling was performed in the autumn; under CT it was done after cropping of wheat and 10 days after shallow cultivation to a depth of approximately 0.10 m. In order to avoid soil damage, the cylinders were slowly pressed into the soil with a hydraulic jack. This method of sampling is suggested for laboratory studies for a realistic assessment of chemical dissipation and movement in soils (Mallants et al.

Table 6.1 Some properties of tilled (CT) and orchard (OR) soil[a]

Particle size (mm) distribution (%, w/w)							
Depth [m]	2–0.02	0.02–0.002	< 0.002	Bulk density [$Mg m^{-3}$]	Soil organic carbon [%]	Organic matter [%]	pH H_2O
CT							
0–0.1	66	28	6	1.38	1.17	2.02	5.91
0.1–0.2	62	29	9	1.62	1.13	1.95	5.80
OR							
0–0.1	71	27	2	1.33	1.77	3.05	6.36
0.1–0.2	70	23	7	1.34	1.59	2.74	6.40

[a]From Siczek et al. (2008a)

1997; Farenhorst et al. 2000). Afterward all columns were saturated with water and allowed to stay for drainage to obtain field water capacity. Forty eight hours later 28 mg (7.85 kg ha^{-1}) atrazine (purity 97.5 %) was suspended in 6.5 mL of distilled water and dripped uniformly onto the surface of each column without plain residue. The rate and way of application of atrazine were similar to those used by Shipitalo et al. (2000). Twenty four hours later all the columns were subjected to watering at an amount of 30 mm (1100 mL) distilled water per each column. Long-term data indicate that on average, rainfalls of such a size occur in the experimental area four times a year (Rejman 2006). Water was applied with manual irrigation system in 100 mL doses to maintain only shallow ponding (a few mm) during infiltration. Successive 100 mL doses started when 1 mm of ponding water remained after the previous dose. To avoid damage of soil structure by water, filter paper was used on the soil surface while on irrigation. Breakthrough times and infiltration rates were recorded as soon as the column began to produce leachate. All the leachates were collected in 50-mL increments from each column separately and the time of percolation was noted. In all cases leaching lasted longer than filtration.

Atrazine in leachates was analyzed by means of HPLC Dionex with a UV–VIS (254 nm) detector and the following operating conditions: the column was Nucleosil CC 250/4 50-5 Macherey Nagel C_{18}, 150 mm long, the mobile phase consisted of a mixture of methanol and water (60:40), and the flow rate was set at 1 ml min^{-1}. Separation was carried out at a constant temperature (30 °C). The correlation coefficient of calibration function was 0.994 (Baran and Oleszczuk 2003).

NH_4^+ NO_3^-, PO_4^{-3} was assayed using ion chromatograph HIC-6A Shimadzu (Japan) equipped with a LP-6A isocratic HPLC pump, conductivity detector CDD-6A, a rotary valve fitted with 20 μl sample loop, protected with a guard column of the same material (25 × 2.3 mm I.D.). The detection was monitored at the range of sensitivity 1 μS. The column was operated at a temperature of 25 °C (Szajdak and Jaskulska 2011). For NH_4^+ determination the mobile phase consisted of 4-mM nitric acid in 7:3 water/methanol ratio and the column of Hamilton PRP X-200 were used. For NO_3^- and PO_4^{-3} measurements, we used the column of Hamilton PRP X-100 and mobile phase 4 mM p-hydroxybenzoic acid in 2.5 % methanol at pH = 8.5 (Szajdak and Gaca 2010).

The contents of inorganic elements including K^{+1}, Mg^{+2}, Mn^{+2}, and Fe^{+3} in the 0–0.2 m depth of soil were determined by ICP-AES using high-pressure system-sealed microwave digestion. The samples of soil were digested with HNO_3 - HF - H_2O_2 acid system. The operation would be simplified and the blank value would be decreased with the above acid systems. The results were proved to be reliable with the relative standard deviation being 0.600–3.683 % for Fe^{+3}, 0.347–2.829 % for K^{+1}, 0.626–2.593 % for Mg^{+2}, and 0.705–4.828 % for Mn^{+2}. The results of determination were proved that the method of sealed microwave digestion ICP-AES was sensitive, precise, easy to operate, and rapid for the determination of inorganic elements in the soil and could satisfy the sample examination request. The methods and results were meaningful to research on the soil mineral system including the contents of mineral elements, the circulation of mineral elements, the interaction, and the application of mineral additive.

Infiltration of water into the soil was determined by the double-ring infiltrometer with a 0.215-m-diameter inner and 0.3-m-diameter outer cylinder inserted 0.14 m into the soil (three replicates). Water entering the soil was measured with a calibrated Mariotte bottle. A constant water head of 0.015 m was maintained in both rings. The measurements were done at the initial soil water content corresponding to approximately field water capacity in all the treatments. This allowed to minimize the effect of different water content. The infiltration data were described according to Philip-type equation (Philip 1957).

Satisfactory precisions based on replicate analyses were ±3.5% for atrazine, ±0.01 for pH measurements, ±5% for bulk density, ±3.5% for TOC, ±3% for NO_3^-, ±3% for NH_4^+, and ±3.8% for PO_4^{-3}. All the chemicals used in this study were of analytical grade.

6.3 Conclusion

The management systems under study revealed very high impact on the physical and chemical properties of soil. CT in comparison with OR soil was less porous as shown by higher bulk density (Table 6.1) and had less biological pores made by soil fauna and/or plant roots (Lipiec et al. 2011).

Content of soil organic matter (SOM) under OR was 1.51 and 1.41 times higher than under CT at the depths of 0–0.1 and 0.1–0.2 m, respectively (Table 6.1). The reduced SOM under CT is due to tillage practices and associated aeration. The SOM amount of the mineral soil with the same constant long-term rotation is in a state of quasi-equilibrium if the plowing frequency and depth are approximately constant. The increase in tillage frequency leads to high SOM losses through intensive soil aeration. The soil under OR was less acidic (pH 6.38) than CT soil (pH 5.85) (Table 6.1).

The OR compared to the CT soil is characterized by lower contents of NO_3^-, PO_4^{-3}, K^{+1}, Mg^{+2}, and Fe and higher concentrations of NH_4^+ and Mn^{+2} (Table 6.2).

Table 6.2 Mean concentrations of the chemicals in the 0–0.02 m soil layer under conventional tillage (CT) and orchard (OR)

Management system	NO_3^- [mg kg^{-1}]	NH_4^+ [mg kg^{-1}]	PO_4^{-3} [mg kg^{-1}]	K^{+1} [g kg^{-1}]	Mg^{+2} [g kg^{-1}]	Fe^{+3} [g kg^{-1}]	Mn^{+2} [g kg^{-1}]
CT	61.1	12.1	2.01	2.31	1.61	10.10	0.35
OR	45.6	14.2	0.71	2.00	1.45	8.10	0.46

6.3.1 Kinetics of Leaching

The percolation of the leachate was considerably faster in orchard (200 min) than tilled soil (800 min). To better understanding the time-dependent behavior of a system of interacting species, a first-order kinetics reaction model was used. The model proceeds at a rate that depends linearly only on one reactant concentration. The differential equation describing first-order kinetics reaction model is

$$-\frac{d[A]}{A} = k[\mathrm{A}] \tag{6.1}$$

$$\frac{d[A]}{A} = -k[A] \tag{6.2}$$

$$\int_{[A]o}^{[A]} \frac{d[A]}{[A]} = \int_{to}^{t} kdt \tag{6.3}$$

$$\int_{[A]o}^{[A]} \frac{1}{[A]} dt = \int_{to}^{t} kdt \tag{6.4}$$

$$\int \frac{1}{x} = \ln(x) \tag{6.5}$$

We can rearrange the equation above to

$$\ln[\mathrm{A}] \quad ln[\mathrm{A}]\mathrm{o} = kt \tag{6.6}$$

$$\ln[\mathrm{A}]=\ln[\mathrm{A}]\mathrm{o} - kt \tag{6.7}$$

Recall from algebra $y = mx + b$ is the equation of a straight line, which $\ln[\mathrm{A}] = -kt + \ln[\mathrm{A}]\mathrm{o}$ demonstrates.

$$y = ax + b \tag{6.8}$$

$$ax = -kt; and\, x = t, a = -k \tag{6.9}$$

$$b = \ln[\mathrm{A}]\mathrm{o} \tag{6.10}$$

Now that we recall the laws of logarithms, we can say that $\frac{\ln[\mathrm{A}]}{\ln \mathrm{Ao}} = -kt$ is the time t with its final concentration of [A] and $[\mathrm{A}]_0$ is at time 0, and it is initial of A and k is rate constant. Since the logarithms of numbers do not have any units, the product will be s^{-1}. Thus, the equation of a straight line is applicable to represent $\ln[\mathrm{A}] = -kt + \ln[\mathrm{A}]_0$.

Rate is the reaction rate (in unites of molar/time) and k is the reaction rate coefficient (in units of 1/time). However, the units can vary with other order reactions. These differential equations are separable, which simplifies the solutions.

The statistical parameters in Table 6.3 indicate that the first-order kinetics reaction model provided an excellent fit for the measured concentration changes vs. time data (Figs. 6.3 and 6.4) with low values of coefficient of variation (CV) ranged from 3.2 to 14.8 %.

The cumulative concentrations of atrazine and inorganic compounds (K^{+1}, Fe^{+3}, Mg^{+2}, Mn^{+2}, NH_4^+, NO_3^-, and PO_4^{-3}) in tilled and orchard silty loam soils as a function of time were characterized by exponential Eq. (6.1) (Figs. 6.3 and 6.4):

$$(c_t) = (c_\infty)(1 - e^{-kt}) \tag{6.11}$$

where c_t = the substance amounts, c_∞ = the maximum of the substance concentrations, k = the first-order rate constant, t = the reaction time, and e = the base of natural logarithm.

Semilogarithmic function of the concentrations measured vs. time showed linear curves that agree well with the first-order kinetics reaction model with respect to all compounds studied (Lasaga et al. 1981; Connors 1990; Reemtsma et al. 1999). Selected examples for atrazine and ammonium are presented in Figs. 6.5 and 6.6:

$$\ln(c_\infty - c_t) = \ln(c_\infty) - kt \tag{6.12}$$

The first-order reaction rate constants (k) were calculated as the slope of the Eq. (6.12) by means of the least squares formulas. The correlation coefficients (r) between the decrease in concentrations of the chemicals ln ($c_\infty - c_t$) and time elapsed varied from −0.807 to −0.993 (Table 6.3).

The slopes of the Eq. (6.2) describe the rates of the chemical concentration changes. The first-order reaction rate constants determined for the studied soils at two management systems (OR and CT) with the parameters of their statistical analysis were shown in Table 6.3. In general, the first-order reaction rate constants for the changes of all chemicals in leachates under OR were shown to be significantly higher than under CT. The first-order reaction rate constants for the leaching of all compounds under CT ranged from 0.2212 10^{-4} s^{-1} for phosphate to 0.7581 10^{-4} s^{-1} for manganese. The corresponding half-life values for these first-order reaction rate constants were 8.70 and 2.54 h, respectively. However, under OR the constants were greater and ranged from 2.9107 10^{-4} s^{-1} for manganese to 8.5867 10^{-4} s^{-1} for atrazine; the respective half-life values were 0.66 h and 0.22 h.

In addition the ratio (k) of the first-order reaction constants under OR and CT for the leaching of all compounds was from 3.8 for manganese to 19.0 times higher for phosphates (Table 6.4). It means that the leaching of all compounds was significantly faster under OR than CT.

Higher first-order reaction rate constants and lower half-life values of the chemicals under OR than CT can be due to both higher total porosity and the contribution of biological pores made by soil fauna and/or plant roots. Additionally, under OR

Table 6.3 Statistical evaluation of the pseudo-first-order reaction rate constants (k) for the leaching of investigated compounds under tilled and orchard soil

Management system	Investigated compounds		$10^{-4}(k \pm \Delta k)$ $[s^{-1}]$	$t_{0.5}$ [h]	r	B	CV [%]
CT	Cations	Atrazine	0.7556 ± 0.0612	2.55	−0.969	6.4545	6.5
		Mn^{+2}	0.7581 ± 0.0682	2.54	−0.966	−2.7468	6.9
		Fe^{+3}	0.6405 ± 0.0575	3.01	−0.985	0.867	5.9
		NH_4^+	0.6325 ± 0.0514	3.04	−0.993	−0.7725	3.2
		Mg^{+2}	0.5386 ± 0.0482	3.58	−0.972	2.9461	5.8
		K^{+1}	0.5263 ± 0.0672	3.66	−0.807	1.4436	14.8
	Anions	NO_3^-	0.5398 ± 0.0431	3.57	−0.967	4.3908	6.8
		PO_4^{-3}	0.2212 ± 0.0183	8.70	−0.958	0.867	7.1
OR	Cations	Atrazine	8.5867 ± 0.7728	0.22	−0.943	6.8612	7.1
		Mn^{+2}	2.9107 ± 0.3784	0.66	−0.873	−3.1774	13.6
		Fe^{+3}	4.9775 ± 0.4281	0.39	−0.952	0.1916	5.8
		NH_4^+	5.3130 ± 0.2883	0.49	−0.947	0.8376	6.8
		Mg^{+2}	3.8483 ± 0.3425	0.50	−0.955	2.9829	6.7
		K^{+1}	3.8972 ± 0.3352	0.49	−0.954	2.1283	7.0
	Anions	NO_3^-	3.9468 ± 0.3552	0.49	−0.955	4.7410	6.9
		PO_4^{-3}	4.2088 ± 0.3787	0.46	−0.910	−1.1346	8.2

$k \pm \Delta k$, confidence limits for pseudo-first-order reaction rate of soil samples at α −0.05 and (n−2) degrees of freedom; $t_{0.5}$, half-life; r, correlation coefficient; b, the constant term, is the y-intercept; CV, coefficient of variation

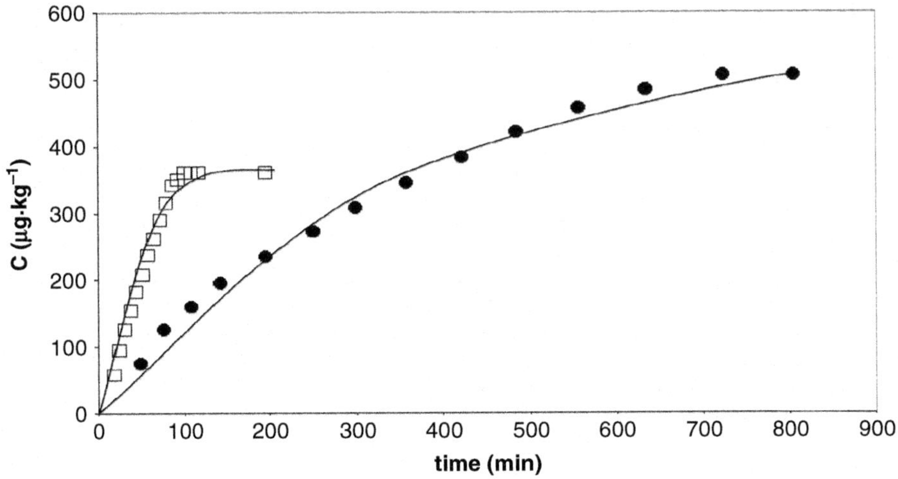

Fig. 6.3 Cumulative concentrations of atrazine in soil under OR (*square*) and CT (*circle*)

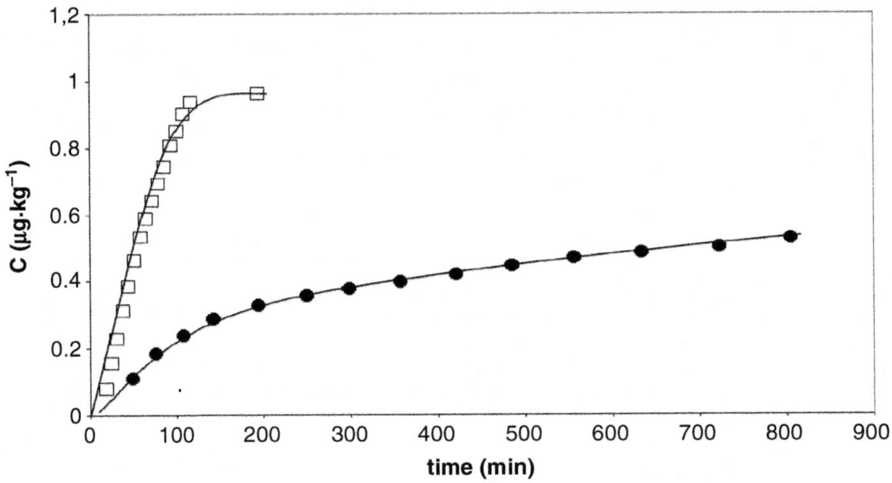

Fig. 6.4 Cumulative concentrations of ammonium in soil under OR (*square*) and CT (*circle*)

the biological pores were surface open. These were reflected in leachate percolation rate. The same quantity of percolate (800 mL) under OR and CT was obtained after 200 and 800 min, respectively. Corresponding breakthrough times were 8 min and 23 min. A greater infiltrability of OR than CT soil was confirmed by field measurements of water infiltration that was described by Philip-type Eq. (6.13) (Philip 1957):

$$I = At + St^{0.5} \tag{6.13}$$

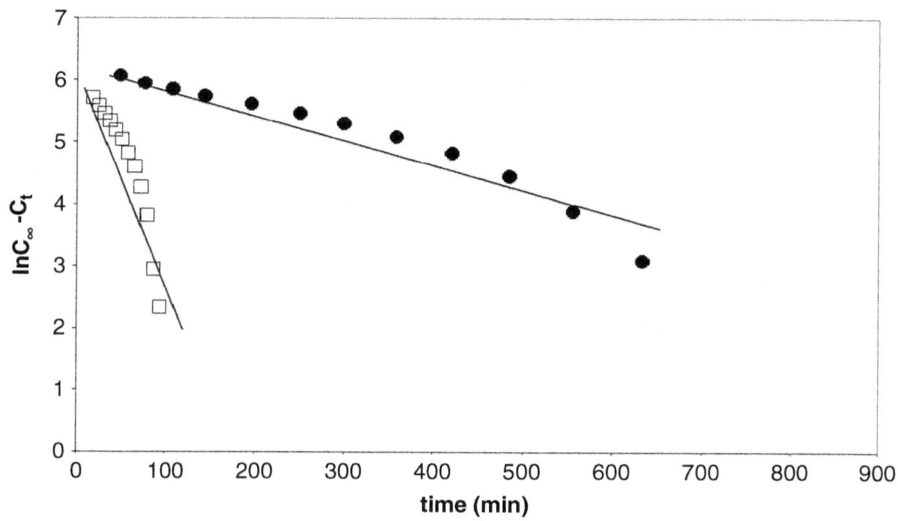

Fig. 6.5 Semilogarithmic function of the atrazine changes in soil under OR (*square*) and CT (*circle*)

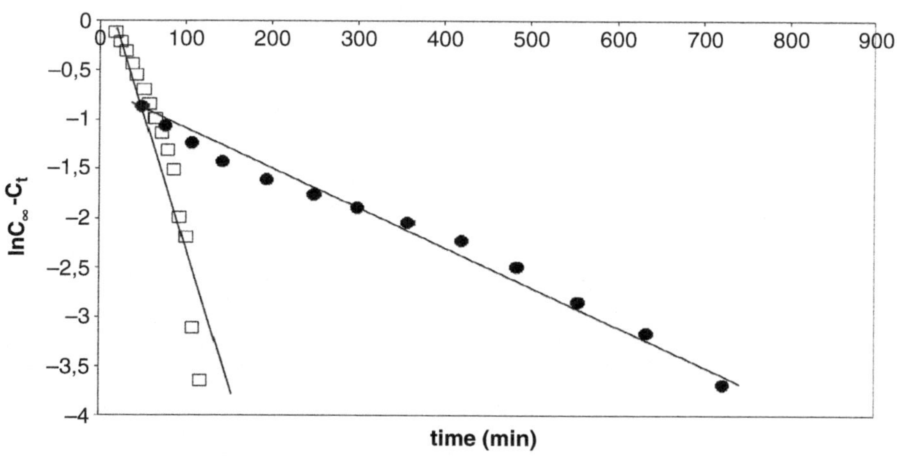

Fig. 6.6 Semilogarithmic function of the ammonium in soil under OR (*square*) and CT (*circle*)

where I = cumulative infiltration (cm), $A = a$ coefficient related to saturated soil conductivity, and S = soil sorptivity.

The value of A is proportional to the saturated soil conductivity, multiplied by coefficient varying from 0.2 to 0.67 (Philip 1987; Kutilek and Niielsen 1994). Both A and S were greater under OR than CT; A calculated for OR soil was about 20 times higher than for CT; differences for S was much smaller (Figs. 6.5 and 6.7) (Table 6.5).

Table 6.4 Ratios (*K*) of the first-order reaction rate constant for the leaching of chemicals under orchard soil and tilled soil

Cations						Anions	
Atrazine	NH_4^+	Fe^{+3}	K^{+1}	Mg^{+2}	Mn^{+2}	PO_4^{-3}	NO_3^-
11.4	8.4	7.8	7.4	7.2	3.8	19.0	7.3

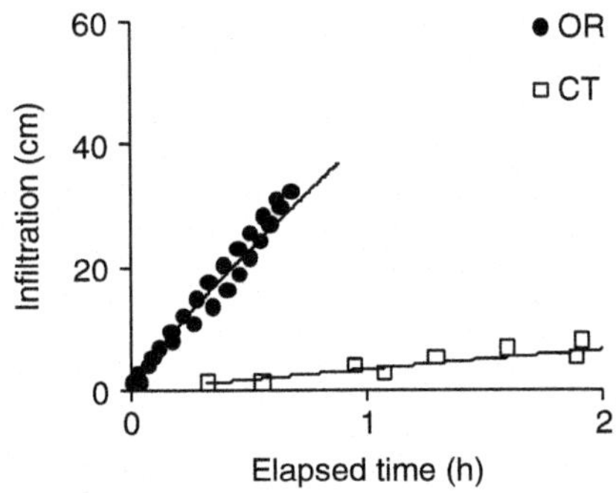

Fig. 6.7 Philip-type model fitted to infiltration data measured in the OR and CT under field conditions. Parameters of the model are presented in Table 6.5

Table 6.5 Parameters and coefficient of determination *R* of a Philip-type Eq. (6.3) for CT and OR soil

Management system	$A\ [\mathrm{cm\,h^{-1}}]$	$S\ [\mathrm{cm\,h^{-0.5}}]$	r
CT	2.44	1.49	0.97
OR	44.55	1.71	0.99

A = coefficient related to saturated soil conductivity, S = describes soil sorptivity, r = correlation coefficient

Both factors were related to water infiltration, the proportion of the soil, through which the water passes and the residence time of the water in the soil, affect process of chemicalsorption (Addiscot and Thomas 2000; Mohsen and Zahra 2006). A greater infiltrability of OR than CT soil can be largely a result of higher contribution of large and more continuous and conductive pores under OR where they are not disturbed by tillage. The differences in pore structure are in line with all values of the first-order reaction rate constants to be higher in OR than in CT soils (Table 6.4). Besides the effects of pore structure and associated infiltrability, the first-order reaction rate constants under OR and CT can be influenced by different acidity and SOM content (Table 6.1). Therefore the first-order reaction constants (Table 6.3) were in line with the parameters of Philip-type equation (Table 6.5).

In less acidic OR than CT soil, the first-order reaction rate constants of Fe^{+3}, Mg^{+2}, and Mn^{+2} cations can be indirectly influenced by a greater conversion into insoluble hydroxides $Fe(OH)_2$, $Fe(OH)_3$, $Mg(OH)_2$, and $Mn(OH)_2$. As for atrazine, the constants can be modified by greater sorption of atrazine under OR than CT due to a greater SOM content in the former (Table 6.1). Sparks and Swift (2002) postulated that total organic matter content is more important for the sorption process of the atrazine than the composition of the organic matter. However, Kulikova and Perminova (2002) observed that the extent of atrazine sorption to humic substances was strongly correlated to the aromaticity of the organic matter. Although Lima et al. (2010) pointed out that carboxyl- and aromatic-rich organic matters are the most efficient binding structures for atrazine. In addition, a charge transfer mechanism is also postulated for the interaction of atrazine and soil organic matter (Müller-Wegener and Ziechman 1980; Müller-Wegener 1988; Piccolo et al. 1992).

The knowledge of the first-order reaction constant for the leaching of chemical compound calculated in every field condition has practical meaning. It enables the researcher to perform the actual concentration of the compound calculations to predict the effectiveness action of the substance in soil condition.

The first-order kinetics reaction model is adequate to describe process of leaching of the agricultural chemicals under different soil management systems. The first-order rate constants reflected different leaching potential of the chemicals from orchard and tilled soil. In general, the first-order reaction rate constants for the changes of all chemicals in leachates under OR were shown to be significantly higher than under CT. The contrary was showed for half-lifes that were lower in orchard soil. The ratios of the first-order reaction constants between OR and CT for the leaching of cations or anions ranged from 3.8 to 8.4 and from 7.3 to 19, respectively.

Higher first-order reaction constants and lower half-life values of the compounds under OR than CT were in line with both higher total porosity and the contribution of biological pores as an effect of soil fauna and plant roots. Surface-open biological pores under OR reflected in percolation time and greater infiltrability, where they were not disturbed by tillage. The leaching includes behavior of chemicals through soil upon soil adsorption, hydrodynamic dispersion and diffusion, adsorption dynamics, and evapotranspiration.

An understanding of the effect of simultaneous leaching and degradation is needed to properly understand the movement of chemicals into groundwater. Elaboration of the knowledge of the dynamics of herbicides will lead not only to a better understanding of the fate and behavior of chemicals in the environment, but will point the development of improvement of our technology of the use of these compounds. This can mean safer, less environmentally contaminating practices and greater effectiveness, both which should contribute to the production of food and the protection of man's health.

References

Addiscott TM, Thomas D (2000) Tillage, mineralization and leaching: phosphate. Soil Till Res 53:255–273

Amellal N, Burtin G, Bartoli F, Heulin T (1998) Colonization of wheat roots by an exopolysaccharide-producing Pantoea agglomerans strain and its effect on rhizosphere soil aggregation. Appl Environ Microbiol 64:3740–3747

Bailey R, Lebel G, Lawrence JF (1978) Gas-liquid chromatography of triazine herbicides as heptafluorobutyryl derivatives and some applications to analysis in foods. J Chromatogr A161:251–257

Bandeen JD, Stephenson GK, Cowett ER (1982) Discovery and distribution of herbicide-resistant weeds in North America. In: LeBaron HM, Gressel J (eds) Herbicide resistance in plants. Willey-Interscience, New York, pp 9–30

Baran S, Oleszczuk P (2003) Determination of S-triazine herbicides in soils of different organic matter content. Chem Anal 48:817–827

Biradar DP, Rayburn AL (1995) Chromosomal damage induced by herbicide contamination at concentration observed in public water supplies. J Environ Q 24:1232–1235

Bouwer H (1986) Intake rate: cylinder infiltrometer. In: Klute A (ed) Methods of soil analysis, part 1 physical and mineralogical properties, Monograph 9. ASA, Madison, pp 825–843

Brye KR, Norman JM (2004) Land-use effects on anion associated cation leaching in response to above-normal precipitation. Acta Hydrochim Hydrobiol 32(3):235–248

Connors KA (1990) Chemical kinetics. The study of reaction rates in solution. VCH Publishers Inc., New York, pp 17–309

Dobrzański B, Zawadzki S (1951) The soils of the experimental station at Felin. Annales Universitatis Mariae Curie Skłodowska, Sectio E. VI. pp 231–252 (in Polish with English summary)

Drury CF, McKenney DJ, Findlay WI, Gaynor JD (1993) Influence of tillage on nitrate loss in surface runoff and tile drainage. Soil Sci Soc Am J 57:797–802

Farenhorst A, Topp E, Bowman BT, Tomlin AD (2000) Earthworm burrowing and feeding activity and the potential for atrazine transport by preferential flow. Soil Biol Biochem 32:479–488

Feiza V, Feiziene D, Riley HCF (2003) Soil available P and P off take responses to different tillage and fertilization systems in the hilly morainic landscape of western Lithuania. Soil Tillage Res 74:3–14

Gajda A (2010) M. Microbial activity and particulate organic matter content in soils with different tillage system use. Int Agrophys 24:129–137

Glotfelty DE, Seiber JN, Liljedahl LA (1987) Pesticides in fog. Nature 325:602–605

Grant FR (1997) Changes in soil organic matter under different tillage and rotations: mathematical modeling in ecosystems. Soil Sci Soc Am J 61:1159–1175

Grapelli A, Rossi W (1979) The effect herbicides, respiration and IAA biosynthesis in Arthrobacter sp. Chemosphere 26:377–382

Graymore M, Stagnitti F, Allinson G (2001) Impacts of atrazine in aquatic ecosystems. Environ Int 26:483–495

Habecker MA (1989) Environmental contamination at Wisconsin pesticide mixing/loading facilities: case study, investigation and remediation action evaluation. Department of Agriculture Trade and Consumer Prot. Ag. Resource Mgmt. Division, Madison, pp 1–80

Hajnos M, Lipiec J, Świeboda R, Sokołowska Z, Witkowska-Walczak B (2006) Complete characterization of pore size distribution of tillied and orchard soil using water retention curve, mercury porosimetry, nitrogen adsorption, and water desorption methods. Geoderma 135:307–314

Holland JM (2004) The environmental consequences of adopting conservation tillage in Europe: reviewing the evidence. Agric Ecos Envir 103:1–25

Jalali M, Rowell DL (2003) The role of calcite and gypsum in the leaching of potassium in a sandy soil. Exp Agric 39:379–394

Kulikova NA, Perminova IV (2002) Binding of atrazine to humic substances from soil, peat, and coal related to their structure. Environ Sci Tech 36:3720–3724

Kutilek M, Niielsen DR (1994) Soil hydrology. Catena – Verlag, Cremlingen, pp 1–370

Lasaga AC, Berner RA, Fisher GW, Anderson DE, Kirkpatrick JR (1981) Kinetics of geochemical processes. In: Lasaga AC, Kirkpatrick RJ (eds) Review in mineralogy. Mineralogy Society of America Mineralogy Society of America, Washington, DC, pp 1–68

Liebl RA, Worsham AD (1983) Inhibition of pitted morning glory (*Ipomea lacunosa*) and certain other weed species by phytotoxic components of wheat (*Triticum aestivum L.*). Straw. J Chem Ecol 9(8):1027–1043

Lima DLD, Schneider RJ, Scherer HW, Duarte AC, Santos EBH, Esteves VI (2010) Sorption-desorption of atrazine on soil subjected to different organic long-term amendments. J Agric Food Chem 58:3101–3106

Lipiec J, Stępniewski W (1995) Effects of soil compaction and tillage systems on uptake and losses of nutrients. Soil Tillage Res 35:37–52

Lipiec J, Nosalewicz A, Siczek A, Kotowska U (2011) Effects of land use on leaching of nitrate-N, ammonium-N and phosphate in relation to stained surface area. Int Agrophys 25:149–154

Long T (1987) Groundwater contamination in the vicinity of agrochemical mixing and loading facilities. Proceedings of the 16th ENR annual conference, Illinois, Department of Energy and Natural Resources. Environmental significance of atrazine in aqueous systems, p 173

Mallants D, Mohanty BP, Vervoort A, Feyer J (1997) Spatial analyses of saturated hydraulic conductivity in a soil with macropores. Soil Tech 10:115–131

McDaniel PA, Buol SW (1991) Manganese distribution in acid soils of the North Carolina Piedmont. Soil Sci Soc Am J 55:152–158

Mohsen J, Zahra K (2006) Groundwater quality in an irrigated, agricultural area of northern Malayer, western Iran. Nutr Cycling Agroecosyst 80(1):95–104

Müller-Wegener U (1988) Interaction of humic substances with biota. In: Frimmel FH, Christman RF (eds) Humic substances and their role in the environment. Wiley, New York, pp 179–192

Müller-Wegener U, Ziechmann W (1980) Elektronen-Donator-Akzeptor-Komplexe zwischen aromatischen Stickstoffheterocyclen und Huminsaure. Zeit Pflanzen Bodenk 143:247–249

Pervanchon F, Bockstaller C, Amiaud B, Peigne J, Bernard PY, Vertes F, Fiorelli JL, Plantureux S (2005) A novel indicator of environmental risks due to nitrogen management on grasslands. Agric Eco Environ 105:1–16

Philip JR (1957) The infiltration joining problem. Soil Sci 84:329–339

Philip JR (1987) The infiltration joining problem. Water Resour Res 23:2239–2245

Piccolo A, Gelano G, De Simone C (1992) Interactions of atrazine with humic substances of different origins and their hydrolysed products. Sci Tot Environ 117/118:403–412

Pranab KG, Ligy P (2006) Environmental significance of atrazine in aqueous systems and its removal by biological processes: an overview. Glob NEST J 8(2):159–178

Puustinen M, Koskiaho J, Peltonen K (2005) Influence of cultivation methods on suspended solids and phosphorus concentrations in surface runoff on clayey sloped fields in boreal climate. Agric Ecos Environ 105:565–579

Reemtsma TA, Bredov A, Gehring M (1999) The nature and kinetics of organic matter release from soil by salt solutions. Eur J Soil Sci 50:53–64

Rejman J (2006) Effect of water and tillage erosion on transformation of soils and loess slopes. Acta Agrophys 3:5–90 (in Polish, with English summary)

Ritter WF, Scarborough RW, Chirnside AEM (1994) Contamination of ground water by triazines, metolachlor, and alachlor. J Contam Hydrol 15:73–92

Rossi W, Grapelli A, Pietrosantu W (1984) Phytohormones in soil after atrazine application. Folia Microbiol 29:325–329

Sharpley AN, Chapra SC, Wedopohl R, Sims JT, Danie TC, Reddy KR (1994) Managing agricultural phosphorus for protection of surface waters. J Environ Q 23:437–451

Shilling DG, Liebl RA, Worsham DA (1985) Rye (*secale cornutum L.*) and wheat (*Triticum aestivum L.*) mulch: the suppression of certain broadleaved weeds and the isolation and identification of phytotoxins. In: Thompson AC (ed) Chemistry of allelopathy. American Chemical Society, Symposium series no. 268. American Chemical Society, Washington. D.C, pp 243–271

Shipitalo MJ, Edwards WM, Dick WD, Owens LB (1990) Initial storm effect on macropore transport of surface-applied chemicals in no-till soil. Soil Sci Soc Am J 54:1530–1536

Shipitalo MJ, Dick WA, Edwards WM (2000) Conservation tillage and macropore factors that affect water movement and the fate of chemicals. Soil Tillage Res 53:167–183

Siczek A, Kotowska U, Lipiec J, Nosalewicz A (2008a) Leaching of potassium, magnesium, manganese and iron in relation to porosity of tilled and orchard loamy soil. Acta Agric Scand Sect B-Soil Plant Sci 58:60–65

Siczek A, Kotowska U, Lipiec J, Nosalewicz A (2008b) Macro-porosity and leaching of atrazine in tilled and orchard loamy soils. Chemosphere 70:1973–1978

Six J, Bossuyt H, Degryz S, Denef K (2004) A history of research on the link between (micro) aggregates, soil biota, and soil organic matter dynamics. Soil Tillage Res 79:7–31

Słowińska-Jurkiewicz A, Świca M, Domżał H, Bielińska E (2001) Effect of management practices on structure of cherry orchard soil. Acta Agrophys 56:271–281

Sparks KM, Swift RS (2002) Effect of soil composition and dissolved organic matter on pesticide sorption. Sci Tot Environ 298:147–161

Stehouwer RC, Dick WA, Traina SJJ (1994) Sorption and retention of herbicides in vertically oriented earthworm and artificial burrows. J Environ Q 23:286–292

Stout WL, Fales SL, Muller LD, Schnabel RR, Weaver SR (2000) Water quality implications of nitrate leaching from intensively grazed pasture swards in the northeast US. Agric Eco Environ 77:203–210

Szajdak LW, Gaca W (2010) Nitrate reductase activity in soil under shelterbelt and an adjoining cultivated field. Chem Ecol 26(4):123–134

Szajdak LW, Jaskulska R (2011) Biochemical and chemical characterization of soils under shelterbelts and adjoining cultivated fields. In: Szajdak LW (ed) Shelterbelts: efficient element of the landscape. Chemical and biochemical investigations of ground water and soil. LAP Lambert Academic Publishing, Saarbrücken, pp 21–32

Szajdak L, Maryganova M (2007) Occurrence of auxin in some organic soils. Agron Res 5(2):175–187

Tindall JA, Vencill WK (1995) Transport of atrazine, 2,4-D and dicamba through preferential flowpaths in an unsaturated clayen soil near Centralia Missouri. J Hydrol 166:37–59

Triplett GB, Lytle GD (1972) Control and ecology of continuous corn grown without tillage. Weed Sci 20:453–457

Trotter DM, Baril A, Wong MP, Kent RA (1990) Canadian water quality guidelines for atrazine. Can Sci Ser 168:106

Chapter 7
Impact of Long-Term Agricultural Management and Native Forest Ecosystem on the Chemical and Biochemical Properties of Retisols' Organic Matter

Alar Astover, Lech Wojciech Szajdak, and Raimo Kõlli

Abstract The objective of the chapter is to assess the impact of long-term fertilization on biologically active substances in arable *Fragic Glossic Retisols* and to compare received results with corresponding native forest soil. The analysis of interactions of chemical and biochemical compounds in soils was based on soil samples taken from the long-term field experiment (crop rotation, as cultivated ecosystem, with application of mineral and organic fertilizers: potato, spring wheat, spring barley) and from the forest soil (as native ecosystem). The relationship between quantities of chemical and biochemical compounds in arable soils' organic matter and in the yield of cultivated plants was greatly influenced by the use of mineral and organic fertilizers. The contents of phytohormone indole-3-acetic acid (IAA), participating in nitrogen cycle enzymes (urease and nitrate reductase), and different forms of nitrogen and organic carbon in soils have a marked effect on the crop yield. Arable soil in rotation without organic fertilizers treated with mineral fertilizer (120 kg N ha^{-1}) revealed the highest increase of nitrate reductase activity, and the highest concentration of IAA, however, the lowest activity of urease. Therefore, mineral fertilizer in comparison with farmyard manure and alternative organic fertilizers created the most suitable conditions for the crop yield increase. The highest activity of urease in forest soil is closely associated with the concentration of biochemically available (dissolvable in water) organic carbon.

Keywords Biochemical properties • Enzymes • Nitrogen • Organic carbon • Phytohormone • Retisols' organic matter • Long-term experiment • Organic and

---

A. Astover (✉) • R. Kõlli
Institute of Agricultural and Environmental Sciences, Estonian University of Life Sciences, Fr.R. Kreutzwaldi Str. 1A, 51014 Tartu, Estonia
e-mail: Alar.Astover@emu.ee; Raimo.Kolli@emu.ee

L.W. Szajdak
Institute for Agricultural and Forest Environment, Polish Academy of Sciences, Str. Bukowska 19, 60-809 Poznań, Poland
e-mail: lech.szajdak@isrl.poznan.pl

L.W. Szajdak (ed.), *Bioactive Compounds in Agricultural Soils*,
DOI 10.1007/978-3-319-43107-9_7

mineral fertilization • Forms of nitrogen • Activity of urease and nitrate reductase • Auxin • Crop yields

7.1 Introduction

The phenomenon of allelopathy is defined as biochemical interaction between all types of plants, including microorganisms. Plants release chemical compounds into the environment through root exudation, leaching by dews and rains from plant surfaces, decaying plant tissues, and volatilization. Most of the allelopathic compounds of the plants can be characterized as secondary metabolites like phenolic acids, flavonoids, cyanogenic glycosides, tannins, isothiocyanates, and others. Soil organic matter (SOM) contains these substances or their metabolites and degradation products in its structure (Honermeier 2007). Allelochemicals as secondary metabolites partially bred out of crop plants because of undesirable qualities, such as bad taste or reduced yield. Most allelochemicals are relatively nonphytotoxic compared with synthetic herbicides, sometimes antagonistic in combinations, and may be inactivated by soil (Duke and Lydon 1987). All these substances or their products of the degradation and their metabolites and biochemical pathways are also incorporated into the SOM.

SOM is considered to be as the most important factor in soil forming, development, and continuous functioning (Paustian et al. 1997; Smith et al. 1997; West and Post 2002). Sequestrated into SOM, organic carbon concentrations, pools, and residence time in soil as well as acting intensity of interconnected with SOM edaphon are soil-type specific or characteristic to certain soil types (Kern et al. 1997; Percival et al. 2000). Soil moisture regime, calcereousness, clay content, soil organic carbon (SOC)-retaining capacity and its verival distribution pattern for each soil is characteristic. However, land-use change (crop rotation, continuous cropping, no-tillage, melioration, rewetting) has the greatest influence mainly on fabric of epipedon and biological functions of soil cover.

It is well documented that SOM may influence soil fertility by supplying micronutrients to the roots, improve soil structure, increase the soil microbial population, increase the exchange capacity and the pH buffering capacity of the soil, supply defined biochemical compounds to plant roots, as well as supply humic substances, which serve as carriers of micronutrients or growth factors (Chen et al. 2004). Nitrogen included in SOM structures is more constant and stable than that of mineral forms or presented in free low-molecular-weight organic substances (Szponar and Szajdak 2003). N mineralization includes a biotic and abiotic processes producing NH_4^+ from N organic compounds which may range from low-molecular-weight (amino acids) to high-molecular weight (e.g., proteins and peptizes). Schimel (1986) indicated that nitrogen immobilization was lower in cropland than in native grassland suggesting that microbial activity in the cropland was limited by carbon substrate availability. Degradation of available carbon, in the form of high-molecular-weight humic substances, suggests the more rapid nitrogen turnover and losses than in forest or native grassland.

SOM plays also significant function on regulating root metabolism of cultivated plants (Voughan and Malcolm 1985). However, low-molecular-weight humic compounds may be released from organic and mineral–organic complexes in the process by low-molecular-weight carboxylic acid presented in root exudates (Albuzio and Ferrari 1989). These compounds enter the plant root and impact on plant metabolism, by inducing or inhibiting the mechanism of protein, sugars, and fat synthesis. These processes led to the changes of plant root cells and result in morpho-functional properties of plants (Ladd and Butler 1971; Voughan and Malcolm 1979; Piccolo et al. 1992).

Besides the scientific interest, the knowledge of organic substances in soil may lead to a better understanding of reactions during plant nutrition, soil productivity, and soil genesis. Many of these substances has no effect on metabolism, but some are also energy source for the soil organisms (Flaig 1971).

The differences between *Fragic Glossic Retisols* and its close neighbors (*Luvisols*) are well documented in our previous works (Kõlli and Kanal 1995; Kanal and Kõlli 1996). Although, depending on variation of local site peculiarities, the characteristics of soil-type humus status indices may vary to a considerable extent, they may be used as benchmarks in estimation of the whole ecosystem status and in arrangement of sustainable land use.

Depending on local ecological conditions, each soil has a specific character of SOC flow (input => transformation and sequestration => output) throughout the soil cover (Körchens et al. 1998; Neill et al. 1998; Yakimenko 1998; Janzen 2006). The SOC flow in the composition of organic matter begins with litter falling on or into the soil and continues with its disintegration, its transformation into humus and accumulation, and its ultimate disappearance, via consumption by soil organisms, complete mineralization or illuviation into the subsoil, or eluviation out of the soil cover. The investigation of SOM flux pathways in connection with its composition and structure had an utmost importance in SOM or humus studies during the last half century (Athertnon et al. 1967; Olsen 1986; Hart et al. 1994).

At first decades of named period in humus quality studies, the priority was done to chemical investigations (first of all the determination of humus fractional composition). Later the leading role of biological features in SOM transformation was accentuated. But only in the last three decades, large perspectives of biochemical studies in researches of SOM were understood.

The objective of the study was to assess the relationship between different kinds of fertilizers and the biologically active substances in arable soil, to investigate the impact of different kinds of chemical and biochemical compounds in soils on the crop yield of cultivated plants, and to compare the amounts of biologically active substances in arable soil with results obtained from forest soil.

7.2 Characterization of Research Conditions and Objects

7.2.1 General Methodological Principles

The general schema of field studies on soil chemical and biochemical properties was the following. In the first level, the comparative research of general and pedoecological characteristics, as well as the functioning of representative to soil-type ecosystems, was studied in natural and cultivated conditions. This kind of comparative research enables to elucidate and to prognosticate the changes taking place in soil fabric and functioning in connection with land-use change.

In the second level of the research, more profound investigations into biochemical aspects (pathways) of SOM transformation and composition were conducted. For these more specific researches, two research areas (one on arable land and another in forest) were selected. The base of our complex pedoecological characterization of *Fragic Glossic Retisol* (IUSS WG WRB 2015) was from one-side long-term field trial as cultivated ecosystem and from the other prematured *Oxalis* site-type spruce forest as native ecosystem. The representatives of these areas were proven by comparing their chemical and pedoecological characteristics, which are typical to *Fragic Glossic Retisols.*

7.2.2 Climatic and Meteorological Conditions

Results of our research are typical for the cool temperate subhumid regions or for *frigid-udic* and *frigid-aquic* pedo-climatic conditions of Estonia. The annual average air temperature is in limits from 4.4 to 6.6 °C, and the annual precipitation varies between 550 and 800 mm (Ahas et al. 2002). Average monthly air temperatures in Tartu range from −7.1 °C in January to 16.5 °C in July. The average annual amount of precipitation is 585 mm, with 270 mm during the main growing season (EMHI 2009). The average air temperature of 20 years (1990–2009) at experimental area in the growing season (from May to August) was 15.2 °C and the average amount of precipitation 290 mm (Table 7.1). The vegetation period of 2002 was extremely dry and relatively warm. In 2008–2009 the precipitation was significantly higher (420–475 mm) than the average of many years.

7.2.3 Soils and Their Humus Status

Formed on two-layered parent material (loamy sand on sandy loam or on loam), *Fragic Glossic Retisols* form from Estonian soil cover totally 9.5 % but 21.3 % from arable and 3.6 % from forest land (Kokk 1995). In South Estonia the red–brown weakly calcareous or somewhat noncalcareous loamy till and in Central Estonia the

Table 7.1 The weather conditions in the growing period (May–August) at the Eerika experimental station

Years	Month				
	May	June	July	August	May–August
	Average temperature, °C				
1990–2009	11.2	15.2	17.7	16.5	15.2
2007	13.9	16.6	20.1	19.2	17.4
2008	11.6	12.9	19.4	15.3	14.8
2009	12.1	13.4	16.4	17.0	14.7
	Sum of precipitation, mm				
1990–2009	57	82	72	79	290
2007	15	81	45	22	163
2008	112	71	104	133	420
2009	34	211	113	116	475

yellow–gray moderate (somewhere weakly) calcareous loamy till are covered with the deposits of postglacial sandy loam or loamy sand. Under the humus layer, the illuvial horizon enriched with amorphous ferric oxide and followed by the horizon, where *albeluvic glossae* (interfingering of albic material into the argillic horizon), can be found. On the boundary of two-membered parent material layers, the seasonal (in spring and autumn) stagnation of water and consequently the temporal reducing conditions may occur. This produces visible redoximorphic features in the contact area, which is reflected by presence of *stagnic* or/and *gleyic* color pattern. The name of this soil characterizes well the presence of *albeluvic glossae* (special case of retic properties) and the formation of *fragic–argic* horizon (IUSS WG WRB 2015).

The general pedoecological characteristics of arable and forest *Fragic Glossic Retisols* presented in Table 7.2 originate from soil profile horizons' database PEDON (Kõlli et al. 2009). The database PEDON was formed mainly during 1967–1985 but was updated in 1986–1995 and 1999–2002. The comparative analysis demonstrates clearly these changes which take place in soil cover fabric and properties in connection with land-use change, from arable land to forest land and vice versa.

As *Fragic Glossic Retisols* are largely distributed at Tartu County (they form here more than half from arable soils), the establishment of long-term field trial and forest research area in these regions for biochemical analysis of *Fragic Glossic Retisols'* organic matter is in all respects justified (Rowel 1994; Reeuwijk 1995). The quantitative characteristics of *Fragic Glossic Retisols'* humus status originate as well from database PEDON, which contains data of 13 research areas (RA) founded on arable and of 18 RA on forest soils (Table 7.3). In this work the data on SOC and total nitrogen (N_{tot}) content, as well as bulk density and particle size distribution of forest floor and humus, eluvial, and illuvial horizons, was used. The concentrations (g kg^{-1}) of SOC and N_{tot} and bulk density were determined by soil horizons (or in some cases by 5 cm layers). The pools (Mg ha^{-1}) of SOC and N_{tot}

Table 7.2 General pedoecological characterization of *Fragic Glossic Retisols*

Characteristic	Arable land	Forest land
pH_{KCl} of humus horizon	5.1–5.9a[a]	3.5–4.4b
Prevailing soil texture: epipedon	Loamy sand	Loamy sand
Subsoil	Sandy loam	Sandy loam
Substratum	Loam	Loam
Type of epipedon	Eluvic low humous	Fresh (moist) moder
Prevailing plant species	Barley, rye, potato, ley	Spruce, pine, birch
Relative quality of soils as compared with the best soils of the same land-use status, %	80–85a	95–100b
Phytoproductivity, Mg DM[b] ha^{-1} $year^{-1}$	6.5–10.2	10.2–14.4
Annual influx DM into soil, Mg DM ha^{-1} $year^{-1}$	4.0–9.0[c]	7.0–8.5
Exchangeable acidity in EP[d], cmol kg^{-1}/kmol ha^{-1}	0.3/9.0a	1.2/26.1b
Extractable Al in EP, mg kg^{-1}/kg ha^{-1}	1.4/54a	9.4/176b
Hydrolytical acidity in EP, cmol kg^{-1}/kmol ha^{-1}	2.9/105a	5.9/258b
Exchangeable basic cations in EP, cmol kg^{-1}/kmol ha^{-1}	11.4/446a	5.6/105b
Cation exchange capacity in EP, cmol kg^{-1}/kmol ha^{-1}	14.2/551	14.5/363
Base saturation stage, % in FF[e]/EP/solum	–/81/85	41/36/67

[a]Limits mean ± standard error at $p < 0.05$, different letters denote significant differences ($p < 0.05$) between arable and forest soil

[b]DM – dry matter

[c]barley – 4–4.5, rye – 5–5.5, and ley 8–9 Mg DM ha^{-1} $year^{-1}$

[d]EP – epipedon

[e]FF – forest floor

Table 7.3 Comparative results of humus status characteristics of *Fragic Glossic Retisols* functioned in cultivated and native conditions

	Arable land	Forest land
Characteristic, unit	$n - 13$	$n - 18$
Depth of epipedon, cm	22.3–29.7a[a]	15.5–20.5b
Among this forest floor	0a	2.7–3.5b
Depth of solum, cm	88.1–97.9a	82.8–101.2a
Pools of SOC in epipedon, Mg ha^{-1}	46.7–51.3a	40.0–44.1b
Among this forest floor	0a	6.7–7.6b
Pools of SOC in solum, Mg ha^{-1}	66.6–71.4a	61.9–66.0a
Pools of N_{tot} in epipedon, Mg ha^{-1}	3.9–5.5a	2.0–3.1b
Among this forest floor	0a	0.26b
Pools of N_{tot} in solum, Mg ha^{-1}	6.0–8.2a	3.8–5.4b
Mean content of SOC in humus horizon, g kg^{-1}	13.2	21.1
Mean content of N_{tot} in humus horizon, g kg^{-1}	1.3	1.5
C/N in humus horizon	10.1	14.5
C/N in epipedon	10.1	16.9

[a]Limits—mean ± standard error at $p < 0.05$; different letters denote significant differences ($p < 0.05$) between arable and forest soil

were estimated on forest areas in three soil layers: (i) in forest floor, (ii) in humus horizon, and (iii) in subsoil, but on arable areas in two layers (i) in plow layer or humus horizon and (ii) in subsoil. For comparative analysis, the data were calculated as well for epipedon and for solum as a whole. On arable areas epipedon consists only humus horizon, but on forest areas the forest floor is added. The soil cover or solum, whose depth reaches from the surface to the unchanged parent material, composed therefore from epipedon and subsoil.

7.2.4 Field Experiment

In 1989, an international long-term experiment on the organic nitrogen or IOSDV (Internationale Organische Stickstoffdauerdiingungsversuche) with three-field crop rotation (potato, spring wheat, spring barley) was established at Eerika near Tartu (58°22.5′ N; 26°39.8′ E) on *Fragic Glossic Retisol*. The main aims of this experiment were to determine the long-term effects of cropping systems on soil properties and productivity. The design of this field experiment is similar to other European network of IOSDV experiments (Boguslawski 1995). Before the establishment of this experiment in 1989, it was in set-aside state for 5–6 years as field-grass fallow. It was used as arable land in 1957–1983.

Average agrochemical characteristics of the plow horizon of soil in the year of establishment were the following: humus content 17.1 g kg^{-1}, N_{tot} concentration 0.9 g kg^{-1}, C/N ratio 11, and pH_{KCl} 6.3. Double lactate-soluble phosphorus amount was 44 mg kg^{-1} and potassium was 133 mg kg^{-1}. Quantities of magnesium (48 mg kg^{-1}) and calcium (1090 mg kg^{-1}) were determined by ammonium acetate–lactate method (Kuldkepp et al. 1995).

Each field of the experiment was divided according to three treatments of organic fertilizers as follows: (i) without any organic fertilizers (WOF), (ii) farmyard manure 40 Mg ha^{-1} for potato (FYM), and (iii) alternative organic fertilizers in autumn before plowing (beet leaves until 1995, since 1996 straw) (RS).

Two first organic treatments have been unchangeable since the beginning of the experiment. In the treatment of alternative organic fertilizers in autumn 2007, recultivation substance (RS-oil shale semicoke mixed with sphagnum peat at voluminous ratio of 1:1) of 20–60 Mg ha^{-1} was applied. In autumn 2008, compost (40 Mg ha^{-1}) was applied formed from RS and solid fraction of pig slurry for potato or created from RS with town waste sludge for spring barley. RS is with alkaline reaction and has high content of potassium, calcium, and magnesium, but the content of heavy metals is close to their average value in Estonian mineral soils (Raave et al. 2004).

Each treatment of organic manure (three in total) was divided into five mineral fertilizer treatments by nitrogen amount (0, 40, 80, 120, and 160 kg N ha^{-1}), from which in current study two variants of mineral fertilizer (0 and 120 kg ha^{-1}) were used in three replications. The combined mineral fertilizers $N_{120}P_{23}K_{57}$ for spring cereals and $N_{120}P_{55}K_{185}$ for potato were applied in spring before sowing/plantation.

The soil materials from different fertilizer treatments of arable soils (S1 – S6) and from forest (S7) were sampled in October 2003 (Szajdak et al. 2006). Soil samples are marked using the numeration and the nomination as follows:

S_1 (WOF/N-0)—without any organic and mineral fertilizers
S_2 (FYM/N-0)—direct or aftereffect of farmyard manure without additional mineral fertilizer
S_3 (RS/N-0)—aftereffect of RS or direct effect of compost made from RS with solid fraction of pig slurry or town wastewater sludge without additional mineral fertilizer
S_4 (WOF/N-120)—without organic fertilizer, but using mineral fertilizer 120 kg N ha^{-1}
S_5 (FYM/N-120)—farmyard manure with mineral fertilizer 120 kg N ha^{-1}
S_6 (RS/N-120)—composts from RS with mineral fertilizer 120 kg N ha^{-1}
S_7—sample of forest soil humus horizon taken from the profile of *Fragic Glossic Retisol* located in prematured (75–85 years) *Oxalis* site-type spruce forest (situated very close to Eerika field trial at Tiksoja)

7.3 Methods of Laboratory Analyses

Ten soil subsamples from each arable land or forest were pooled together to give the so-called average mixed sample, from which the roots and stones were removed. Samples were air-dried and crushed to pass a 1-mm-mesh sieve. These "mean samples" were used for the potentiometric determination of pH (in 1M KCl) and, for the measurements of total organic carbon (TOC), dissolved organic carbon (DOC) and N_{tot} and its mineral forms, indole-3-acetic acid (IAA), and activity of enzymes. The contents of TOC were determined by TOC 5050A facilities (Shimadzu, Japan). The Kjeldahl method was introduced for the determination of N_{tot} in soils.

For the estimation of DOC, air-dried soil samples in deionized water were heated at 100 °C by 2 h under reflux condenser. Extracts were separated by the mean filter paper and analyzed on facility TOC 5050A (Smolander and Kitunen 2002).

7.3.1 Ammonium

Ammonium ions were measured on ion chromatograph Waters 1515 (USA) equipped with a 1515 Isocratic HPLC pump, conductivity detector Waters 432, a rotary valve fitted with 20 μl sample loop, and column PRP-X200 (150 × 4.1 mm ID) from Hamilton, protected with a guard column of the same material (25 × 2.3 mm ID). The detection was monitored at the range of sensitivity 10 μS. The column was operated at a temperature of 25 °C. The mobile phase consisted of 4 mM HNO_3 in the water and methanol (70:30, v/v) at a flow rate of 1 ml/min.

Table 7.4 Parameters for the determination of $N–NH_4^+$, $N–NO_3^-$, urease activity, and nitrites (for measurements of nitrate reductase activity)

Compounds	Slope, L mol^{-1} cm^{-1} ε +/−95 % confidence interval	Correlation coefficient (*r*)
$N–NH_4^+$	$1.47\ 10^6 \pm 2.06\ 10^4$	0.999
$N–NO_3^-$	$3.66\ 10^3 \pm 2.20\ 10^2$	0.996
	Molar absorption coefficient (ε), L mol^{-1} cm^{-1} ε +/−95 % confidence interval	
Urease activity	8326 ± 5	0.996
Nitrite (for measurements of nitrate reductase activity)	21,115 ± 1	0.999

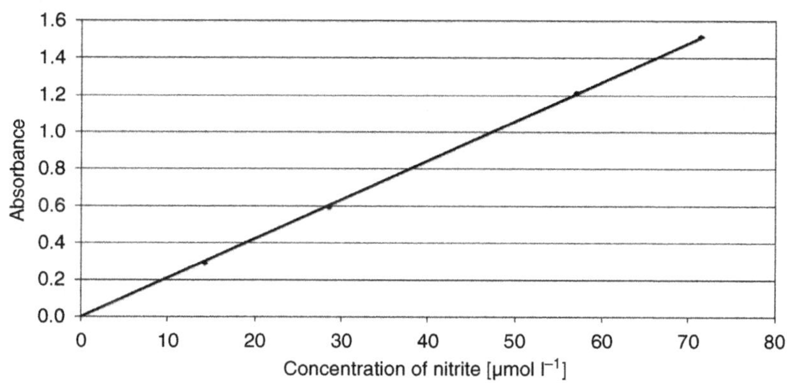

Fig. 7.1 Analytical curve of the concentrations of nitrite ions

Ammonium in soils was calculated from the early-prepared analytical curve by means of the least squares formulas (7.1) (Table 7.4, Fig. 7.1).

$$A = \varepsilon \cdot c \cdot l \tag{7.1}$$

where A is absorbance, ε slope [1 mol^{-1} cm^{-1}], *c* concentration [mol l^{-1}], and *l* thickness of layer [1 cm].

In order to prepare ammonium standard stock solution, 3.8167 g NH_4Cl was dissolved in deionized water and the volume diluted to 1000 ml with deionized water in a volumetric flask. From this solution working standard solution was prepared, where the concentration of $N–NH_4^+$ was equal to 10 mg $N–NH_4^+$ l^{-1}. The calibration standards were constructed by adding of 5.0-, 10.0-, 50.0-, 70.0-, and 90.0-ml working standard solution in five 100-ml volumetric flasks. All standards were analyzed six times.

The soil extracts were prepared by placing 10 g of air-dried soil in a 250-ml beaker and adding 30-ml deionized water. The samples were extracted for 30 min in a shaker at room temperature. Next, the mixture was centrifuged and filtered by Whatman filter GF/C. Ammonium as nitrogen and nitrate as nitrogen were determined in these extracts.

7.3.2 Nitrates

In order to obtain the standard stock solution, nitrate was prepared by dissolving 0.6067-g $NaNO_3$ in deionized water and the volume diluted to 1000 ml with deionized water in the volumetric flask. The concentration of $N–NO_3^-$ was equal to 100-mg $N–NO_3^-$ l^{-1}. The calibration standards were constructed by adding of 2.5-, 5.0-, 7.0-, 9.0-, and 15.0-ml working standard solution in five 100-ml volumetric flasks. All standards were analyzed six times. Nitrate ions were measured by ion chromatograph HIC-6A Shimadzu (Japan) equipped with a LP-6A Isocratic HPLC pump, a conductivity detector CDD-6A, a rotary valve fitted with 20-μl sample loop, and a PRP-X100 (150×4.1-mm ID) column from Hamilton, protected with a guard column of the same material (25×2.3-mm ID). The detection was monitored at the range of sensitivity of 1 μS. The column was operated at a temperature of 25 °C. The mobile phase consisted of 4 mM p-hydroxybenzoic acid with 2.5% methanol (pH = 8.4) at a flow rate of 1.5 ml/min. Nitrate in soils was calculated from the early-prepared analytical curve by means of the least squares formulas (7.1). The calculation of analytical curve for nitrate was similar to ammonium (Table 7.4).

The soil extracts were prepared by placing 10 g of air-dried soil in a 250-ml beaker and adding 30-ml deionized water. The samples were extracted for 30 min in a shaker at room temperature. Next, the mixture was centrifuged and filtered by Whatman filter GF/C. Ammonium as nitrogen and nitrate as nitrogen were determined in these extracts.

7.3.3 Urease

The urease [EC 3.5.1.5] is an enzyme participating in the hydrolytic degradation of urea with release of ammonium and carbon dioxide. Ammonium ions created in these processes are included in fast conversion in soils such as nitrification, and it is strongly bond to clay minerals, and in this way it is easily accessible to plants. Seasonal changes of urease activity depend mainly on moisture, temperature, the pH values of soil, and microbial activity (Bremner and Mulvaney 1978; Szajdak et al. 2002; Makoi and Ndakidemi 2008). Urease activity in soils was determined by Hoffmann and Teicher method (Wyczółkowski and Dąbek-Szreniawska 2005).

In order to prepare standard stock solution, 0.0472 g of ammonium sulfate was dissolved in deionized water in a measuring flask of 100 ml in volume, where 1 ml of this solution contains 10 μg of nitrogen. The analytical curve was constructed by adding of 0.2-, 0.5-, 1.0-, and 1.5-ml working standard solution in four 50-ml volumetric flasks. 9.0 ml of deionized water, 4.0 ml of 1.32 M sodium phenolate, and 1.0 ml of 0.3 M NaClO were added for each standard for colorimetric analysis. The absorbance of the reaction was measured colorimetrically at $\lambda_{max} = 630$ nm using a UV–VIS spectrophotometer Beckman DU®-68, USA. Urease activity in soils was calculated from the early-prepared analytical curve according to the Beer–Walter light absorption law by means of the least squares formulas (7.1). The calculation of analytical curve for urease activity was similar to ammonium (Table 7.4).

According to the procedure, 10 g of air-dried soil was placed in a measuring flask of 100 ml in volume. Next, 1.5-ml toluene was added. All samples were mixed and allowed to stand for 15 min at 20 °C. After this time, 10 ml of 10 % urea solution and 20 ml of 0.88 M citric buffer at pH = 6.7 were added. All the samples were placed at temperature of 37 °C for a period of 3 h. After the incubation, the volumetric flasks were made up deionized water. Next, samples were centrifuged at 4000 rpm for 20 min, and soil solutions were filtered. The urease activity assay was as follows: 1 ml of soil extract, 9.0 ml of deionized water, 4 ml of 1.32 M sodium phenolate, and 1 ml of 0.3 M sodium hypochlorite solution which were combined in a 50-ml measuring flask. After 20 min the volumetric flasks were made up deionized water. The ingredients were mixed and an absorbance of the reaction mixture was measured colorimetrically at $\lambda_{max} = 630$ nm using a UV–VIS spectrophotometer Beckman DU®-68. The deionized water served as control.

7.3.4 *Nitrate Reductase*

Nitrate reductase [EC 1.6.6.3] is an enzyme involved in the process of denitrification. During low content of oxygen, nitrate ions are reduced to nitrite and this process is catalyzed by nitrate reductase. Nitrogen present in the structure of this enzyme acts as a terminal acceptor of electrons instead of molecular O_2 by bacteria and is irreversible once NO is formed (Abdelmagid and Tabatabai 1987). Prosthetic group of this enzyme creates a flavoprotein containing molybdenum. Nitrate reductase was determined by Kandeler method (Kandeler 1996; Szajdak and Gaca 2010).

Color reagent: the dye of reagent was obtained by dissolving 2 g of sulphanilamide and 0.1 g of N-(1-naphtyl)-ethylenediamine hydrochloride in 150 ml of deionized water and 20 ml of concentration phosphoric acid. The mixtures were instantaneously cooled to the 20 °C and diluted to the volume of 200 ml with deionized water in a volumetric flask. Solution has to be colorless and should be prepared daily.

The standard stock solution was prepared by dissolving 4.9257-g $NaNO_2$ in deionized water and the volume diluted to 1000 ml with deionized water in a volumetric flask. From this solution, working standard was prepared, where the concentration of $N\text{–}NO_2^-$ was equal to 10 μg $N\text{–}NO_2^-$ ml^{-1}. The calibration standards were constructed by adding of 0 (blank reagent)-, 2.0-, 4.0-, 8.0-, and 10.0-ml working standard in five 100-ml volumetric flasks. 50 ml of potassium chloride solution was added and the volume was made up with deionized water. For colorimetric analysis, 5 ml of each standard, 3 ml of ammonium chloride buffer at pH = 8.5, and 2 ml of color reagent were added. All calibration standards were mixed and allowed to stand for 15 min at 20 °C. Nitrite ions revealed purple complex with sulphanilamide and N-(1-naphtyl)-ethylenediamine hydrochloride determined colorimetrically at wavelength $\lambda_{max} = 520$ nm using a UV–VIS spectrophotometer Beckman DU®-68.

Activity of nitrate reductase in soils was calculated from the early-prepared analytical curve according to the Beer–Walter light absorption law by means of the least squares formulas (7.1) (Table 7.4, Fig. 7.1).

KNO_3 is used as a substrate for the measurement of activity of nitrate reductase in field-moist soil samples. 5 g-soil samples were incubated for 24 h at 25 °C with 1 ml of 25 mM KNO_3 solution and 4 ml of 0.9 mM 2.4-dinitrophenol solution and 5 ml of deionized water. The controls were incubated for 24 h at −20 °C. After incubation the controls were thawed at 20 °C. Nitrite reductase is inhibited by the addition of 2.4-dinitrophenol. Nitrate is released as a result of incubation extracted with 10 ml of 4 M potassium chloride. Next, samples were centrifuged at 4000 rpm for 10 min, and soil solutions were filtered. For colorimetric analysis 5 ml of each soil extracts, 3 ml of ammonium chloride buffer (0.19 M at pH = 8.5), and 2 ml of color reagent were added. All samples were mixed and allowed to stand for 15 min at 20 °C. The concentration of nitrite is determined colorimetrically at $\lambda_{max} = 520$ nm from the early-prepared analytical curve according to the Beer–Walter light absorption law by means of the least squares formulas (7.1). The calculation of analytical curve for nitrite was similar to ammonium (Table 7.4). The deionized water served as control.

7.3.5 Indole-3-Acetic Acid (IAA)

IAA concentrations were determined by fluorimetric method. 2 g of soil was added to 10-ml 0.1 M NaOH. The mixture was shaken for 5.0 h vigorously. Then, the suspension was allowed to stand overnight. The following day, the sample was centrifuged for 20 min (15,000 *g*). 4 ml of supernatant was taken and added to 4 ml of n-pentanol. The mixture was shaken for 30 min. After centrifugation, the mixture was centrifuged by 20 min (15,000 *g*). 3 ml of top layer was taken and added to 3 ml 0.1 M phosphate buffer at pH = 7.0. The mixture was shaken for 1.0 h and next centrifuged for 20 min (15,000 *g*). Organic layer was washed two times by 2-ml 0.1 M HCl for the removal of tryptophane. IAA in the resulting soil extraction was measured fluorometrically in bottom layer at $\lambda_{excitation} = 290$ nm and $\lambda_{emission} = 368$ nm. The concentration of IAA was calculated from analytical curve, ranged from 50 to 300 ng/ml, and prepared similar to investigated soil samples. The mean fluorescence of IAA concentrations is linear. The correlation coefficient of calibration curve, which crosses its origin, is high ($r = 0.998$). The slopes of IAA of the following conditions: (i) before extraction, (ii) without any washing with 0.1 M HCl, and (iii) after washing with 0.1 M HCl, were used for the calculations of IAA in soil samples.

Satisfactory precision based on replicate analyses was ±0.01 for pH measurements, ±3 % for TOC, ±3 % for DOC, ±4 % for N_{tot}, ±3 % for $N–NO_3^-$, ±3 % for $N–NH_4^+$, ±4 % for nitrate reductase activity, ±4 % for urease activity, and ±4 % for IAA. All the experiments were performed in five replicates, and the results averaged. All the chemicals used in this study were of analytical grade.

7.4 Results and Discussion

SOM consists of a mixture of plant and animal products in various stages of decomposition together with substances synthesized from breakdown products. A large number of relatively simple compounds of known structures and belonging to well-known groups such as carbohydrates, acids, esters, amino acids, alkaloids, phenolic compounds, glycosides, vitamins, enzymes, and bases have been isolated (Athertnon et al. 1967).

With land-use change from natural (forest) into arable status, the rate of SOM mineralization and the stage of soil saturation are increased, but biological diversity is decreased. Because a new epipedon type is formed and the leading role in soil functioning goes to the anthropogenic factor (tillage technology, fertilizers), land-use change causes the biggest alterations in biological and pedoecological characteristics; some little changes may be obtained in physical characteristics also, but there are no changes in substratum properties.

We would like to accentuate that soil cover is not a passive component in ecosystems as it influences in different manner to the environmental status of the area, depending mainly on soil cover composition (soil types, texture, calcareousness), regimes (moisture conditions), soil edaphon activity, and on land-use peculiarities (tillage technology). The soils should be treated as an active superficial layer of landscape, which, thanks for its multiple functions, acts as determining environmental status of area natural body. Understanding more profoundly these processes and arrangement of sustainable use of soil the biochemical indices are with utmost importance. In addition, crop residue management affects the biological, chemical, and biochemical processes that govern the conversion of C and N to SOM and the residual availability of N to succeeding crops (Bird et al. 2003; Talgre et al. 2012). Consequently, an enhanced knowledge of the effect of long-term soil incorporation of straw and winter flooding on crop residue C and N cycling would enable improved utilization of fertilizer and N of crop residue.

The importance of organic matter widely presented in arable agricultural soils by the participation in biogeochemical conversions and pathways is well established. Its dynamic is affected by humification, mineralization, immobilization, leaching, and plant uptake and also root exudates. The chemical parameters of different forms of soil nitrogen and carbon, IAA, activity of urease, and nitrate reductase were determined in the samples taken in autumn after harvesting of crops (Table 7.5).

The N_{tot} concentration in different fertilizer treatments of field experiment ranged from 0.90 to 1.02 g kg^{-1}. However, the application of FYM/N-120 did not significantly decrease the concentrations of N_{tot} from 0.102 to 0.099 %. The application of RS/N-120 did not reveal any significant effect on the content of N_{tot}, while RS/N-0 is poor in nitrogen. Mineral fertilizer (WOF/N-120) increased the N_{tot} content only in comparison with RS/N-0. The application of mineral fertilizer WOF/N-120 kg N per ha in comparison with WOF/N-0 decreased the content of N_{tot}. Moreover, the application of WOF/N-120 irrespective of cultivated plants, potato, spring barley, and spring wheat showed the highest increase in the crop yield (Table 7.6).

Table 7.5 Agrochemical parameters of soils sampled in autumn 2003 (average of nine samples with standard deviation—LSD_{95})

Parameter, unit	Arable land						Forest land
	WOF/ N-0 S_1	WOF/ N-120 S_4	FYM/ N-0 S_2	FYM/ N-120 S_5	RS/N-0 S_3	RS/N-120 S_6	Tiksoja S_7
N_{tot}, %	0.090	0.094	0.102	0.099	0.090	0.098	0.174
N_{min}, mg kg^{-1}	35.1	16.1	22.2	21.3	13.4	20.5	10.6
N–NH_4, mg kg^{-1}	17.5	8.4	10.8	9.0	9.1	9.4	7.3
N–NO_3^-, mg kg^{-1}	17.6	7.7	11.4	12.2	4.3	11.2	3.3
TOC, %	0.93	0.82	1.12	1.11	0.98	1.06	2.37
DOC, %	0.04	0.04	0.045	0.045	0.04	0.045	0.13
C/N	10.1	10.1	10.7	11.6	12.0	11.0	13.6
pH_{KCl}	6.5	6.4	6.6	6.5	7.1	7.1	4.2
Activity of urease, μmol urea g^{-1} DM h^{-1}	0.88	0.94	1.13	1.47	0.97	1.75	5.15
+/–, in %		***+6.8 %***		***+30.1***		***+79.4 %***	
Activity of nitrate reductase, μg N g^{-1} DM 24 h^{-1}	0.0732	0.1255	0.1934	0.1690	0.2507	0.2410	0.1805
+/–, in %		***+71.4 %***		***−12.6 %***		***−3.8 %***	
IAA, μg kg^{-1} DM	25.95	20.48	26.96	24.97	18.97	22.47	39.17
+/–, in %		***−21.1 %***		***−7.4 %***		***+18.5 %***	

Table 7.6 The yields of potato, spring wheat, and spring barley as an average of crop rotation (in 2002–2004) under different kinds of fertilization (mean ± 95 % confidence intervals, Mg ha^{-1})

Organic treatments	Rate of mineral N, kg ha^{-1}	Sample No	Potato	Spring wheat	Spring barley
WOF—without organic fertilizer	0	S_1	13.21 ± 4.00b[a]	1.93 ± 0.52c	1.77 ± 0.24a
	120	S_4	21.29 ± 6.21a	4.05 ± 0.38a	4.81 ± 0.75b
			61.2 %	*109.8 %*	*172.80 %*
FYM—farmyard manure	0	S_2	17.29 ± 5.21ab	2.41 ± 0.57b	2.19 ± 0.36a
	120	S_5	23.13 ± 6.58a	4.03 ± 0.40a	4.84 ± 0.80b
			33.8 %	*70.8 %*	*121.0 %*
RS—recultivation substance and compost	0	S_3	16.07 ± 3.79ab	2.33 ± 0.45b	2.24 ± 0.59a
	120	S_6	22.71 ± 5.49a	3.98 ± 0.31a	5.02 ± 0.76b
			41.3 %	*70.8 %*	*124.1 %*

[a]Different letters denote significant differences ($p < 0.05$) between fertilizer variants

The lowest content of TOC was measured in treatment of WOF/N-0 where in 14 years the mineral N and organic fertilizers were not used. The application of farmyard manure during investigated period led to the increase of the content of TOC. Thus, the present results support the previous findings from IOSDV (Kuldkepp 1997). Moreover, the highest mineral fertilizer WOF/N-120 increased the content of TOC only in comparison with RS/N-0.

The hot water extractable carbon is well known as a sensitive indicator of the changes, transformations, and challenges in soil management and also of agricultural practices. DOC can contribute significantly to cycling of soil nutrients. DOC seems to be a major vehicle for leaching of many elements from the litter. The primary sources of DOC are considered leaching substances from fresh litter and the products of plant residue decomposition (Qualls and Haines 1991; Smolander and Kitunen 2002). The concentrations of DOC in these investigations ranged from 0.040 to 0.045 g kg^{-1}. The lowest concentrations of DOC were measured in the treatment without organic manure and also using RS/N-0 and were equal to 0.04 g kg^{-1}. The highest content of DOC revealed the forest soil equaled to 0.13 g kg^{-1}. The labile nature of dissolved C and reasonable function of its pool (usually constitutes of 3–6 % of TOC in soils) give DOC measurements an early indication possibility of organic matter loss (Ghani et al. 2003). Moreover, DOC exhibits better relationship with mineral fertilization and N_{tot} and also with mineral N (N_{min}) forms than TOC in soil. There was also a negative correlation between the content of different forms of organic carbon and $N–NH_4^+$.

Significant source of N_{min} in soil is the degradation of organic substances. Its dynamics are affected by mineralization, immobilization, leaching, root exudates, and plant uptake. Sufficient N_{min} supply for the crop growth is crucial for organic farming to be economically sustainable. On the other hand, minimization of inorganic nitrogen leaching is a criterion for any farming system to be considered environmentally sound. Any supply of inorganic nitrogen through fertilizers or mineralization from incorporated organic matter are only partly available to the growing crop, the rest being prone to immobilization and losses (Korsaeth et al. 2002). N_{min} in cultivated soils revealed very low concentrations ranged from 13.4 to 35.1 mg kg^{-1} in arable soil. The lowest content of N_{min} was measured in forest soils and equaled to 10.6 mg kg^{-1}. However, the content of $N–NH_4^+$ and $N–NO_3^{-1}$ was almost similar in comparison with the background (WOF/N-0). The contents of $N–NH_4^+$ was twice higher than the concentrations of $N–NO_3^{-1}$ in soils fertilized with RS/N-0. Similar in soil fertilized with RS/N-0, the content of $N–NH_4^+$ was two times higher than the concentrations of $N–NO_3^{-1}$ in very acidic forest soil. In this investigation the contents of $N–NH_4^+$ in the fields ranged from 8.4 to 17.5 mg kg^{-1}. Similar dates were expressed by Hannolainen (1970), 25–39 mg kg^{-1}.

The total amount of the concentrations of $N–NH_4^+$ and $N–NO_3^{-1}$ in soil without organic manure (WOF/N-0) was two times higher than in the treatment, where 120 kg N per ha was applied with mineral fertilizer. Organic nitrogen fertilizer included in the form of RS/N-120 increased 2.6 times the content of $N–NO_3^{-1}$ in comparison with RS/N-0 (Table 7.5).

In natural ecosystems, decomposition and mineralization of plants and animal detritus regulate the availability of nutrients for plant uptake or loss from systems (Swift et al. 1979). It has been clearly documented that plowing increases rates of organic matter decomposition and nutrient mineralization in soils (Hendrix et al. 1986). The arable soils of experimental plots were from slightly acid to neutral. Humus horizon of Tiksoja forest soil revealed the highest acidity. The relationship between different agrochemical parameters of soil fertility and experimental factors is moderate. However, the content of mineral forms of nitrogen and N_{tot} was negatively correlated with the treatments of fertilization. However, the total and easily mineralizable carbon (TOC and DOC) exhibited positive correlation with organic manure application. The negative impact of mineral nitrogen fertilizers on accumulation of $N–NH_4^+$ in soil was mentioned earlier by Kuldkepp and Suitso (1997). The reason of that is supposed to be in enhanced attack of nitrifiers on ammonium. The $N–NO_3^{-1}$ used to be higher in spring compared with autumn.

The reason of low level of N_{min} probably is associated with time of sampling in late autumn. Therefore, low temperature and lack of substrate for microorganism were the reason of low N_{min}. Niklinska et al. (1999) have shown that the rate of the nitrogen mineralization increases significantly above 15 % in autumn. The study showed that the DOC represented negative correlation with mineral forms of nitrogen as $N–NH_4^+$ and $N–NO_3^{-1}$ as well as TOC and IAA. The highest linear correlation was between DOC and IAA ($r = 0.90$). The correlation between DOC and TOC and between DOC and N_{tol} was negative ($r = –0.48$ and $r = –0.33$, respectively).

In addition, our investigations showed that soil treated with organic manure (FYM/N-120) characterized high microbial activity represented by high activity of enzymes. The activity of urease and nitrate reductase in soils fertilized with FYM/N0 and FYM/N-120 and also RS/N-0 and RS/N-120 was higher than in soil treated only with chemical fertilizers. The similar findings presented by our researches were observed also in the case of using dairy effluent (Zaman et al. 2002). The synergism of mineral and organic fertilizers resulted always in higher activity of microorganisms in comparison with manure FYM/N-0 or FYM/120 treatment.

Urease participates in the hydrolytic decomposition of urea. Ammonia produced during this process is strongly absorbed by the soil, which makes it safer in the case of larger nitrogen losses, being at the same time easily accessible to plants (7.2).

$$\text{peptides} \longrightarrow CO(NH_2)_2 + H_2O \xrightarrow{\text{urease}} CO_2 + 2\,NH_3 \tag{7.2}$$

The investigation of urease in soils revealed wide range of the activity of this enzyme from 0.88 to 5.15 μg N g^{-1} DM 24 h^{-1} in soils of investigated plots (Table 7.5). The lowest activity of this enzyme was measured in soil without any organic fertilizer and the highest in Tiksoja forest soil. Due to this combination of organic and mineral fertilizers, the increase of the activity of urease was observed. As a result of this, applying the highest increase equal to 79.4 % was measured in soil, applying the composts from RS with mineral fertilizer in the rate of 120 kg N ha^{-1}

and the smallest (6.8 %) using mineral fertilizer 120 kg ha^{-1} without organic fertilizer (Table 7.5).

Denitrification is defined as the "microbial reduction of nitrate or nitrite coupled to electron transport phosphorylation resulting in gaseous N either as molecular N_2 or as an oxide of N" (Martens 2005; Szajdak and Gaca 2010). Nitrate reductase is an enzyme involved in the process of denitrification. Nitrogen present in the structure of this enzyme acts as a terminal acceptor of electrons instead of molecular O_2 by bacteria and is irreversible once NO is formed (Abdelmagid and Tabatabai 1987). During low content of oxygen, nitrate ions are reduced to nitrite and this process is catalyzed by nitrate reductase.

The fertilizers WOF/N-120, FYM/N-120, and RS/N-120 impacted on the changes of activity of nitrate reductase in soils (Table 7.5). WOF/N-120 has resulted in 71.4 % increase of the activity of this enzyme than WOM/N-0. The highest increase of activity of this enzyme in soil with WOF/N-120 suggests the highest loss of nitrogen in the form of volatile compounds such as N_2O and N_2 from all applied fertilizers. WOF/N-120, FYM/N-120 and RS/N-120 in comparison with FYM/N-0 and RS/N-0 decreased the activity of nitrate reductase and the loss of nitrogen from soils.

The high value of activity of nitrate reductase in forest soil equal to 0.1805 μg N g^{-1} DM 24 h^{-1} and the highest content of TOC equal to 2.37 % and also the highest concentration of N_{tot} are responsible for the optimal conditions for denitrification. Mazur (1991) noted that high content of moisture and organic matter and also neutral and basic pH favors the denitrification.

Allelopathic interactions occur between crops and weeds, between two crops, from decomposing crop and weed residues, and from crop and weed exudates (Anaya 1999). Nonpathogenic allelopathic bacteria can produce plant-inhibiting compounds (Barazini and Friedman 1999). However, crop rotation can be used to alleviate the allelopathic or autopathic effects a crop plant might have on itself. In addition, long-term continuous cropping encourages proliferation of allelopathic bacteria (Barazini and Friedman 1999). Continuous cropping led to the changes in the soil community, which increased pathogen load and reduced barley growth with that by grain in multiple rotation (Olsson and Gerhardson 1992). Continuous cropping of wheat, however, can lead to suppression of the take-all pathogen bacteria that produce the antibiotics phenazine and phloroglucinol (Mazzola et al. 1995). By crop rotation, it is possible to lessen the negative effects a crop might have on itself and on subsequent crop. In addition, crop rotation can influence root colonization by mycorrhiza (Rice 1995).

Humic substances can indirectly or directly affect the physiological processes of plant growth (Tan 1998). Indirectly, they provide minerals (Rauthan and Schnitzer 1981), increase the microorganism population (Visser 1985), provide low-molecular-weight metabolites (Vaughan and Malcolm 1979), and carry trace elements and growth-related regulators (Chen and Schnitzer 1978). Directly, they influence microalgae growth (Heil 2005), seed germination (Muscolo et al. 2002)

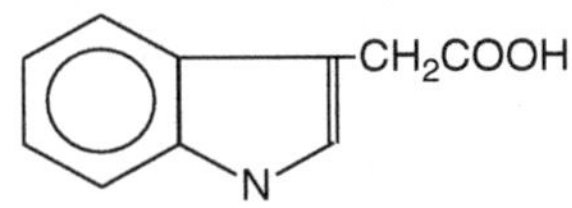

Fig. 7.2 Indole-3-acetic acid

and plant growth (Chen and Aviad 1990; Muscolo et al. 2000; Arancon et al. 2006), modify some metabolic processes such as respiration (Varanini and Pinton 2001) and nutrient uptake (Varanini and Pinton 2001; Panuccio et al. 2001), and induce macrofunctional changes in the root architecture (Canellas et al. 2002). However in many cases, these effects depend on the molecular weight and the concentration of humic substances (Nardi et al. 2002; Muscalo et al. 2007; Muscolo and Sidari 2009).

Plant growth hormones (phytohormones), particularly growth retardants, may maintain internal hormonal balance and efficient sink–source relationship and thus enhance crop productivity. IAA seems to play an important function in nature as result to its influence in regulation of plant growth and development (Fig. 7.2).

A principal feature of IAA is its ability to affect growth, development, and health of plants. This compound affects root morphology and metabolic changes in the host plant. The physiological impact of this substance is involved in cell elongation, apical dominance, root initiation, parthenocarpy, abscission, callus formation, and the respiration (Tena et al. 1986; Strzelczyk et al. 1992). Significantly higher concentrations of IAA were determined in soil under *Robinia pseudoacacia* than under *Quercus petraea*, *Q. robur*, *Larix deciduas*, *Pinus silvestris*, *Sorbus aucuparia*, *S. intermedia*, and *Tilia cordata* during entire vegetation's season (Szajdak and Maryganova 2009).

The concentrations of IAA in cultivated fields ranged from 18.97 to 26.96 μg kg^{-1} DM. The lowest concentration of IAA was in soil fertilized with RS/N-0 (Table 7.5). The use of RS/N-120 increased the highest the concentration of IAA (18.5 %). WOF/N-120 and FYM/N-120 decreased the concetration of IAA in cultivated fields 21.1% and 25.0 %, respectively. The content of IAA in forest soils was 30 % higher than in cultivated fields.

Applying mineral fertilizer 120 N kg ha^{-1} in rotation without organic fertilizers increased crop yields of potato, spring wheat, and spring barley (61.2 %, 109.8 %, and 172.8 %, respectively) more than organic fertilizers, farmyard manure (FYM/N-120), or RS and composts (RS/N-120) (Table 7.6). Compost amendment of soil is an attractive way to add organic matter to soil as it has optimal C/N ratio, is devoid of weed seeds, is suppressive to soil-borne diseases (Hoitink et al. 1993), and is richer than manure in humic substances, thereby facilitating plant growth enhancement (Inbar et al. 1990).

In this experiment RS and composts (RS/N-120) increased the crop yields of potato 41.3 %, spring wheat 78.8 %, and spring barley 124.1 %, but farmyard manure (FYM/N-120) 33.8 %, 70.8 %, and 121.8 %, respectively. Such findings are in agreement with conclusions of Ryszkowski et al. (1998) and Szajdak et al. (2004). They have shown that during long-term study on biologically active substances in soils under crop rotation and continuous cropping of rye mineral fertilizer, NPK increased higher content of free and bound amino acids, total acidity, and phenolic and carboxylic groups in humic acids than manure.

7.5 Conclusions

The chemical and biochemical composition of different kinds of fertilizers (mineral or organic) greatly influenced on the relationship between the quantities of chemical and biochemical compounds in arable soil and the crop yield of cultivated plants. The study has shown that the content of compound of well-known structure, such as indole-3-acetic acid, representing phytohormone, the concentrations of different forms of nitrogen, and organic carbon, and enzymes participating in nitrogen cycle such as urease and nitrate reductase in soils, has a marked effect on the crop yield of cultivated plants. Arable soil treated WOM/N-120 in comparison with other fertilizers revealed the highest increase of the activity of nitrate reductase, and the highest concentration of IAA, however the lowest activity of urease. Therefore, WOM/N-120 in comparison with FYM and RS created the most suitable conditions for the crop yield of cultivated plants. This fertilizer caused the highest increased crop yield of potato, spring wheat, and spring barley, 61 %, 110 %, and 173 %, respectively, in comparison with soil without any organic and mineral fertilizers. Moreover, the highest amounts of IAA, TOC, DOC, and N_{tot} and the highest C/N ratio as well as the highest activity of urease in forest soil have been found to be closely associated with the highest concentrations of DOC, representing organic form of carbon available for biochemical pathways.

References

Abdelmagid HM, Tabatabai MA (1987) Nitrate reductase activity of soils. Soil Biol Biochem 19:421–427

Ahas R, Jaagus J, Russak V (2002) Kliima. In: Eesti Entsüklopeedia, vol 11. Eesti Entsüklopeediakirjastus, Tallinn, pp 112–118

Albuzio A, Ferrari G (1989) Modulation of the molecular size of humic substances by organic acids of the root exudates. Plant Soil 113:237–241

Anaya AL (1999) Allelopathy as a tool in the management of biotic resources in agroecosystems. Crit Rev Plant Sci 18:697–739

Arancon QA, Edwards CA, Lee S, Byrne R (2006) Effects of humic acids from vermicomposts on plant growth. Eur J Soil Biol 42:565–569

Athertnon NM, Cranwell AJ, Haworth RD (1967) Humic acid – I. ESR spectra of humic acids. Tetrahedron 23:1653–1667

Barazini O, Friedman J (1999) Allelopathic bacteria and their impact on higher plants. Crit Rev Plant Sci 18:741–755

Bird JA, van Kessel C, Horwath WR (2003) Stabilization of ^{13}C-carbon and immobilization of ^{15}N-nitrogen from rice straw in humic fractions. Soil Sci Soc Am J 67:806–816

Boguslawski E von (1995) Das Zusammenwirken der mineralischen Düngung mit verschiedenen Formen der organischen Düngung. In: Boden und Düngung. EAU, Tartu, pp 9–22

Bremner JM, Mulvaney RL (1978) Urease activity in soils. In: Burns RG (ed) Soil enzymes. Academic, London, pp 149–196
Canellas LP, Olivares FL, Okorokova-Facanha AL, Facanha AR (2002) Humic acids isolated from earthworm compost enhance root elongation, lateral root emergence, and plasma membrane H+ATPase activity in maize roots. Plant Physiol 130:1951–1957
Chen Y, Aviad T (1990) Effects of humic substances on plant growth. In: MacCarthy et al (eds) Humic substances in soils and crop science: selected readings. SSSA, Madison, pp 161–186
Chen Y, Schnitzer M (1978) The surface tension of aqueous solutions of soil humic substances by E_4/E_6 ratios. Soil Sci 125:7–15
Chen Y, De Nobili M, Aviad T (2004) Stimulatory effects of humic substances on plant growth. In: Magdoff F, Weil RR (eds) Soil organic matter in sustainable agriculture. CRC Press, Boca Raton, pp 103–129
Duke SO, Lydon J (1987) Herbicides from natural compounds. Weed Technol 1:122–128
EMHI (2009) Estonian Meteorological and Hydrological Institute. http://www.emhi.ee/index.php?ide=6. Accessed 15 Dec 2009
Flaig W (1971) Organic compounds in soil. Soil Sci 111(1):19–33
Ghani A, Deuter M, Perrot KW (2003) Hot water extractable carbon in soils: a sensitive measurement for determining impacts of fertilization, grazing and cultivation. Soil Biol Biochem 35(9):1231–1243
Hart SC, Nason G, Myrold D, Perry D (1994) Dynamics of gross nitrogen transformation in an old-growth forest: the carbon connection. Ecology 75:880–891
Heil CA (2005) Influence of humic acid, fulvic and hydrophilic acids on the growth, photosynthesis and respiration of the dinoflagellate Prorocentrum minimum (Pavillard) Schiller. Harmful Algae 4:525–531
Hendrix PF, Parmelee RW, Crossley DA Jr, Coleman DC, Odum EP, Groffman PM (1986) Detritus food webs in conventional and no-tillage agroecosystems. Bioscience 36(6):374–380
Hoitink HAJ, Boehm MJ, Hadar Y (1993) Mechanisms of suppression of soil borne plant pathogens in compost-amended substrates. In: Hoitink HAJ (ed) Science engineering of composting. Renaissance Publication, Worthington, pp 601–622
Honermeier B (2007) Diversity in crop production systems. In: Benckiser G, Schnell S (eds) Biodiversity in agricultural production systems. CRC Press/Taylor & Francis, Boca Raton, pp 1–19
Hannolainen E (1970) Content of mobile nitrogen in humus layer of arable lands of Estonian SSR. EMMTUI Proc 20:172–183
Inbar Y, Chen Y, Hadar Y (1990) Humic substances formed during the composting of organic matter. Soil Sci Soc Am J 54:1316–1323
IUSS Working Group WRB (2015) World reference base for soil resources 2014, update 2015. International soil classification system for naming soils and creating legends for soil maps. World Soil Resources Reports No 106. FAO, Rome, p 182
Janzen HH (2006) The soil carbon dilemma: shall we hoard it or use it? Soil Biol Biochem 38:419–424
Kanal A, Kõlli R (1996) Influence of cropping on the content, composition and dynamics of organic residue in the soil of the plough layer. Biol Fertil Soils 23:153–160
Kandeler E (1996) Nitrate reductase activity. In: Schinner F, Öhlinger R, Kandeler E, Margesin R (eds) Methods in soil biology. Springer, Berlin/Heidelberg, pp 176–179
Kern JS, Turner DP, Dodson RF (1997) Spatial patterns of soil organic carbon pool size in the Northwestern United States. In: Lal R, Kimble JM, Follett RF, Stewart BA (eds) Soil processes and the carbon cycle. CRC Press, Boca Raton, pp 29–43
Kokk R (1995) Distribution and properties of soils. In: Raukas A (ed) Estonia. Nature. Valgus, Tallinn, pp 430–439
Kõlli R, Kanal A (1995) Humus status of arable Podzoluvisols and the role of annual litterfall in its formation. Z Pflanzenernähr Bodenkd 158:235–241

Kõlli R, Ellermäe O, Köster T, Lemetti I, Asi E, Kauer K (2009) Stocks of organic carbon in Estonian soils. Est J Earth Sci 58(2):95–108
Körchens M, Weigel A, Shulz E (1998) Turnover of soil organic matter (SOM) and long-term balances – tools for evaluating sustainable productivity of soils. Z Pflanzenernähr Bodenkd 161:409–424
Korsaeth A, Henriksen TM, Bakken LR (2002) Temporal changes in mineralization of N during degradation of plant material: implications for the plant N supply and nitrogen losses. Soil Biol Biochem 34:789–799
Kuldkepp P (1997) Wirkung und Wechselwirkung unterschiedlicher mineralischer und organischer N- Düngung auf Ertrag und Bodeneigenshaften im IOSDV Tartu (Estland) nach 6 Jahren. Arch Agron Soil Sci 42:21–32
Kuldkepp P, Suitso A (1997) The dynamics of nitrate nitrogen content in sandy loam soil during the vegetation period as depending on fertilization. In: Proceedings of NorFA seminar, Jelgava, pp 59–63
Kuldkepp P, Liiva I, Teesalu T (1995) Zum Einfluss organischer Düngung, verschiedener Stickstoffdüngungsstifen und Witterungsfaktoren auf den Ertrag von Kartoffel, Sommerweizen und Sommergerste. In: Boden und Düngung. EAU, Tartu, pp 28–39
Ladd JN, Butler JHA (1971) Inhibition by soil humic acids of native and acetylated proteolytic enzymes. Soil Biol Biochem 3:157–160
Makoi JHJR, Ndakidemi PA (2008) Selected soil enzymes: examples of their potential roles in the ecosystem. Afr J Biotechnol 7(3):181–191
Martens DA (2005) Denitrification. In: Hatfield JL, Powlson DS, Crosenzweig DS, Scow KM, Singer MJ, Sparks DL (eds) Encyclopedia of soils in the environment. Academic, Amsterdam, pp 378–382
Mazur T (1991) The conversions of nitrogen compounds in soil. In: Mazur T (ed) Nitrogen in agricultural soils. PWN, Warszawa, pp 66–98
Mazzola M, Fujimoto DK, Thomashow SS, Cook RJ (1995) Variation in sensitivity of Gaeumannomyces graminis to antibiotics produced by fluorescent Pseudomonas spp., and effect on biological control of take-all of wheat. Appl Environ Microbiol 61:2554–2559
Muscolo A, Sadiri M (2009) Carboxyl and phenolic humic fractions affect Pinus Nigra callus growth and metabolism. Soil Sci Soc Am J 73(4):119–1129
Muscolo A, Panuccio MR, Sidari M, Nardi S (2000) Effect of two humic substances on some glycolytic enzymes in callus culture of Pinus laricio. Humic Subst Environ 2:19–24
Muscolo A, Panuccio MR, Sidari M, Sessi E, Nardi S (2002) Alteration of amino acids metabolism by humic substances during germination of Pinus laricio seeds. Seed Sci Technol 30:205–210
Muscolo A, Sidari M, Francioso O, Tugnoli V, Nardi S (2007) The auxin-like activity of humic substances is related to membrane interaction in carrot cell cultures. J Chem Ecol 33:115–129
Nardi S, Pizzeghello D, Muscolo A, Vianello A (2002) Physiological effects of humic substances on higher plants. Soil Biol Biochem 32:1527–1536
Neill C, Cerri CC, Melillo JM, Feigl BJ, Steudler PA, Moraes JFL, Piccolo MC (1998) Stocks and dynamics of soil carbon following deforestation for pasture in Rondonia. In: Lal R, Kimble JM, Follett RF, Stewart BA (eds) Soil processes and the carbon cycle. CRC Press, Boca Raton, pp 9–28
Niklinska M, Mariański M, Laskowski R (1999) Effect of temperature on humus respiration rate and nitrogen mineralization: implication for global; climate change. Biochemistry 44:239–257
Olsen SR (1986) The role of organic matter and ammonium in producing high corn yields. In: Chen Y, Avnimelech Y (eds) The role of organic matter in modern agriculture. Martinus Nijhoff, Dordrecht, pp 29–70
Olsson S, Gerhardson B (1992) Effects of long-term barley monoculture on plant-affecting soil microbiota. Plant Soil 143:99–108
Panuccio MR, Muscolo A, Nardi S (2001) Effect of humic substances on nitrogen uptake and assimilation in two species of Pinus. J Plant Nutr 24:693–704

Paustian K, Collins HP, Paul EA (1997) Management controls on soil carbon. In: Paul EA, Paustian K, Elliott ET, Cole VV (eds) Soil organic matter in temperate agroecosystems. Long-term experiments in North America. CRC Press, Boca Raton, pp 15–49

Percival HJ, Parfitt RL, Scott NA (2000) Factors controlling soil carbon levels in New Zealand grasslands: is clay content important. Soil Sci Soc Am J 64:1623–1630

Piccolo A, Nardi S, Concheri G (1992) Structural characteristics of humic substances as related to nitrate uptake and growth regulation in plant systems. Soil Biol Biochem 4:373–380

Qualls RG, Haines BL (1991) Geochemistry of dissolved organic nutrients in water percolation through a forest ecosystem. Soil Sci Soc Am J 55:1112–1123

Raave H, Kuldkepp P, Leedu E, Merivee A (2004) Recultivation substance and composts produced from semi-coke: the effect on soil characteristics, the yield of field crops and the environment. Oil Shale 21:59–73

Rauthan BS, Schnitzer M (1981) Effect of soil fulvic acid on the growth and nutrient content of cucumber (Cucumis sativus). Plant Sci 63:491–495

Reeuwijk LP Van (ed) (1995) Procedures for soil analysis. Technical Paper 9. ISRIC, Wageningen

Rice EL (1995) Biological control of weeds and plant diseases: advances in applied allelopathy. University of Oklahoma Press, Norman

Rowel DL (1994) Soil science: methods and applications. Prentice Hall, Harlow

Ryszkowski L, Szajdak L, Karg J (1998) Effects of continuous cropping of rye on soils biota and biochemistry. Crit Rev Plant Sci 38(1):225–244

Schimel DS (1986) Carbon and nitrogen turnover in adjacent grassland and cropland ecosystems. Biogeochemistry 2:345–357

Smith P, Powlson DS, Glendining MJ, Smith JU (1997) Opportunities and limitations for C sequestration in European agricultural soils through changes in management. In: Lal R, Kimble JM, Follett RF, Stewart BA (eds) Management of carbon sequestration in soil. CRC Press, Boca Raton, pp 143–152

Smolander A, Kitunen V (2002) Soil microbial activities and characteristics of dissolved organic C and N in relation to tree species. Soil Biol Biochem 34:651–660

Strzelczyk E, Pokojska A, Kampert M (1992) The effect of pH on production of plant growth regulators by mycorrhizal fungi. Symbiosis 13:201–215

Swift MJ, Heal OW, Anderson JM (1979) Decomposition in terrestrial ecosystems. University of California Press, Berkeley

Szajdak LW, Gaca W (2010) Nitrate reductase activity in soil under shelterbelt and an adjoining cultivated field. Chem Ecol 26(4):123–134

Szajdak L, Maryganova V (2009) Impact of age and composition of shelterbelts plants on IAA content as allelochemical in soils. Allelopath J 23(2):461–468

Szajdak L, Maryganova V, Meysner T, Tychinskaja L (2002) Effect of shelterbelt on two kinds of soils on the transformation of organic matter. Environ Int 28:383–392

Szajdak L, Życzyńska-Bałoniak I, Meysner T, Blecharczyk A (2004) Bound amino acids in humic acids from arable cropping systems. J Plant Nutr Soil Sci 167:562–567

Szajdak L, Kuldkepp P, Leedu E, Teesalu T, Toomsoo A, Kõlli R (2006) Effect of different management on biochemical properties of organic matter in Fragi-Stagnic Albeluvisols. Arch Agron Soil Sci 52(2):127–137

Szponar A, Szajdak L (2003) Amino acids in quaternary soil horizons from southwest Poland. Eur J Soil Sci 54:237–242

Talgre L, Lauringson E, Roostalu H, Astover A, Makke A (2012) Green manure as a nutrient source for succeeding crops. Plant Soil Environ 58:275–281

Tan KH (1998) Principles of soil chemistry, 3rd edn. Marcel Dekker, New York

Tena M, Pinilla JA, Magallanes M (1986) L-phenylalanine deaminating activity in soil. Soil Biol Biochem 18(3):321–325

Varanini Z, Pinton R (2001) Direct versus indirect effect of soil humic substances on plant growth and nutrition. In: Pinton et al. (eds) The rhizosphere. Marcel Dekker, New York, pp 415–421

Vaughan D, Malcolm ER (1979) Effect of humic acids on invertase synthesis of higher plants. Soil Biol Biochem 11:247–252
Vaughan D, Malcolm ER (1985) Influence of humic substances on growth and physiological processes. In: Vaughan D, Malcolm ER (eds) Soil organic matter and biological activity. Martinus Nijhoff Publisher, Dordrecht, pp 37–76
Visser SA (1985) Effect of humic acid on numbers and activities of microorganisms within physiological groups. Org Geochem 8:81–85
West TO, Post WM (2002) Soil organic carbon sequestration rates by tillage and crop rotation: a global data analysis. Soil Sci Soc Am J 66:1930–1946
Wyczółkowski AI, Dąbek-Szreniawska M (2005) Enzymes participating in organic nitrogen mineralization. In: Russel S and Wyczółkowski AI (eds) Methods of the determination of enzymes activity in soil. Acta Agrophysica 120(3):37–61
Yakimenko EY (1998) Soil comparative evolution under grasslands and woodlands in the forest zone of Russia. In: Lal R, Kimble JM, Follett RF, Stewart BA (eds) Management of carbon sequestration in soil. CRC Press, Boca Raton, pp 391–420
Zaman M, Cameron KC, Di HJ, Inubushi K (2002) Changes in mineral N, microbial biomass and enzyme activities in different soil depths after surface application of dairy shed effluent and chemical fertilizer. Nutr Cycl Agroecosys 63:275–290

Part II
The Physiological Importance of Biologically Active Substances

Chapter 8
Auxin, One Major Plant Hormone, in Soil

Junichi Ueda, Marian Saniewski, and Kensuke Miyamoto

Abstract This chapter describes "auxin, one major plant hormone, in soil," especially focusing on "outline of plant hormones" including the brief history of auxin, "auxins in soil," "polar movement of auxin" including recent findings of its molecular mechanisms, "auxin polar transport inhibitors," and "important role of auxin polar transport and its inhibitors in plant growth and development." Unlike other plant hormones, auxin shows a specific movement in plant tissues recognized as auxin polar transport. The process of auxin polar transport generates auxin maxima and gradients within tissues that are instrumental in the diverse regulation of various plant developmental processes. Some novel naturally occurring inhibitors of auxin polar transport such as dehydrocostus lactone (decahydro-3,6,9-tris-methylene-azulenol(4,5-b)furan-2(3H)-one) and artabolide (3-hydroxy-4,6,7(H)-germacra-1(10),11(13)-dien-6,12-olide) with α-methylene-γ-lactone moiety and 4-hydroxy-β-thujone (4-hydroxy-4-methyl-1-(1-methylethyl)-bicyclo[3.1.0]hexan-3-one), as well as potent synthetic products such as 2,3,5-triiodobenzoic acid (TIBA), 1-*N*-naphthylphthalamic acid (NPA), methyl 2-chloro-9-hydroxyfluorene-9-carboxylate (morphactin IT 3456), and 9-hydroxyfluorene-9-carboxylic acid (HFCA) are intensively described. Auxin polar transport inhibitors interfere with basipetal polar transport of auxin and change auxin distribution, and in consequence perturb plant growth and development. NPA, TIBA, morphactin IT 3456, and HFCA have been expected to show various effects on plant growth and development. Generally, various effects of auxin polar transport inhibitors have been proposed to be caused, at molecular level, by interference with the auxin efflux carrier and/or facilitator(s) located in the polar side of plasma membrane, but detailed mechanism

J. Ueda (✉)
Graduate School of Science, Osaka Prefecture University,
1-1 Gakuen-cho, Naka-ku, Sakai, Osaka 599-8531, Japan
e-mail: ueda@b.s.osakafu-u.ac.jp

M. Saniewski
Research Institute of Horticulture, Konstytucji 3 Maja 1/3, 96-100 Skierniewice, Poland
e-mail: marian.saniewski@inhort.pl

K. Miyamoto
Faculty of Liberal Arts and Sciences, Osaka Prefecture University,
1-1 Gakuen-cho, Naka-ku, Sakai, Osaka 599-8531, Japan
e-mail: miyamoto@las.osakafu-u.ac.jp

L.W. Szajdak (ed.), *Bioactive Compounds in Agricultural Soils*,
DOI 10.1007/978-3-319-43107-9_8

remains still unclear whether or how blockage of auxin efflux by the inhibitors is involved in the effects on physiological and developmental processes in plants.

Keywords Auxin • Auxin polar transport • Auxin polar transport inhibitors • 4-Hydroxy-β-thujone • Dehydrocostus lactone • Artabolide • 2,3,5-Triiodobenzoic acid (TIBA) • 1-*N*-Naphthylphthalamic acid (NPA) • Morphactin IT 3456 • 9-Hydroxyfluorene-9-carboxylic acid (HFCA) • Auxin efflux carrier and/or facilitator(s) • Plant hormones • Cholodny–Went theory • Auxin in soil • Polar movement of auxin • AUX1 protein • PIN proteins • MDR/ABC (ABCB)/PGP proteins • Tropic growth (tropism) • Phototropism • Gravitropism • Cambial activity • Xylem differentiation • Stem growth • Morphogenetic processes • Organogenesis • Embryogenesis • *Arabidopsis thaliana* • Rooting • Root formation • Crassulaceae • *Bryophyllum daigremontianum* • *Bryophyllum calycinum* • *Kalanchoe blossfeldiana* • *Kalanchoe tubiflora*

8.1 Outline of Plant Hormones

Plant hormones (synonyms: phytohormones) and plant growth substances play an important role in controlling growth and development at extremely low concentrations in plants. Seven major kinds of endogenous plant growth substances in plants, auxins, gibberellins, cytokinins, abscisic acid, ethylene, brassinosteroids which is a class of steroid hormones regulating various plant development and physiology (Adam and Petzold 1994; Bishop and Koncz 2002), and jasmonates (jasmonic acid and its derivatives) playing important roles in regulating senescence and defense response against insect and fungi (Wasternack 2007; Wasternack and Hause 2013), have been recognized as plant hormones. In addition, polyamines which are organic compounds with molecules composed of at least two amino groups such as the diamine putrescine, the triamine spermidine, and the tetraamine spermine have been considered to be new types of growth substances similar to plant hormones (Galston and Sawhney 1990; Alcázar et al. 2010).

The discovery of first plant hormone, auxin, has been described in most of the textbooks of plant hormones and plant physiology. The importance of plant hormones was stated by the success of isolation of "Wuchsstoff" from the tip of *Avena* coleoptiles into agar blocks by Went (1928a, b) and in his famous pronouncement, "Ohne Wuchsstoff, kein Wachstum (Without growth substance, no growth)." Though it is thought not to be required further description, nomenclature of plant hormone, auxin is described briefly from the historical aspect in this article.

Plant hormone, auxin, was similarly thought to produce a growth response at a distance from its site of synthesis and thus to fit the definition of a transported chemical messenger. Auxin plays a crucial role in many aspects and development in plants (Berleth and Sachs 2001). The term "hormone" in plant physiology was derived from the mammalian hormone concept (Bayliss and Starling 1904) which originally comes from the Greek ὁρμή, "impetus," and is used in animal physiology to denote a chemical messenger. Mammalian hormone concept involves a localized site of synthesis, transport in the bloodstream to a target tissue, and the control of a physiological response in a target tissue via the concentration of the hormone. According to the nomenclature published by the Committee of the American Society of Plant Physiologists in 1954 (Tukey et al. 1954), plant hormones are also defined as follows: "Plant hormones are regulators produced by plants, which in low concentrations regulate plant physiological processes. Hormones usually move within the plant from a site of production to a site of action." On the other hand, "plant regulators" are defined as organic compounds, other than nutrients, which in small amounts promote, inhibit, or otherwise modify any physiological processes in plants. The term "hormone" is restricted to naturally occurring plant products. In contrast, the term "regulator" has very wide boundaries, being able to apply to any, natural, and synthetic, compounds that modify the physiological process in plants. It is now clear that plant hormones do not fully fit the requirements of a hormone in the mammalian sense.

The hormone concept in plants was developed from three discoveries in the nineteenth century (Went and Thimann 1937; Thimann 1977), namely, "organ-forming substances" by Julius Sachs (1882, 1887), the idea of "growth enzymes" from the study of galls by Beijerinck (1888, 1897), and "the influence of light from the study of phototropism" by Charles Darwin (1880). Sachs interpreted the morphological differences among plant organs by corresponding differences in "organ-forming substances" which move in different directions through the plant and direct development in minute amounts. For example, when stem segments excised from plants were kept vertically or horizontally, shoot and roots were regenerated at the top and the bottom ends, respectively, since the substances suitable for the formation of shoots which are already present in it will, as heretofore, move in the acropetal direction, and those which form roots, on the contrary, in the basipetal one. He also suggested that polar distribution of these substances is modified by external force such as light and gravity. On the other hand, Beijerinck (1888) proposed the "growth enzyme," a protein being different from ordinary proteins but resembles enzyme, for considering the developmental mechanism of *Nematus capreae* gall on *Salix*. Later, he extended his view to the development of organisms in general that form is determined by liquid substances, which move freely through considerable numbers of cells in growing tissues (Beijerinck 1897; Went and Thimann 1937). At about same period, the existence of polarity in correlation phenomena, tropisms, was realized and emphasized by Charles Darwin (1880). The study of phototropism using *Pharalis* and *Avena* coleoptiles by Darwin was the first clue to the isolation of "Wuchsstoff" or auxin. He concluded when seedlings are exposed to a lateral light, some "influence" is transmitted from the upper to the lower part where the influence reacts, resulting in the bending. Although at first Darwin's statement met with much opposition, Rothert's work on 1894 of phototropism in shoots substantially confirmed the separation between the perceive and the reaction zones (Rothert 1894). In gravitropism (also known as geotropism) of plant roots, first Ciesielski's work (1872) should be highlighted. In his conclusion, "transmitted influence" present in the tip of plant roots was clearly affecting growth differentially as Darwin's findings of phototropism in shoots (1880).

Darwin's work concerning with "influence" was followed by the elaborated experiments by Boysen Jensen (1910). He demonstrated that when the excised tip was replaced with gelatin inserted between it and the stump, the coleoptile regained its ability to respond to the light, indicating clearly that phototropic stimulus is transmitted across a wound gap. In 1919, Paál demonstrated that when the excised tip is replaced on one side of the *Avena* coleoptile stump, accelerated growth beneath the tip results in curvature similar to the curvature caused by light, and gave a name to this "influence" from the tip to base as "correlation carrier." He also suggested that this correlation carrier, which under normal conditions continually moves downward from the tip along all sides, is upon illumination of the tip, either interfered within its formation, photochemically inactivated, or inhibited in its downward movement, through some changes in the protoplasm, the effect being greater on the lighted side (Paál 1919).

In 1928, success in isolating this "influence" into the agar block as a substance was finally achieved by Went. He placed the excised tips of *Avena* coleoptiles on agar blocks for a couple of hours, and then applied agar blocks containing "influence," asymmetrically on the stumps of decapitated coleoptiles, resulting in a curvature away from the agar block. The curvature was proportional, within limits, to the concentration of the active substances. Using this so-called *Avena* curvature test, Went (1928a, b) demonstrated that an asymmetrical distribution of the growth substance occurs when unilateral light falls upon an excised *Avena* coleoptile tip; this being the cause of the curvature. This finding led to "lateral transport" as used in the Cholodny–Went theory for tropism, although Yamada et al. (2000) have recently proposed it is against considerations.

Within few years after Went's investigation, Kögl and his colleagues, Haagen-Smit and Erxleben (1933, 1934a, b) isolated and identified three highly active substances causing the specific growth reaction which is conveniently measurable by the curvature of *Avena* coleoptiles, auxins a and b from plant materials and indole-3-acetic acid (IAA) from human urine. They first proposed the term "auxin" from the Greek "αύξάνω"(increase or grow) for these plant growth substances in plants (Kögl and Haagen-Smit 1931), and heteroauxin for IAA. The discovery of auxins a and b was a real mystery, since no one succeeded to find auxins a and b from plants after the first discovery. On the contrary, IAA has been isolated from yeast plasmolysate (Kögl and Kostermans 1934), from culture of fungus *Rhizopus suinus* (Thimann 1935), and finally from plant materials, corn meal (Haagen-Smit et al. 1942) and the endosperm of immature corn grain (Haagen-Smit et al. 1946); IAA becoming to be realized to be one of the widespread auxin in higher plants. Thus the discovery of IAA from human urine was certainly a milestone in plant hormone physiology.

As definition of auxin mentioned above, the term of "auxin" does not refer to IAA alone; rather, it represents a group of compound with physiological activity such as cell elongation of the *Avena* coleoptiles, although IAA is the most abundant naturally occurring auxin. Plant produces active IAA both by de novo synthesis and by releasing IAA by hydrolysis from IAA conjugates such as IAA-amino acids, IAA-*myo*-inositol, and IAA-glucose. Intensive studies of biosynthesis of IAA chemically and genetically have proposed two major pathways, tryptophan-dependent and tryptophan-independent pathway. In the former, four pathways, indole-3-acetamide (IAM) pathway, indole-3-pyruvic acid (IPA) pathway, the tryptamine (TAM) pathway, and the indole-3-acetaldoxime (IAOX) pathway, have been postulated in plants (Bartel 1997; Mano and Nemoto 2012). In the latter, IAA has been suggested to be biosynthesized via possible intermediates, indole-3-acetonitrile (IAN) or IPA, possibly from indole-3-glycerol phosphate (IGP) or indole (Bartel 1997). Besides IAA, 4-chloro-3-indoleacetic acid (4-Cl-IAA) and its methyl ester were identified as naturally occurring chlorinated auxin in immature seeds of various leguminous plants (Marumo et al. 1968a, b; Gandar and Nitsch 1967; Engvild et al. 1980) and *Pinus sylvestris* (Ernstsen and Sandberg 1986). Indole-3-butyric acid (IBA) synthesized from IAA has been also shown to have auxin activity (Blommaert 1954; Bayer 1969; Schneider et al. 1985), indicating that IBA is substantially one of the natural auxins. However, it has not been clear whether IBA itself acts as natural auxin, since IBA is possible to be converted to IAA again. A lot of precursors and metabolites of IAA and 4-Cl-IAA have been known to be widespread in the plant kingdom as mentioned above.

On the other hand, many compounds have been synthesized in order to examine the physiological activity and the mode of action of auxins. The representative compounds with auxin activities are chlorinated phenoxy and benzoic acids; 2,4-dichlorophenoxy acetic acid (2,4-D), naphthaleneacetic acid (NAA), *cis*-cinnamic acid, and phenylacetic acid (PAA) having relatively high auxin activity. On the contrary, some substitute synthetic compounds or isomers such as 2,4,6-trichlorophenoxyacetic acid (McRae and Bonner 1953), *trans*-cinnamic acid (van Overbeek et al. 1951), 4-chlorophenoxyisobutyric acid (PCIB) (Burström 1950; Frenkel and Haard 1973), 5,7-dichloroindole-3-isobutyric acid (Hatano et al. 1989), and others have been reported to act as antiauxins which competitively inhibit auxin action. Furthermore, nagilactone A~F and inumakilactone A, and argophyllin-A and B, have been isolated as naturally occurring antiauxin from *Podocarpus nagi* (Hayashi and Sakan 1974) and *Helianthus argophyllus* (Watanabe et al. 1982), respectively. Synthetic and naturally occurring inhibitors of auxin polar transport are intensively described in the below part as well.

8.2 Auxin in Soil

Green algae living in freshwater habitats have evolved into land plants. To grow in soils, land plants take and utilize some important materials from soils. Many soils contain various kinds of inorganic and organic compounds. Organic materials in soil are important resources of the nutrients for plant growth and development (Power 1994). Among them, humic substances have been recognized as one of important materials for promoting plant growth (Szajdak and Maryganova 2007).

Some review papers have also reported the promoting effects of humic substances on plant growth (Chen and Aviad 1990; Nardi et al. 1996; Musculo and Nardi 1997; Cesco et al. 2002). Some organic compounds in soil have been known to show strong auxin-like activity (Parker-Rhodes 1940; Stewart and Anderson 1942; Hamence 1946; Whitehead 1963; Sheldrake 1971). IAA is considered to be one of the major auxin produced in soil. Synthesizing abilities of IAA in soil depend on the fertility status and the content of organic materials (Stewart and Anderson 1942; Hamence 1946; Chandramohan and Mahadevan 1968).

Microorganisms and/or microflora in soil have been considered to be important in the production of IAA and other auxin-like substances (Strzelczyk et al. 1973; Purushothaman et al. 1974). Heterotrophic microflora and microorganisms living in soil have been responsible for the production of IAA, and a change of the bacterial population has been reported to activate or inactivate directly plant growth (Cooper 1959), suggesting that the microbial synthesis of plant growth regulators including IAA is an important factor in soil fertility (Kampert et al. 1975). Microorganisms isolated from the rhizosphere and rhizoplane of various crops are more active in producing auxins (Roberts and Roberts 1939; Dvornikov et al. 1970; Brown 1972; Purushothaman et al. 1974; Kampert et al. 1975; Strzelczyk et al. 1977; Strzelczyk and Pokojska-Burdziej 1984). Higher amounts of IAA were observed in the rhizosphere environments compared with non-rhizosphere ones (Narayanaswami and Veerraju 1969; Rossi et al. 1984), indicating that the production of auxin in rhizosphere soil is most likely due to abundance of substrates and microorganisms. Since plants uptake auxin released from microorganisms in rhizosphere soil, plant growth is positively affected as a result (Arshad and Frankenberger 1991).

IAA is biosynthesized from the physiological precursor, L-tryptophan (L-TRP), in microorganisms in soil (Frankenberger et al. 1990) as well as in plant cells. It is possible for L-TRP to affect plant growth and development when it is applied to soil. Frankenberger et al. (1990) have reported that the 3.0 mg/kg of L-TRP applied to soil resulted in the optimum radish growth in greenhouse. Interestingly, the root yield by 1.31-fold over the control was observed. In this case, the root/shoot ratio was increased by 1.10-fold, and one L-TRP application was sufficient to promote growth. Application time of L-TRP also affects plant growth and development. Similar observations have been reported in the growth of maize when L-TRP was applied to the field as a nutrient (Ahmad et al. 2008). The best time of the application of L-TRP was at the onset of seedling emergence. The application of L-TRP was the most effective on promoting radish yield comparable to those plants treated with IAA, indole-3-acetamide and indole-3-lactic acid. The mechanism of L-TRP to promote plant growth and development has not been clear yet, but it could be attributed to (i) substrate-dependent auxin production in soil by the indigenous microflora, (ii) uptake directly by plant roots followed by metabolism within their tissues, and/or (iii) a change in the balance of rhizosphere microflora affecting plant growth (Frankenberger et al. 1990). On the other hand, foliar application of L-TRP has been reported not to be effective in the root and shoot dry weight. This fact also suggests that the appropriate times and concentrations of applied L-TRP are important in radish yield (Frankenberger et al. 1990).

Auxins produced in soil are incorporated into plants and function to regulate the growth and development due to their chemical and biological characteristics as described below.

8.3 Polar Movement of Auxin

8.3.1 Auxin Polar Transport

Auxin moves from the shoot apex to the root apex between plant cells through a combination of membrane diffusion and carrier-mediated transport, and different internal and external signals have been shown to modulate both auxin biosynthesis and auxin polar transport (Kramer and Bennett 2006; Roberts and Friml 2009). As intensively described in the below subheading, many theories concern the mechanism of cellular and molecular aspects of auxin polar transport and its regulation in plants (Zazimalova and Napier 2003; Kramer and Bennett 2006; Vieten et al. 2007; Kuppusamy et al. 2009; Roberts and Friml 2009). The auxin polar transport generates auxin maxima and gradients within tissues that are instrumental in the diverse regulation of various plant developmental processes, including elongation growth, organogenesis, vascular tissue formation, embryogenesis, tropisms, and many other processes (Roberts and Friml 2009).

The overall morphology of a plant is largely determined by developmental decisions taken within or near the terminally positioned apical meristems of shoots and roots, the spatial separation of these apical meristems emphasizing the need for long-distance signaling. Auxin moves from the shoot apex to the root apex between plant cells through a combination of membrane diffusion and carrier-mediated transport as a coordinating signal in plants and also as patterning signal in meristem organization and embryos (Berleth and Sachs 2001; Friml 2003; Friml et al. 2003).

The concept of the polarity of auxin transport was established by the 1930s. As mentioned above, Paál has suggested that transport of the "correlation carrier" in the coleoptile might itself be polar (1919). Based on the results of the extended experiments of Boysen Jensen and Nielsen (1925), Beyer (1928) confirmed the fact that movement of phototropic stimulus was strictly polar. He interposed a cylinder of coleoptiles in normal position or in inverted position between the tip and the base of *Avena* coleoptile, demonstrating when the tip was illuminated, the stimulus was transmitted to the base through the normally inserted section only. Such characteristic movement of auxin has been known as auxin polar transport. The nature of this polar transport was intensively studied by van der Weij (1932, 1934), and later found to be a widespread feature of shoot and root tissues. In shoot tissues, the polarity of auxin transport is preferentially basipetal from morphological apices to more basal region with a velocity of 5 ~ 20 mm/h. In root, it continues in the same physical direction as in shoots, but is now described as acropetal, since it is directed toward the root tip. In immature growing portions of the root, basipetal transport away from the root tip appears to occur.

Polarity of auxin transport has been considered to be dependent on tissue polarity which probably reflects an underlying asymmetry of individual cells. Thus, auxin movement across membranes has been extensively studied at the cellular level. Such works lead to the "chemiosmotic" interpretation of auxin polar transport (chemiosmotic hypothesis), which was independently proposed by Rubery and Sheldrake (1974) and Raven (1975) (see review by Estelle 1998; Friml 2010). This model posits that auxin polar transport occurs through the action of cellular auxin influx and efflux carriers located in the plasma membrane. In support of this aspect of the model, basal localization of a putative efflux carrier was suggested by using immunological approach (Jacobs and Gilbert 1983).

The fundamental transport process of principle auxin, IAA, is summarized as follows: IAA can cross membranes by both diffusive and carrier-mediated routes, responding to transmembrane pH (ΔpH) and electrical potential gradient ($\Delta\psi$). The relatively alkaline cytoplasm (pH 7 ~ 7.4) can accumulate IAA from more acidic compartments such as the cell wall because of (i) the high diffusive membrane permeability of undissociated IAA molecules (pK = 4.7) relative to IAA anions (IAA^-) and (ii) a high-affinity, saturable uptake carrier which may operate by electroimpelled $IAA^-/2H^+$ cotransport. There is also a carrier, probably for IAA anions, catalyzing efflux down the electrochemical gradient set up by the accumulative uptake process. The driving forces of polar transport are envisaged as metabolically maintained pH and electrical potential gradients, while the polarity is most simply obtained by preferential localization of the IAA anion carrier at the basal ends of cells in the transport pathway. Molecular mechanisms of auxin polar transport are described in the following subheading.

8.3.2 A Brief Outline of Molecular Mechanisms of Auxin Polar Transport

Auxin polar transport has been recognized to be regulated by several functional proteins located in plasma membrane. Recent genetic and molecular biological approaches in the model system *Arabidopsis thaliana* have contributed fundamentally to our understanding of the molecular mechanisms of auxin polar transport. In the chemiosmotic model (see Friml 2010), several proteins function as plasma membrane-based influx and efflux carriers, and/or facilitators of auxin transport. Such proteins designate as auxin transport proteins and are divided into three groups as follows. Figure 8.1 shows the outline of molecular mechanisms of auxin polar transport.

The first class of auxin transporters is permease-like protein. The amino acid permease-like protein AUX1 has been isolated and characterized as an auxin carrier and/or facilitator located in an anti-polar side of the plasma membrane catalyzing

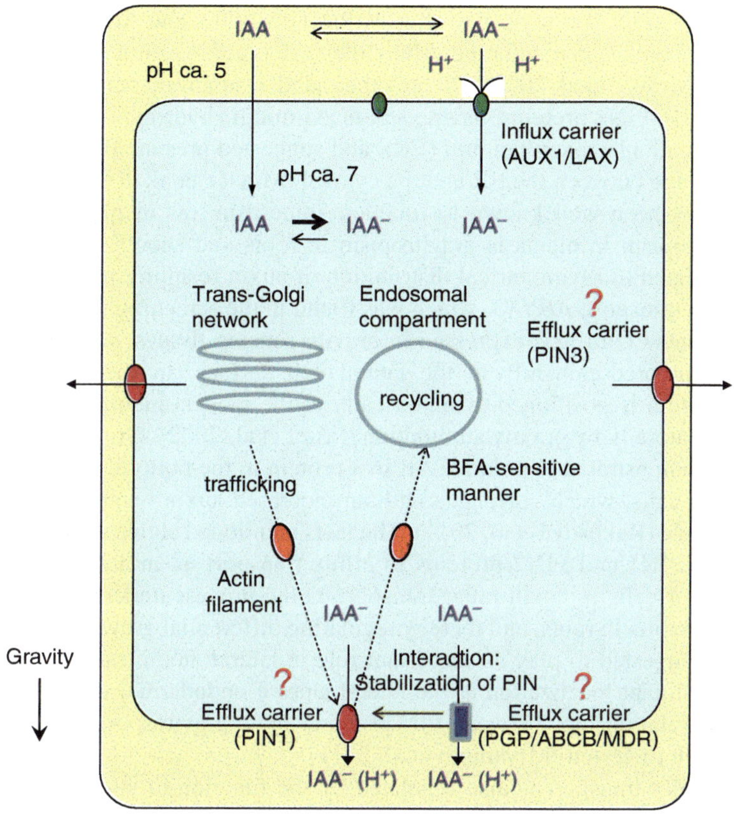

Fig. 8.1 Schema of molecular mechanisms of auxin polar transport (Adopted from Miyamoto et al. (2011) with some modifications)

the uptake of protonated auxin, IAA (IAAH), into cells (Marchant et al. 1999; Swarup et al. 2001). Auxin is cotransported into cells with two protons through AUX1 protein.

The second class of auxin transporters is PIN proteins which have been suggested from the investigation of flower formation using *Arabidopsis thaliana pin-formed* or *PIN1* mutant (Okada et al. 1991). Strenuous effort to isolate PIN proteins has been made, resulting in the successful isolation of AtPIN1 protein using a tagging system in *Arabidopsis thaliana* in 1998 (Gälweiler et al. 1998). In addition, AGR/EIR1/PIN/WAV6 (Chen and Masson 2005) membrane proteins have been found in a polar side of the plasma membrane and suggested to act as efflux carriers and/or facilitators of an anion form of IAA (IAA^-) out of cells (Gälweiler et al. 1998). These facts indicate the presence of a specific system consisting with an asymmetric distribution of these proteins in plant cells to regulate auxin polar transport. Until now, eight members of the PIN protein family referred to as AtPIN1 to AtPIN 8 have been isolated in *Arabidopsis thaliana*. Members of PIN families possess two transmembrane domains, separated by a large hydrophilic linker region.

AtPIN4 and AtPIN7 proteins as well as AtPIN1, AtPIN2, and AtPIN3 proteins are localized at the plasma membrane, where they act as auxin efflux carriers (Petrášek et al. 2006; Mravec et al. 2008). On the other hand, a subgroup comprising AtPIN5, AtPIN6, and AtPIN8 proteins, having a reduced-middle hydrophilic loop, are localized on the endoplasmic reticulum (ER), and suggested presumably to regulate the auxin exchange between the ER and the cytosol (Mravec et al. 2009).

Auxin has been well known to function some tropisms in plants. One of the important tropism in plants is gravitropism in roots and shoots. A graviresponse might be related to asymmetrical distribution of auxin resulting from auxin movement and/or transport. *AtPIN3*, which was found in the screening for new members of the *PIN* gene family, is expressed in gravity-sensing tissues, with PIN3 protein accumulating predominantly at the lateral cell surface. In the root columella, AtPIN3 protein is positioned symmetrically at the plasma membrane but rapidly relocalizes laterally by gravity stimulation (Friml et al. 2002). Gravistimulation has also been demonstrated to polarize AtPIN3 protein to the bottom side of hypocotyl endodermal cells, which correlates with an increased auxin response at the lower hypocotyl side (Rakusová et al. 2011). The facts mentioned above support an important role of PIN1 and PIN2 proteins as efflux transport of an anion form of IAA (IAA^-) out of cells in auxin polar transport as the process underlying differential auxin distribution in roots, and thereby regulating differential growth, while AtPIN3 protein is suggested to play an important role in lateral auxin transport system by their predominant localization at the lateral side of endodermal cells (Friml et al. 2002). The relationships between PIN proteins and a graviresponse in plants have recently been reviewed (Miyamoto et al. 2011).

Auxin polar transport is also regulated by the function of some other proteins instead of PIN proteins. The *PINOID* gene encodes a protein-serine/threonine (Ser/Thr) kinase (Friml et al. 2004). PINOID has been shown to function as a positive regulator of auxin polar transport. In addition, PINOID is considered to involve in the fine changing of auxin polar transport during organ formation in response to local auxin concentration (Benjamins et al. 2001). The Ser/Thr protein kinase PINOID-dependent binary switch, direct phosphorylation of the hydrophilic loop of PIN proteins, is considered to control polarity of PIN proteins and to mediate changes in auxin flow to create local gradients for patterning processes (Friml et al. 2004; Grunewald and Friml 2010). Loss of PINOID activity alters auxin transport and gravitropism without causing an obvious change in cellular polarity (Sukumar et al. 2009). The *Arabidopsis GNOM* gene has also been known to encode an ARF GDP/GTP exchange factor involved in embryonic axis formation and polar localization of the auxin efflux regulator PIN1 (Geldner et al. 2004). In addition, myosins are eukaryotic molecular motors moving along actin filaments, and a loss-of-function mutation for a myosin of plant-specific class XI, *Arabidopsis* myosin XI mutant, has reported to be defective in organelle movement and auxin polar transport (Holweg and Nick 2004). Although auxin polar transport substantially affects a graviresponse in plants as described above, detailed relationships between the function of these proteins and gravistimulation in plants have not been clear yet.

The third class of auxin transporters is phospho-glycoproteins (PGPs) that belong to the ATP-binding cassette protein subfamily B (ABCB), subgroup of the ATP-binding cassette (ABC) transporter superfamily. Recent observations have provide us the fact that *multidrug resistance* (*MDR*)-like gene of *Arabidopsis* required for auxin transport and auxin-mediated development (Noh et al. 2001; Santelia et al. 2005). In addition, *mdr* mutants of *Arabidopsis* have shown enhanced gravi- and phototropism compared to wild type as mislocalizing AtPIN1 protein of auxin efflux carrier and/or facilitator (Noh et al. 2003). Different from PIN proteins, the ABC transporters are uniformly distributed in the plasma membrane and organelle membranes (Noh et al. 2001; Wu et al. 2007, 2010). These facts described above together with the evidence that MDR/ABC (ABCB)/PGP (P-glycoprotein) proteins have been widely found in the plant kingdom (Geisler et al. 2005; Geisler and Murphy 2006; Petrášek et al. 2006) strongly suggest that not only AGR/EIR1/PIN/WAV6 membrane proteins but also MDR/ABC (ABCB)/PGP proteins play an important role in auxin polar transport and/or respond to gravistimulation in plants. Recent review paper focusing on the possible function of MDR/ABC (ABCB)/PGP proteins has been published (Ueda et al. 2011).

It is very surprising that there is no evidence supported by biochemical data that PIN proteins substantially carry auxin molecules across the plasma membrane in plant cells. Coexpression of PIN and PGP transporters in HeLa cells, however, increases substrate specificity, inhibitor sensitivity, and efflux. Significant net efflux of [^{3}H]IAA is observed in HeLa cells expressing PGP1 and PGP19 proteins but not PIN1 protein (Blakeslee et al. 2005). Judging from this finding together with the fact that ABC proteins are really transporter bound to plasma membrane and/or organelle membranes, there is a possibility that ABC proteins substantially carry auxin molecules from inside to out of cells, although PIN proteins are essential molecules in auxin polar transport. Direct transport of auxin by MDR proteins or positive regulation of a PIN-type auxin efflux channel is also possible. Further intensive studies will be necessary to clarify the relationships between PIN and ABC proteins. The important role of ABC proteins for growth and development of plants will be clarified as well.

8.4 Auxin Polar Transport Inhibitors

Auxin polar transport is inhibited by not only artificial but also naturally occurring compounds. Among them, 1-*N*-naphthylphthalamic acid (NPA), 2,3,5-triiodobenzoic acid (TIBA), methyl 2-chloro-9-hydroxyfluorene-9-carboxylate (morphactin IT 3456), and 9-hydroxyfluorene-9-carboxylic acid (HFCA) have been known to be potent inhibitors (Fig. 8.2a).

NPA is known to inhibit auxin polar transport in dicotyledons and monocotyledons by inhibiting active auxin secretion (Depta et al. 1983). NPA has been reported to inhibit auxin polar transport to bind to the specific site, the so-called NPA receptor, in plant cells (Muday et al. 1993), indicating that the NPA-binding site is

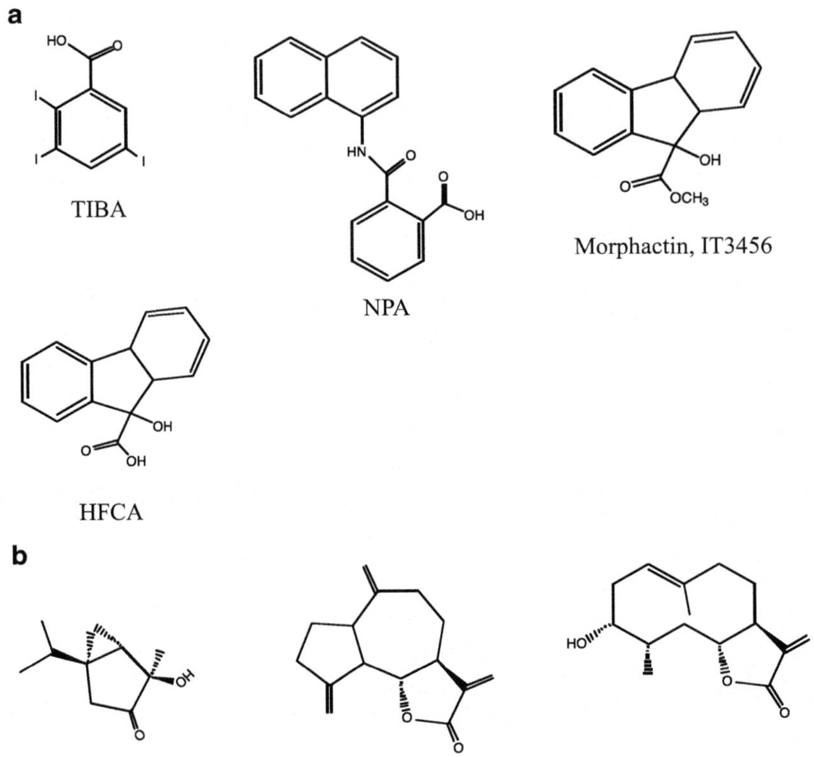

Fig. 8.2 (**a**) Chemical structures of 2,3,5-triiodobenzoic acid (TIBA), 1-*N*-naphthylphthalamic acid (NPA), methyl 2-chloro-9-hydroxyfluorene-9-carboxylate (morphactin IT 3456), and 9-hydroxyfluorene-9-carboxylic acid (HFCA) of potent inhibitors of auxin polar transport. (**b**) Chemical structures of 4-hydroxy-β-thujone, dehydrocostus lactone, and artabolide of novel auxin polar transport inhibitors

important for auxin polar transport (Ruegger et al. 1997). The NPA-binding site, however, has been controversial. Katekar and Geissler (1980) reported that NPA inhibits auxin polar transport by specifically binding to the auxin efflux carrier. The endogenous auxin, IAA, does not compete with NPA for binding site (Lomax et al. 1995). Thein and Michalke (1988) showed that bisulfite interacts with binding sites of NPA and that inhibition of NPA binding by bisulfite is noncompetitive. Bisulfite at 2 or 3×10^{-5} M reduces NPA binding to 50 % of the control for membrane-bound and solubilized binding sites, respectively.

TIBA has been well known to inhibit the basipetal auxin polar transport in plants for many years (Galston 1947; Thimann and Bonner 1948; Niedergang-Kamien and Leopold 1957; Morris et al. 1973). There have been reports showing that TIBA competes for the same binding sites as that of IAA (Thomson et al. 1973; Jablanović and Noodén 1974; Michalke et al. 1992). Moreover, it is suggested that TIBA is transported in a polar basipetal manner, but NPA is not (Thomson et al. 1973). Dhonukshe et al. (2008) showed that TIBA interferes with actin dynamics in plants providing a mechanism by which TIBA disrupts vesicle subcellular trafficking, including that of PIN auxin efflux carriers.

Morphactins interfere with many processes in plant growth and development and have been characterized as specific inhibitors of the polar transport of auxin (Krelle and Libbert 1968; Tognoni and Alpi 1969; Parups 1970; Schneider 1970; Naqvi 1972; Bridges and Wilkins 1973; Kaldewey et al. 1973; Gagianas and Berg 1977; Katekar and Geissler 1980; Tamimi and Firn 1985). The binding site of morphactin has been controversial. It has been also shown that morphactin binds to NPA receptor, suggesting that morphactin inhibits auxin (IAA) polar transport by the same mechanism as NPA (Thomson and Leopold 1974; Sussman and Goldsmith 1981). It is believed that morphactin is translocated in plants basipetally as well as acropetally through both sieve tubes and xylem elements (Neumann et al. 1977; Sundberg et al. 1994), suggesting that morphactin is faster to move in plant tissues than TIBA and NPA. Tamimi and Firn (1985) showed that NPA and morphactin IT 3456 are very effective inhibitors of the basipetal transport also of naphthalene-3-acetic acid (NAA) in hypocotyl sections taken from young *Helianthus annuus* and *Cucurbita pepo* seedlings when they were preincubated in the solution containing either NPA and morphactin or when these compounds were applied in a narrow ring at the apical end of these sections. Other compounds such as 1-naphthoxyacetic acid (1-NOA), 3-chloro-4-hydroxyphenylacetic acid (CHPAA, Parry et al. 2001), and 3,3a-dihydro-2-(*p*-methoxyphenyl)-8H-pyrazolol[5,1-α]isoindol-8-one (DPX-1840, Beyer and Quebedeaux 1974; Paterson 1983) have also been reported to inhibit auxin polar transport.

Recently, several artificial compounds which affect PIN trafficking and then inhibit auxin polar transport have also been screened using maize coleoptile assay, drawing the conclusion that compounds affecting cellular vesicle trafficking systems related to PIN trafficking are potent candidates for novel regulators of auxin polar transport system (Nishimura et al. 2012). Besides such artificially synthesized compounds, some naturally occurring compounds have also been shown to have inhibitory effects on auxin polar transport. Flavonoids have been shown to act as negative regulators of auxin transport in vivo in *Arabidopsis* (Brown et al. 2001), although their functions as modulators or regulators were not confirmed yet (Peer and Murphy 2007). As described above, PIN proteins have been considered to be an essential molecule in auxin polar transport (Chen and Masson 2005; Křeček et al. 2009; Ueda et al. 2011), and asymmetric PIN shift during gravity stimulation was promoted by the application of flavonoids, thus redirecting basipetal auxin streams necessary for root bending (Santelia et al. 2008).

Strenuous efforts to learn the presence and the distribution of naturally occurring auxin polar transport regulators in the plant kingdom resulted in the successful identification of physiologically active compounds in the extracts of several Asteraceae plants using an appropriate bioassay system consisting of radish (*Raphanus sativus* L.) hypocotyl segments. Dehydrocostus lactone (decahydro-3,6,9-tris-methylene-azulenol[4,5-b]furan-2(3H)-one), with α-methylene-γ-lactone moiety, and

4-hydroxy-β-thujone (4-hydroxy-4-methyl-1-(1-methylethyl)-bicyclo[3.1.0]hexan-3-one) have been identified as auxin polar transport inhibitors (Ueda et al. 2013, Fig. 8.2b). Another compound with α-methylene-γ-lactone moiety found in the extract of *Artemisia absinthium* has also been suggested to inhibit auxin polar transport in radish hypocotyls (see Ueda et al. 2013). The chemical structure of artabolide, 3-hydroxy-4,6,7(H)-germacra-1(10),11(13)-dien-6,12-olide, has successfully been determined based on the data of some physicochemical analyses (Arai et al. 2013, Fig. 8.2b).

8.5 Important Role of Auxin Polar Transport and Its Inhibitors in Plant Growth and Development

As described above, auxin moves from the shoot apex to the root apex between plant cells through a combination of membrane diffusion and carrier-mediated transport as a coordinating signal in plants and also as patterning signal in embryo and meristem organization. The process of auxin polar transport generates auxin maxima and gradients within tissues that are instrumental in the diverse regulation of various plant developmental processes (Roberts and Friml 2009). Recently, the molecular nature of the underlying positional information importance has been intensively studied in relation to auxin polar transport.

Auxin polar transport inhibitors interfere with basipetal polar transport of auxin and change auxin distribution, and in consequence perturb plant growth and development. NPA, TIBA, morphactin IT 3456, and HFCA have been expected to show various effects on plant growth and development. Generally, various effects of auxin polar transport inhibitors have been proposed to be caused, at molecular level, by interference with the auxin efflux carrier, but detailed mechanism remains still unclear whether or how blockage of the efflux of auxin by the inhibitors is involved in the effects of these inhibitors on physiological and developmental processes in plants (Fujita and Syōno 1996).

8.5.1 *Tropic Growth (Tropism)*

The tropism is caused by differential cell elongation in flanks between the stimulated and opposite sides. According to the classical Cholodny–Went model, lateral auxin transport gives rise to photo- and gravitropism. Cell elongation is usually promoted proportionally by the amount of IAA presented; however, in roots the usual effect of IAA is to inhibit cell elongation.

To test the idea that unequal auxin polar transport gives rise to phototropism, Gardner et al. (1974) followed radiolabeled auxin that had been asymmetrically applied to *Zea* and *Avena* coleoptiles, further demonstrating that unilateral blue light induced lateral auxin transport. Kang and Burg (1974) also reported that the enhancement of phototropism by red light in pea epicotyl, or another growth promoting hormone, gibberellins, did not correlate with an increase in lateral auxin transport and proposed that the magnitude of the phototropic response is determined by adjusting auxin sensitivity. These studies supported the involvement of auxin in phototropism, but the precise mechanism of how auxin transport and signaling cause phototropism remains unknown from the aspect of physiological approach.

Many more details about how changes in auxin transport influence phototropism are now emerging from research using *Arabidopsis* (Whippo and Hangarter 2006). Possible involvement of auxin transport is suggested by the fact that mutations in *PIN3*, which encodes an auxin efflux carrier involved in lateral auxin transport, can disrupt phototropism (Friml et al. 2002). However, work by Blakeslee et al. (2004) suggests that blue light does not affect PIN3 localization as it does for PIN1, an auxin efflux carrier thought to be important for auxin polar transport. Thus, the relationship between blue light signaling and PIN3 is uncertain.

Light-mediated relocalization of PIN1 appears to play a role in phototropism, as mutants in *MDR1*, a gene encoding a P-glycoprotein ABC transporter, show less PIN1 localization to the basal end of hypocotyl cells and an enhanced phototropic response (Noh et al. 2003). With less PIN1 localized to the basal end of hypocotyl cells, it was proposed that by decreasing auxin polar transport, phototropism may be enhanced by increasing the amount of auxin available for lateral transport (Noh et al. 2003). However, a mutant which has diminished auxin polar transport displays a longer phototropic latent periods, suggesting that some auxin polar transport is important for the normal progression of phototropism (Whippo and Hangarter 2006). More research is needed to determine how changes in both lateral and auxin polar transport occur during phototropism. In particular, it will be important to understand how the dynamics of auxin transport are regulated in the differential growth responses leading to curvature, although our knowledge about transport has advanced significantly at molecular levels.

Recently, the important role of auxin response factor 7 (ARF7) in phototropism has been suggested since the *nonphototropic hypocotyl 4* (*nph4*) locus encodes ARF7, a member of the auxin response factor family (Harper et al. 2000). ARFs function as transcriptional regulators whose activity is inhibited by binding to AUX/IAA proteins, and auxin, facilitates ARF activity by promoting the targeting of AUX/IAA proteins for degradation via the ubiquitin–proteasome pathway (Liscum and Reed 2002). With respect to phototropism, the degradation of IAA19, which binds to and inactivates ARF7, participates in phototropism under low-light conditions (Tatematsu et al. 2004). A possible scenario is that the ARFs promote phototropism by controlling the expression of genes containing auxin response elements (AuxREs) (Liscum and Reed 2002). Recently, it is found that eight genes with differential transcript accumulation across phototropic-stimulated *Brassica oleracea* hypocotyls have one or more AuxREs (Esmon et al. 2005). Interestingly, two of these genes encode expansins, which are involved in cell wall extension. Therefore, this study provides a plausible mechanism linking the differential growth underlying phototropism to the auxin regulation of expansin activity (Esmon et al. 2005).

Generally accepted is the hypothesis that basipetal IAA movements may control root elongation and a graviresponse in plants (Rashotte et al. 2000). Auxin gradient plays a key role in root gravitropism. Reorientation of *Arabidopsis* seedlings induces a rapid, asymmetric release of auxin from gravity-sensing columella cells at the root apex. The resulting lateral auxin gradient is hypothesized to drive differential cell expansion in elongation-zone tissues. Targeted expression of the auxin influx facilitator AUX1 demonstrated that root gravitropism requires auxin to be transported via the lateral root cap to all elongating epidermal cells. A three-dimensional model of the root elongation zone predicted that AUX1 causes the majority of auxin to accumulate in the epidermis. Selectively, disrupting the auxin responsiveness of expanding epidermal cells by expressing a mutant form of the AUX/IAA17 protein, *axr*3-1, abolished root gravitropism. Therefore it is concluded that gravitropic curvature in *Arabidopsis* roots is primarily driven by the differential expansion of epidermal cells in response to an influx carrier-dependent auxin gradient (Swarup et al. 2005).

One of the specific effects of auxin polar transport inhibitors is the abolition of gravitropism. Application of NPA to roots blocked a graviresponse, root waving and root elongation. Immediately after the application of NPA, the root gravity response was completely blocked. Inhibition of basipetal IAA transport by local application of NPA blocked a graviresponse. Thus, reduction in basipetal auxin transport by inhibitors leads to a loss of a graviresponse. In contrast, acropetal IAA transport in roots does not appear to be required for a graviresponse. Thus, auxin synthesized in the root tip may be sufficient to control a graviresponse (Rashotte et al. 2000).

Pretreatment of maize roots with NPA reduced both basipetal auxin transport and gravitropic bending (Young and Evans 1996). TIBA inhibited root graviresponse in tomato (Muday and Haworth 1994). It should be mentioned that TIBA, NPA, and HFCA have been shown to induce automorphosis-like epicotyl bending in etiolated Alaska pea seedlings, which is observed under microgravity conditions in space, connected to reduced auxin polar transport in etiolated epicotyls, although mechanisms of epicotyl bending in etiolated pea seedlings induced by auxin polar transport inhibitors are still unclear (Ueda et al. 1999, 2000; Miyamoto et al. 2005a, b; Hoshino et al. 2006, 2007).

Morphactins have been found to influence photo- and gravitropic responses (Khan 1967; Schneider 1970), in which auxin polar transport is essential for differential distribution of auxin resulting in differential cell elongation growth as well. Morphactins interfere with many processes in plant growth and development and have been characterized as specific inhibitors of the polar transport of auxin (Krelle and Libbert 1968; Tognoni and Alpi 1969; Parups 1970; Schneider 1970; Naqvi 1972; Bridges and Wilkins 1973; Kaldewey et al. 1973; Gagianas and Berg 1977; Katekar and Geissler 1980; Tamimi and Firn 1985). They are responsible for complete abolition of polarity in plants, inhibition of mitosis in apical meristems, and abolition of photo- and gravitropic responses.

When seedlings of different plants were grown in the presence of morphactin IT 3456, on a vertical or a horizontal plane, roots and shoots lost the capacity to respond to gravity or to unilateral light source. This was true for both monocots and dicots. This suggests that basic mechanism(s) of the two tropic responses is the same in roots and shoots of the two plant groups. The site(s) of action of morphactins is unknown. The reaction(s) controlling photo- and gravitropism may be closely related as morphactins affected both gravi- and phototropic responses of the same organ (Khan 1967).

Morphactin IT 3456 rapidly inhibited the normal positive gravitropic response in roots of intact pea seedlings; the elongation zone of the morphactin-treated root showed swelling of the epidermal and cortical cells and many long root hairs (Gaither 1975). It was documented that morphactin IT 3456 inhibition of pea root gravitropism could be due to the disruption of auxin polar transport at the level of auxin binding. Other inhibitors of auxin polar transport, TIBA and NPA, also inhibited gravitropism in roots of intact seedlings of *Pisum sativum* but in a much smaller degree (Gaither and Abeles 1975).

8.5.2 Cambial Activity and Xylem Differentiation: Relevance to a Graviresponse

The vacuolar system of plants is composed of a continuous network of vascular bundles, primarily composed of phloem and xylem tissues that translocate dissolved photoassimilates and mineral, respectively. Auxin is the limiting and controlling factor for both phloem and xylem differentiation. Auxin induces phloem with no xylem at low auxin levels, and xylem differentiation only takes place at higher auxin levels. Auxin also appears to play a central role in the induction of vascular differentiation; parenchymatic cells being able to transdifferentiate into vessel elements upon auxin exposure. In addition, local application of auxin on plant organs can induce the formation of new vascular strands within these organs. This effect is along the polarity. In that vascular differentiation always occurs basal to the auxin source. The explanation for this polar effect is found in another auxin-specific process, auxin polar transport. Auxin is actively transported from sources of production in the shoot toward the root. A hypothesis, canalization of auxin flow hypothesis, explaining why distinct vascular strands rather than a radial cloud of vascular cells differentiate from an auxin source has been proposed (Sachs 2000). In this hypothesis, auxin flow is a major organization mechanism and in which the gradual canalization of the auxin polar transport is responsible for patterning at the cell, tissue, and whole-plant levels.

Similar explanation is possible in leaf vascular patterning. For example, AtPIN1, a crucial member of the AtPIN family of auxin efflux-associated proteins, is expressed prior to pre-procambial and procambial cell fate markers in domains that become restricted toward site of procambium formation. Subcellular AtPIN1 polarity indicates that auxin is directed to distinct "convergence points" in the epidermis from where it defines the positions of major vines (Scarpella et al. 2006).

Trees usually develop reaction wood in response to a gravistimulation. The type of reaction wood formed in gymnosperms is referred to as compression wood.

It develops on the lower side of an inclined shoot or stem, where the tissue seems to be under a "compressive" stress. The most characteristic of compression wood is the rounded shape of the tracheid, resulting in intercellular spaces, and cell walls of compression wood are thicker than those in normal wood (Du and Yamamoto 2007). The application of a high concentration of IAA to vertical stems induced compression wood formation at the application point and suggests that IAA is correlated with compression wood formation (Du and Yamamoto 2007). Compression wood, which is normally produced on the lower side of inclined stems and branches as a response to normal gravitropic conditions as described above, in vertically growing shoots of various coniferous (gymnosperm) species was induced by the application of morphactins. The morphactin-induced compression wood is probably caused by the differential redistribution or accumulation of endogenous auxin (IAA) between the lower and the upper sides of branches in the affected part of the shoot (Westing 1965, 1968; Smoliński et al. 1972, 1973; Phelps et al. 1974, 1977; Yamaguchi et al. 1983), since the application of IAA resulted in compression wood formation. Reaction wood in the arboreal dicotyledonous angiosperms is called tension wood and is formed on the upper side of a leaning stem, where the tissues are held in "tension." The most characteristic of tension wood is the presence of thick inner cell-wall layers consisting highly crystalline cellulose (Du and Yamamoto 2007). It is generally accepted that a gravistimulation causes a difference in auxin levels around the stem and an auxin deficiency on the upper side induces tension wood formation. Application of morphactin to the upper side of inclined shoots of *Aesculus hippocastanum* resulted in a suppression of tension wood formation on this side (Smoliński et al. 1974; Phelps et al. 1975).

Sundberg et al. (1994) showed that ringing an intact vertical *Pinus sylvestris* shoot with lanolin containing morphactin IT 3456 or NPA induced the formation of compression wood above the ring, and both the compounds inhibited the polar transport of IAA; the level of free IAA was dramatically decreased below the ring. However, morphactin IT3456- and NPA-induced wood formation above the application bands were not associated with an increased concentration or turnover IAA in the cambial region (Sundberg et al. 1994). Yamaguchi et al. (1980) also documented that NPA applied as a band in vertical shoot of *Cryptomeria japonica* induced typical compression wood above the treatment as a result of inhibition of IAA transport.

The application of morphactin IT 3456 and NPA to poplar stems changed wood differentiation; IT 3456 treatment causing a decrease in mean wall thickness of vessels and NPA increasing cell wall thickness of fibers compared with untreated controls (Junghans et al. 2004). It has been reported that morphactin increased the cambial activity and irregular differentiation of xylem elements in *Sterculia urens* Roxb. (Malvaceae) (Patel and Setia 1979). Morphactin treatments have been reported to increase stem diameter of guayule (*Parthenium argentatum*) (Dierig and Backhaus 1990), *Pinus radiata*, *Eucalyptus globulus*, and olives (Backhaus et al. 1976; Doss et al. 1977), when it was applied as a bark band, the thickening being considered to be connected mostly with the stimulation of cambial activity.

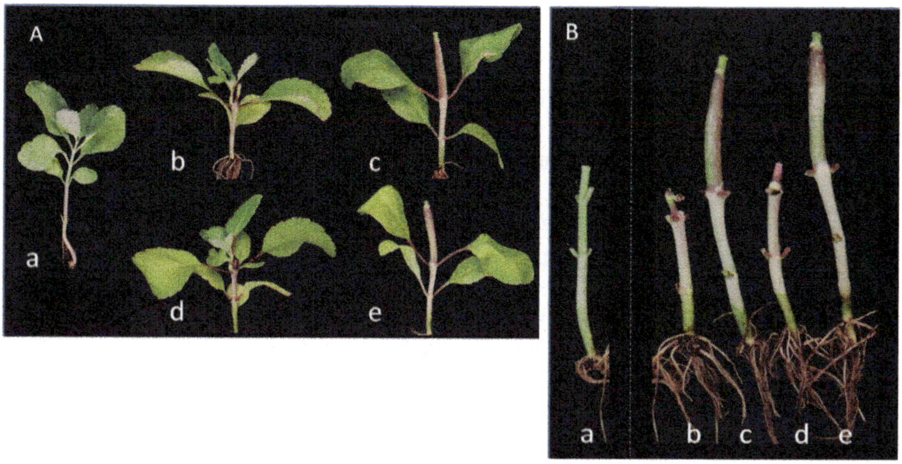

Fig. 8.3 Effects of morphactin IT3456, benzyladenine, and their mixture, all used at a concentration of 0.2 %, applied on the last internode of decapitated shoots of *Bryophyllum calycinum* on the elongation and thickening of treated internodes; treatments made on August 11. Photographs were made on October 15 (Quoted from Miyamoto et al. (2013)).
(**A**) Picture of treatment *a, b, c, d*, and *e*.
(*a*) Initial plants on the day of treatment (August 11);
(*b*) control plants, treated with lanolin only; the upper part of the last internode abscised and sprouting of axillary bud is visible;
(*c*) morphactin; the elongation and thickening of the treated internode can be observed;
(*d*) benzyladenine; the upper part of the last internode abscised and sprouting of axillary bud is visible;
(*e*) morphactin + benzyladenine; the elongation and thickening of the treated internode can be observed.
(**B**) Picture of initial treatment *a*, and treatment *b, c, d*, and *e* after removal of all leaves and axillary buds in the treatments

Morphactin IT 3456 applied at the last internode in decapitated shoot of *Bryophyllum calycinum* substantially stimulated elongation and thickening of the internode (Fig. 8.3). These results suggest that morphactin translocated basipetally from the top of the treated internode inhibits auxin polar transport from the internode, resulting in the accumulation of endogenous auxin for elongation and thickening in the treated internode of decapitated *Bryophyllum calycinum* (Miyamoto et al. 2013). The drastic increase in diameter of internode treated with morphactin in *Bryophyllum calycinum* is due to increase in cambial activity resulting in wider cambial zone and zone of secondary xylem. This suggestion is supported by the observation that HFCA strongly promoted the elongation of internodes and raised IAA level above the application site in the short mutant (*lkb*) of garden pea (*Pisum sativum* L.) which contained two- to threefold less free IAA than those of the wild type of garden pea (McKay et al. 1994).

8.5.3 Stem Growth

One of the main areas of physiological study of auxin is the mechanism of auxin action on cell elongation resulted in stem growth. Numerous reports of auxin physiology relevance to cell elongation have been published (Lyons and Widmer 1984; Kutschera et al. 1987; Yang et al. 1993; McKay et al. 1994; see Jager et al. 2007). Here the effects of auxin and auxin polar transport inhibitors on tulip stem growth are described as an interesting topic.

Tulip bulbs with terminal buds containing a complete flower require a period of 12–14 weeks of low temperature treatment for shoot elongation (De Hertogh 1974). It is well known that elongation of the stem and leaves of tulips is due almost entirely to the elongation of cells produced during earlier developmental stages of flower-bud formation (Gilford and Rees 1973). The leaves and gynoecium provide auxins which control the elongation growth of the stem (Op den Kelder et al. 1971; Hanks and Rees 1977; Saniewski and De Munk 1981; Banasik and Saniewski 1985). Saniewski and De Munk (1981), and Banasik and Saniewski (1985) showed that elongation of all internodes in precooled tulip bulbs is promoted by application of auxins in the place of removed flower buds in the absence of leaves. Okubo and Uemoto (1985) showed that 2,3,5-triiodobenzoic acid (TIBA) treatment at the first internode of sprouting tulip shoot inhibited the dark-induced elongation of the first internode and decreased the amount of diffusible auxin from the upper organs into the first internode but did not affect the gibberellin amount.

As well known, auxin producing in flower buds is an essential for the growth of shoots in nonprecooled and precooled, rooted and derooted tulip bulbs (Saniewski and Okubo 1997, 1998a). They also found that TIBA applied in the middle of the fourth internode (below IAA application) greatly inhibited the growth of flower internodes. Similarly, NPA and morphactin IT 3456 also strongly inhibited the growth of internodes induced by IAA in precooled and rooted tulip bulbs after removal of flower bud and all leaves (Saniewski and Okubo 1998b; Saniewski et al. 1999). Morphactin inhibited also the growth of all internodes in tulips induced by flower bud in the absence of leaves. The inhibitory effects of TIBA, NPA, and morphactin on tulip stem growth induced by IAA were restored by additional application of IAA below at the treatment of auxin polar transport inhibitors. These results clearly indicate that endogenous auxin and its polar transport are substantially responsible for stem growth in plants. This idea is strongly supported by the recent observations (Ueda et al. 2012). Morphactins were applied as single soil drenches 1–2 days after transferring the fully cooled bulbs to the greenhouse. It was found that morphactins at high concentrations were toxic or caused flower-bud blasting and other floral abnormalities. At lower doses most morphactins reduced first internode length, while some increased the length of the upper internodes, including nastic curvature and occasionally altering the flower shape. None of the morphactins reduced the length of the top internode.

8.5.4 Morphogenetic Processes Including Organogenesis and Embryogenesis

In *Arabidopsis*, mutation in *Monopteros* gene has been shown to induce phenotypically rootless seedlings with reduced vascular systems and occasionally fused cotyledons. In the embryos of mutant affected in this gene, early cell divisions are abnormally oriented and do not establish cell files along the apical–basal axis. *Monopteros* gene has been implicated in the relay of an apical–basal auxin signal and encoded an "auxin response" transcriptional factor (Przemeck et al. 1996; Hardtke and Berleth 1998). Recently, mutations in two genes, *auxin resistant 6* (*AXR6*) and *BODENLOS* (*BLD*) have been shown to phenotypically resemble the previously identified *Monopteros* (*mp*) mutant. Therefore it is speculated that all three genes function in auxin signal transduction to promote the alignment of cell differentiation with the apical–basal orientation of auxin flow (Berleth and Sachs 2001). Interestingly, mutations in a single *Arabidopsis* gene, *EMB30/GNOM*, can abolish cell polarity in the entire embryo (Steinmann et al. 1999), and null mutations in another gene, *pin-formed 1* (*PIN1*), result in related but far less severe distortions of embryo symmetry (Gälweiler et al. 1998). Since the PIN1 product is the best characterized member of a family of presumptive auxin efflux carrier proteins, and the *EMB30/GNOM* gene encodes a guanosine exchange factor acting on small G proteins in vesicle transport; auxin transport and coordinate localization of auxin efflux carrier proteins via vesicle transport are suggested to be prerequisites for directional growth in morphogenesis in plants.

Recently, Kumar et al. (2013) showed that NPA treatment of the growing shoots of *Pisum sativum* caused different changes in the stipule structure. NPA disrupts also leaf initiation and formation of leaf margins in maize (Scanlon 2003). DeMason and Chawla (2004) showed many abnormalities in leaves of pea grown both on TIBA and NPA, suggesting that an auxin gradient plays fundamental roles in controlling morphogenesis of leaves of pea, starting from the point of leaf initiation on the shoot apical meristem through the period of leaf expansion. TIBA treatment provoked modification of the structure of the shoot apex in *Lycopersicon esculentum*, inducing the formation of aphyllous structures (Hernández and Driss-École 1989, 1990).

NPA induced fasciation of primordia which sometimes results in changes in the relation vertical spacing of primordia in *Epilobium hirsutum* (Meicenheimer 1981). Somatic embryos of Norway spruce (*Picea abies*) grown on medium containing NPA leads to the formation of embryos with poor shoot apical meristem and fused cotyledons, and to a pin-formed phenotype of the regenerated plantlets (Hakman et al. 2009).

TIBA has been well known to induce the number of leaf abnormalities in different plant species. Different abnormalities induced by TIBA in shoot growth, mostly leaves, have been described in bean and *Kalanchoe blossfeldiana*, respectively (Snyder 1949; van Zeist and Koevoets 1951). Foliar modifications induced by TIBA and HFCA have also been reported in cultured pedicle explants of *Orychophragmus violaceus*, cultured leaf explants of tobacco, and seedlings of tobacco and *Brassica chinensis* (An et al. 1999). These results strongly suggest that auxin polar transport plays an important role in leaf morphogenesis. Schiavone and Cooke (1987) showed that both TIBA and NPA are able to block the ability of different stages of somatic embryos of carrot to undergo morphogenetic transitions to the subsequent stages.

HFCA has been reported to show the unique physiological effect on the flower formation in *Arabidopsis thaliana*. The *pin-formed* and the *pinoid* mutants in *Arabidopsis*, whose flower mutants normally show a unique structure with no flower in the inflorescence axis, have been reported to show reduced activity of auxin polar transport in inflorescence axis (Okada et al. 1991; Ueda et al. 1992; Bennett et al. 1995; Oka et al. 1998) and that the application of HFCA as well as NPA and TIBA to wild-type *Arabidopsis* induced malformations mimicking the *pin-formed* mutant (Okada and Shimura 1994; Oka et al. 1999).

Morphactins have been shown to dispense through shoot apex induced diverse malformations in various species of *Kalanchoe*; the effect being freely transmitted to newly emerging axially branches as well as across leaf lamina to differentiating epiphyllous buds (Sawhney and Mahajan 1995).

It has been demonstrated that morphactin, NPA, and TIBA induced parthenocarpy in cucumber by rapidly blocking the natural outward flow of auxin from the ovary, thereby resulting in an accumulation of auxin within the ovary sufficient to trigger parthenocarpy (Beyer and Quebedeaux 1974). Also Robinson et al. (1971) found earlier that foliar application of morphactin induced parthenocarpy in cucumber (*Cucumis sativus*).

8.5.5 Rooting and Lateral Root Formation

The promoting effect of auxin on root meristem formation is experimentally well established. Monitoring the distribution of perceived auxin in the growing root using a reporter gene expression controlled by a synthetic auxin response element, which is considered to be reflecting auxin distribution, revealed that a sharp-bordered local auxin maximum (auxin peak) just distal to the quiescent center was observed. Furthermore, any shift in the localization of this peak, whether caused by genetic or experimental interference with auxin transport by auxin polar transport inhibitors, was associated with the shift in the pattern of distal cell fates in the root meristem (Berleth and Sachs 2001). Since in normal plant development, the formation of a local auxin peak would probably be dependent on shoot-derived auxin, the peak and its shoot-dependent positioning would reflect both the coordinating and the local-patterning functions of auxin in root meristem formation.

The process of adventitious root formation can be divided into three stages: root induction in which molecular and biochemical changes occur before any cytological event, root initiation when the first anatomical modifications take place, and protrusion, corresponding to emergence of root primordial. Auxin (IAA) is one of the essential endogenous hormones known to play a most important role in the formation of adventitious roots (Casimiro et al. 2001; Han et al. 2009). Auxins are usually synthesized in the stem tip and tender leaves of plants and transported to the basal part of the stem and roots. Auxin polar transport has been described to move

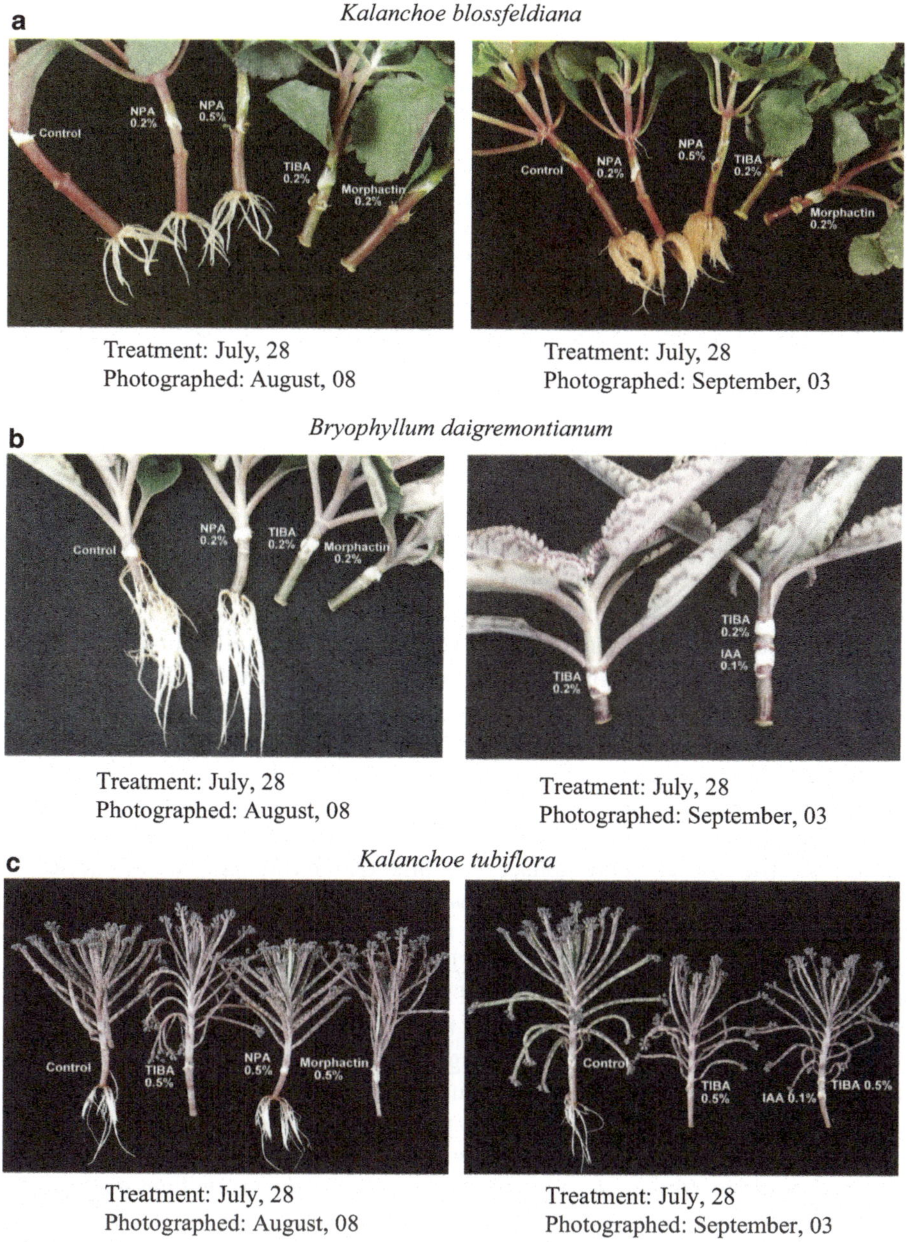

Fig. 8.4 Effects of TIBA, NPA, and morphactin IT 3456 on the rooting of a few species of Crassulaceae, (**a**) *Kalanchoe blossfeldiana*, (**b**) *Bryophyllum daigremontianum*, (**c**) *Kalanchoe tubiflora*, and (**d**) *Bryophyllum calycinum*. TIBA and morphactin completely inhibited root formation, but NPA did not, when TIBA, NPA, and morphactin were used at concentrations of 0.2 and 0.5 % (w/w) in lanolin paste and applied as a ring around the stem under the shoot with leaves (Quoted from Saniewski et al. (2014))

Fig. 8.4 (continued)

IAA to the root apex (acropetal) and toward the root/shoot junction (basipetal) in two different root tissues (Rashotte et al. 2000). Exogenous auxins, IAA, indole-3-butyric acid (IBA) and naphthalene-3-acetic acid (NAA), are widely used in the promotion of rooting of cuttings in a lot of species of plants.

Several studies have pointed to the inhibition of rooting and inhibition of auxin polar transport in many plants after the application of NPA and TIBA (Katsumi et al. 1969; Batten and Goodwin 1981; Guerrero et al. 1999; Liu and Reid 1992; Garrido et al. 2002; Marks et al. 2002). Morphactin IT 3456 also inhibited adventitious root formation in the cuttings of *Pisum sativum* (Khan et al. 1977), hypocotyl cuttings of *Impatiens balsamina* (Nanda et al. 1973), and stem cuttings of *Salix tetrasperma* (Kochhar et al. 1972). Saniewski et al. (2014) intensively studied on the effects of TIBA, NPA, and morphactin IT 3456 on the rooting of a few species of Crassulaceae (*Bryophyllum daigremontianum*, *Bryophyllum calycinum*, *Kalanchoe blossfeldiana*, and *Kalanchoe tubiflora*), resulting in the interesting and particular observations in rooting of these plants. TIBA and morphactin completely inhibited root formation but NPA did not, when TIBA, NPA, and morphactin were used at concentrations of 0.2 and 0.5 % (w/w) in lanolin paste and applied as a ring around the stem under the shoot with leaves (Fig. 8.4a–d). These results strongly suggest that TIBA and morphactin, and NPA, interact with different proteins relevant to auxin polar transport, respectively.

Karabaghli–Degron et al. (1998) found that TIBA inhibits the stimulation of in vitro lateral root formation by Norway spruce (*Picea abies*) seedlings treated with IAA. The ectomycorrhizal fungus *Laccaria bicolor* strain S238 N, which can synthesize IAA in pure culture, also stimulated lateral root formation on spruce seedlings, and the colonization of the taproot cortex by the fungus and lateral rhizogenesis was inhibited by TIBA. Authors suggest that TIBA inhibited the transport of both exogenous IAA and fungal IAA in the root of spruce seedlings and in this way also inhibited lateral root induction. Thus, auxin produced by ectomycorrhizal fungus *Laccaria bicolor* plays a central role in the formation of roots in the *Picea abies*.

NPA, similar to TIBA, counteracted the effect of exogenous IAA on lateral rhizogenesis of spruce seedlings. Although TIBA completely inhibited the stimulatory effect of *Laccaria bicolor* on lateral root formation, NPA inhibited it only partially (Rincón et al. 2003). NPA arrests lateral root development by blocking the first transverse divisions. NPA causes IAA to accumulate in the root apex while reducing levels in basal tissues critical for lateral root initiation. This pattern of IAA redistribution is consistent with NPA blocking basipetal IAA movement from the root tip. NPA did not affect acropetal transport of IAA but inhibited basipetal transport (Casimiro et al. 2001). Reed et al. (1998) also suggested that inhibition of auxin movement from the shoot into the root (acropetal transport) inhibits lateral root development in *Arabidopsis*.

References

Adam G, Petzold U (1994) Brassinosteroide – eine neue Phytohormon-Gruppe? Naturwissenschaften 81:210–217

Ahmad R, Khalid M, Naveed M, Shahazad SM, Zahir ZA, Khokhar SN (2008) Comparative efficiency of auxin and its precursor applied through compost for improving growth and yield of maize. Pak J Bot 40:1703–1710

Alcázar R, Altabella T, Marco F, Bortolotti C, Reymond M, Koncz C, Carrasco P, Tiburcio AF (2010) Polyamines: molecules with regulatory functions in plant abiotic stress tolerance. Planta 231:1237–1249

An ND, Wang LJ, Xu ZH, Xia ZA (1999) Foliar modifications induced by inhibition of polar transport of auxin. Cell Res 9:27–35

Arai T, Toda Y, Kato K, Miyamoto K, Hasegawa T, Yamada K, Ueda J, Hasegawa K, Inoue T, Shigemori H (2013) Artabolide, a novel polar auxin transport inhibitor isolated from *Artemisia absinthium*. Tetrahedron 69:7001–7005

Arshad M, Frankenberger WT Jr (1991) Microbial production of plant hormones. Plant Soil 133:1–8

Backhaus RA, Sachs RM, Hield HZ (1976) Tree growth inhibition by bark application of morphactins. HortSci 11:578–580

Banasik L, Saniewski M (1985) The effect of different auxins on tulip stalk elongation. Acta Hortic 167:193–204

Bartel B (1997) Auxin biosynthesis. Annu Rev Plant Physiol Plant Mol Biol 48:51–66

Batten DJ, Goodwin PB (1981) Auxin transport inhibitors and the rooting of hypocotyl cuttings from etiolated mung-bean *Vigna radiata* (L.) Wilczek seedlings. Ann Bot 47:497–503

Bayer MH (1969) Gas chromatographic analysis of acidic indole auxins in *Nicotiana*. Plant Physiol 44:267–271

Bayliss WM, Starling E (1904) The chemical regulation of the secretory process. Proc R Soc Lond (Ser B) 73:310–322 (cited from Plant Growth Substances in Agriculture, Weaver RJ, 1972)

Beijerinck MW (1888) Über das Cecidium von *Nematus capreae* auf *Salix amygdalina*. Botanische Zeitung 46:17–27

Beijerinck MW (1897) Sur la cécidiogénèse et la génération alternante chez le *Cynips calicia*. Verz. Geschr III: 199–232

Benjamins R, Quint A, Weijers D, Hooykaas P, Offringa R (2001) The PINOID protein kinase regulates organ development in *Arabidopsis* by enhancing polar auxin transport. Development 128:4057–4067

Bennett SRM, Alvarez J, Bossinger G, Smyth DR (1995) Morphogenesis in *pinoid* mutants of *Arabidopsis thaliana*. Plant J 8:505–520

Berleth T, Sachs T (2001) Plant morphogenesis: long-distance coordination and local patterning. Curr Opin Plant Biol 4:57–62

Beyer A (1928) Beiträge zum Problem der Reizleitung. Zeitschrift für Botanik 20:321–417

Beyer EM Jr, Quebedeaux B (1974) Parthenocarpy in cucumber: Mechanism of action of auxin transport inhibitors. J Am Soc Hortic Sci 99:385–390

Bishop GJ, Koncz C (2002) Brassinosteroids and plant steroid hormone signaling. Plant Cell 14(Suppl):S97–S110

Blakeslee JJ, Bandyopadhyay A, Peer WA, Makam SN, Murphy AS (2004) Relocalization of the PIN1 auxin efflux facilitator plays a role in phototropic responses. Plant Physiol 134:28–31

Blakeslee JJ, Peer WA, Murphy AS (2005) Auxin transport. Curr Opin Plant Biol 8:494–500

Blommaert KLJ (1954) Growth- and inhibiting-substances in relation to the rest period of the potato tuber. Nature 174:970–972

Boysen Jensen P (1910) Über die Leitung des phototropischen Reizes in *Avena*-keimpflanzen. Ber Deut Bot Ges 28:118–120

Boysen Jensen P, Nielsen N (1925) Studien über die hormonalen Beziehungen zwischen Spitze und Basis der Avenacoleoptile. Planta 1:321–331

Bridges IG, Wilkins MB (1973) Effects of morphactin on indol-3yl-acetic acid transport, growth, and geotropic response in cereal coleoptiles. J Exp Bot 24:711–723

Brown ME (1972) Plant growth substances produced by micro-organisms of soil and rhizosphere. J Appl Bacteriol 35:443–451

Brown DE, Rashotte AM, Murphy AS, Normanly J, Tague BW, Peer WA, Taiz L, Muday GK (2001) Flavonoids act as negative regulators of auxin transport *in vivo* in *Arabidopsis*. Plant Physiol 126:524–535

Burström H (1950) Studies on growth and metabolism of roots. IV. Positive and negative auxin effects on cell elongation. Physiol Plant 3:277–292

Casimiro I, Marchant A, Bhalerao RP, Beeckman T, Dhooge S, Swarup R, Graham N, Inzé D, Sandberg G, Casero PJ, Bennett M (2001) Auxin transport promotes *Arabidopsis* lateral root initiation. Plant Cell 13:843–852

Cesco S, Nikolic M, Römheld V, Varanini Z, Pinton R (2002) Uptake of ^{59}Fe from soluble ^{59}Fe-humate complexes by cucumber and barley plants. Plant Soil 241:121–128

Chandramohan D, Mahadevan A (1968) Indole acetic acid metabolism in soils. Curr Sci 37:112–113

Chen Y, Aviad T (1990) Effects of humic substances on plant growth. In: MacCarthy D, Clapp CE, Malcolm RL, Bloom PR (eds) Humic substances in soil and crop sciences. Selected reading. Amer Soc Agron Inc, and Soil Sci Soc America Inc, Madison, WI, USA pp 161–186

Chen R, Masson PH (2005) Auxin transport and recycling of PIN proteins in plants. In: Šamaja J, Baluška F, Menzel D (eds) Plant endocytosis. Springer, Berlin, pp 139–157

Ciesielski T (1872) Untersuchungen über die Abwärtskrümmung der Wurzel. Beiträge zur Biologie der Pflanzen 1:1–30

Cooper R (1959) Bacterial fertilizers in the Soviet Union. Soils Fertil 22:327–333

Darwin C (1880) The Power of Movements in Plants. Murray J, London

De Hertogh A (1974) Principles for forcing tulips, hyacinths, daffodils, Easter lilies and Dutch irises. Sci Hortic Amsterdam 2:313–355

DeMason DA, Chawla R (2004) Roles for auxin during morphogenesis of the compound leaves of pea (*Pisum sativum*). Planta 218:435–448

Depta H, Eisele K-H, Hertel R (1983) Specific inhibitors of auxin transport: action on tissue segments and *in vitro* binding to membranes from maize coleoptiles. Plant Sci Lett 31:181–192

Dhonukshe P, Grigoriev I, Fischer R, Tominaga M, Robinson DG, Hašek J, Paciorek T, Petrášek J, Seifertová D, Tejos R, Meisel LA, Zažimalová E, Gadella TWJ Jr, Stierhof Y-D, Ueda T, Oiwa K, Akhmanova A, Brock R, Spang A, Friml J (2008) Auxin transport inhibitors impair vesicle motility and actin cytoskeleton dynamics in diverse eukaryotes. PNAS 105:4489–4494

Dierig DA, Backhaus RA (1990) Effects of morphactin and DCPTA on stem growth and bioinduction of rubber in guayule. HortSci 25:531–533

Doss RP, Neumann PM, Backhaus RA, Sachs RM (1977) Bark banding of *Pinus radiata* with morphactin (IT3456): influence on trunk anatomy assimilate transport and growth. Physiol Plant 39:280–284

Du S, Yamamoto F (2007) An overview of the biology of reaction wood formation. J Integr Plant Biol 49:131–143

Dvornikova TP, Skriabin GK, Suvorov NN (1970) Enzymatic transformation of tryptamine by fungi. Mikrobiologiia 39:42–46

Engvild KC, Egsgaard H, Larsen E (1980) Determination of 4-chloroindole-3-acetic acid methyl ester in *Lathyrus*, *Vicia* and *Pisum* by gas chromatography–mass spectrometry. Physiol Plant 53:499–503

Ernstsen A, Sandberg G (1986) Identification of 4-chloroindole-3-acetic acid and indole-3-aldehyde in seeds of *Pinus sylvestris*. Physiol Plant 68:511–518

Esmon CA, Tinsley AG, Ljung K, Sandberg G, Hearne LB, Liscum E (2005) A gradient of auxin and auxin-dependent transcription precedes tropic growth responses. PNAS 103:236–241

Estelle M (1998) Polar auxin transport: new support for an old model. Plant Cell 10:1775–1778

Frankenberger WT Jr, Chang AC, Arshad M (1990) Response of *Raphanus sativus* to the auxin precursor, L-tryptophan applied to soil. Plant Soil 129:235–241

Frenkel C, Haard NF (1973) Initiation of ripening in bartlett pear with an antiauxin α(*p*-chlorophenoxy)isobutyric acid. Plant Physiol 52:380–384

Friml J (2003) Auxin transport-shaping the plant. Curr Opin Plant Biol 6:7–12

Friml J (2010) Subcellular trafficking of PIN auxin efflux carriers in auxin transport. Eur J Cell Biol 89:231–235

Friml J, Wiśniewska J, Benková E, Mendgen K, Palme K (2002) Lateral relocation of auxin efflux regulator PIN3 mediates tropism in *Arabidopsis*. Nature 415:806–809

Friml J, Vieten A, Sauer M, Weijers D, Schwarz H, Hamann T, Offringa R, Jürgens G (2003) Efflux-dependent auxin gradients establish the apical-basal axis of *Arabidopsis*. Nature 426:147–153

Friml J, Yang X, Michniewicz M, Weijers D, Quint A, Tietz O, Benjamins R, Ouwerkerk PBF, Ljung K, Sandberg G, Hooykaas PJJ, Palme K, Offringa R (2004) A PINOID-dependent binary switch in apical-basal PIN polar targeting directs auxin efflux. Science 306:862–865

Fujita H, Syōno K (1996) Genetic analysis of the effects of polar auxin transport inhibitors on root growth in *Arabidopsis thaliana*. Plant Cell Physiol 37:1094–1101

Gagianas AA, Berg AR (1977) The effect of morphactin (methyl 2-chloro-9-hydroxyfluorene-9-carboxylate) on basipetal transport of indol-3-ylacetic acid in hypocotyl sections of *Phaseoulus vulgaris* L. Ann Bot 41:1135–1148

Gaither DH (1975) Auxin and the response of pea roots to auxin transport inhibitors: morphactin. Plant Physiol 55:1082–1086

Gaither DH, Abeles FB (1975) Sites of auxin action. Regulation of geotropism, growth, and ethylene production by inhibitors of auxin transport. Plant Physiol 56:404–409

Galston AW (1947) The effect of 2,3,5-triiodobenzoic acid on the growth and flowering of soybeans. Am J Bot 34:356–360

Galston AW, Sawhney RK (1990) Polyamines in plant physiology. Plant Physiol 94:406–410

Gälweiler L, Guan C, Müller A, Wisman E, Mendgen K, Yephremov A, Palme K (1998) Regulation of polar auxin transport by AtPIN1 in *Arabidopsis* vascular tissue. Science 282:2226–2230

Gandar JC, Nitsch C (1967) Isolement de l'ester méthylique d'un acide chloro-3-indolyacétique à partir de graines immatures de pois, *Pisum sativum* L. C. R. Acad Sci (Paris) Sér D 265:1795–1798

Gardner G, Shaw S, Wilkins MB (1974) IAA transport during the phototropic responses of intact *Zea* and *Avena* coleoptiles. Planta 121:237–251

Garrido G, Guerrero JR, Cano EA, Acosta M, Sánchez-Bravo J (2002) Origin and basipetal transport of the IAA responsible for rooting of carnation cuttings. Physiol Plant 114:303–312

Geisler M, Murphy AS (2006) The ABC of auxin transport: the role of p-glycoproteins in plant development. FEBS Lett 580:1094–1102

Geisler M, Blakeslee JJ, Bouchard R, Lee OR, Vincenzetti V, Bandyopadhyay A, Titapiwatanakun B, Peer WA, Bailly A, Richards EL, Ejendal KFK, Smith AP, Baroux C, Grossniklaus U, Müller A, Hrycyna CA, Dudler R, Murphy AS, Martinoia E (2005) Cellular efflux of auxin catalyzed by the Arabidopsis MDR/PGP transporter AtPGP1. Plant J 44:179–194

Geldner N, Richter S, Vieten A, Marquardt S, Torres-Ruiz RA, Mayer U, Jürgens G (2004) Partial loss-of-function alleles reveal a role for *GNOM* in auxin transport-related, post-embryonic development of *Arabidopsis*. Development 131:389–400

Gilford JM, Rees AR (1973) Growth of the tulip shoot. Sci Hortic Amsterdam 1:143–156

Grunewald W, Friml J (2010) The march of the PINs: developmental plasticity by dynamic polar targeting in plant cells. EMBO J 29:2700–2714

Guerrero JR, Garrido G, Acosta M, Sánchez-Bravo J (1999) Influence of 2,3,5-triiodobenzoic acid and 1-*N*-naphthylphthalamic acid on indoleacetic acid transport in carnation cuttings: relationship with rooting. J Plant Growth Regul 18:183–190

Haagen-Smit AJ, Leech WD, Bergren WR (1942) The estimation, isolation, and identification of auxins in plant materials. Am J Bot 29:500–506

Haagen-Smit AJ, Dandliker WB, Wittwer SH, Murneek AE (1946) Isolation of 3-indoleacetic acid from immature corn kernels. Am J Bot 33:118–120

Hakman I, Hallberg H, Palovaara J (2009) The polar auxin transport inhibitor NPA impairs embryo morphology and increases the expression of an auxin efflux facilitator protein PIN during *Picea abies* somatic embryo development. Tree Physiol 29:483–496

Hamence JH (1946) The determination of auxins in soils: including a note on synthetic growth substances. Analyst 71:111–116

Han H, Zhang S, Sun X (2009) A review on the molecular mechanism of plants rooting modulated by auxin. Afr J Biotechnol 8:348–353

Hanks GR, Rees AR (1977) Stem elongation in tulip and narcissus: the influence of floral organs and growth regulators. New Phytol 78:579–591

Hardtke CS, Berleth T (1998) The *Arabidopsis* gene *MONOPTEROS* encodes a transcription factor mediating embryo axis formation and vascular development. EMBO J 17:1405–1411

Harper RM, Stowe-Evans EL, Luesse DR, Muto H, Tatematsu K, Watahiki MK, Yamamoto K, Liscum E (2000) The *NPH4* locus encodes the auxin response factor ARF7, a conditional regulator of differential growth in aerial Arabidopsis tissue. Plant Cell 12:757–770

Hatano T, Kato Y, Katayama M, Marumo S (1989) A new indolic antiauxin, α-(5,7-dichloroindole-3-)isobutyric acid: its chemical synthesis and biological activity. Experientia 45:400–402

Hayashi Y, Sakan T (1974) Nagilactones, plant growth regulators with an antiauxin activity. In: Plant Growth Substances 1973, Proceedings of the 8th international conference on Plant Growth Substances, Hirokowa Publishing, Tokyo, Japan, pp 525–532

Hernández MN, Driss-École D (1989) Interaction of auxin-TIBA in *Lycopersicon esculentum*. J Exp Bot 40:1279–1284

Hernández MN, Driss-École D (1990) Acide, 2,3,5-triiodobenzoïque (TIBA) et initiation foliare chez *Lycopersicon esculentum*. Can J Bot 68:167–173

Holweg C, Nick P (2004) *Arabidopsis* myosin XI mutant is defective in organelle movement and polar auxin transport. PNAS 101:10488–10493

Hoshino T, Miyamoto K, Ueda J (2006) Requirement for the gravity-controlled transport of auxin for a negative gravitropic response of epicotyls in the early growth stage of etiolated pea seedlings. Plant Cell Physiol 47:1496–1508

Hoshino T, Miyamoto K, Ueda J (2007) Gravity-controlled asymmetrical transport of auxin regulates a gravitropic response in the early growth stage of etiolated pea (*Pisum sativum*) epicotyls: studies using simulated microgravity conditions on a three-dimensional clinostat and using an agravitropic mutant, *ageotropum*. J Plant Res 120:619–628

Jablanović M, Noodén LD (1974) Changes in competible IAA binding in relation to bud development in pea seedlings. Plant Cell Physiol 15:687–692

Jacobs M, Gilbert SF (1983) Basal localization of the presumptive auxin transport carrier in pea stem cells. Science 220:1297–1300

Jager CE, Symons GM, Glancy NE, Reid JB, Ross JJ (2007) Evidence that the mature leaves contribute auxin to the immature tissues of pea (*Pisum sativum* L.). Planta 226:361–368

Junghans U, Langenfeld-Heyser R, Polle A, Teichmann T (2004) Effect of auxin transport inhibitors and ethylene on the wood anatomy of poplar. Plant Biol 6:22–29

Kaldewey H, Ginkel U, Lehmann I, Seiwert R (1973) Transport and immobilization of indoleacetic acid as affected by morphactins. I. Time course of auxin transport in sections excised from different hypocotyl regions of light-grown seedlings of *Citrullus edulis*. Proc Res Inst Pomol Skierniewice Poland Ser E 3:215–226

Kampert M, Strzelczyk E, Pokojska A (1975) Production of auxins by bacteria isolated from the roots of pine seedlings *Pinus sylvestris* L. Acta Microbiol Polonica 7:135–143

Kang BG, Burg SP (1974) Red Light enhancement of the phototropic response of etiolated pea stems. Plant Physiol 53:445–448

Karabaghli-Degron C, Sotta B, Bonnet M, Gay G, Le Tacon F (1998) The auxin transport inhibitor 2,3,5-triiodobenzoic acid (TIBA) inhibits the stimulation of *in vitro* lateral root formation and the colonization of the tap-root cortex of Norway spruce (*Picea abies*) seedlings by the ectomycorrhizal fungus *Laccaria bicolor*. New Phytol 140:723–733

Katekar GF, Geissler AE (1980) Auxin transport inhibitors. IV. Evidence of a common mode of action for a proposed class of auxin transport inhibitors: the phytotropins. Plant Physiol 66:1190–1195

Katsumi M, Chiba Y, Fukuyama M (1969) The roles of the cotyledons and auxin in the adventitious root formation of hypocotyl cuttings of light-grown cucumber seedlings. Physiol Plant 22:993–1000

Khan AA (1967) Physiology and morphactins: effect on gravi- and photo-response. Physiol Plant 20:306–313

Khan AR, Andersen AS, Hansen J (1977) Morphactin and adventitious root formation in pea cuttings. Physiol Plant 39:97–100

Kochhar VK, Anand VK, Nanda KK (1972) Effect of morphactin on rooting and sprouting of buds on stem cuttings of *Salix tetrasperma*. Bot Gaz 133:361–368

Kögl F, Haagen-Smit AJ (1931) Über die Chemie des Wuchsstoffs. Proc Kon Akad Wetensch Amsterdam 34:1411–1416

Kögl F, Kostermans DGFR (1934) Hetero-auxin als Stoffwechselprodukt niederer pflanzlicher Organismen. Isolierung aus Hefe. XIII. Mitteilung. Zeit Physiol Chem 228:113–121

Kögl F, Erxleben H, Haagen Smit AJ (1933) Über ein Phytohormone der Zellstreckung. Zur Chemie des krystallisierten Auxins. V. Mitteilung. Zeit Physiol Chem 216:31–44

Kögl F, Erxleben H, Haagen-Smit AJ (1934a) Über die Isolierung der Auxine a und b aus pflanzlichen Materialien, IX Mitteilung. Zeit Physiol Chem 225:215–229

Kögl F, Haagen-Smit AJ, Erxleben H (1934b) Über ein neues Auxin ("Hetero-auxin") aus Harn. 11. Mitteilung über pflanziche Wachstumsstoffe. Zeit Physiol Chem 228:90–103

Kramer EM, Bennett MJ (2006) Auxin transport: a field in flux. Trends Plant Sci 11:382–386

Křeček P, Skůpa P, Libus J, Naramoto S, Tejos R, Friml J, Zažímalová E (2009) The PIN-FORMED (PIN) protein family of auxin transporters. Genome Biol 10:249. doi:10.1186/gb-2009-10-12-249

Krelle E, Libbert E (1968) Inhibition of the polar auxin transport by a morphactin. Planta 80:317–320

Kumar A, Sharma V, Khan M, Tripathi BN, Kumar S (2013) *Pisum sativum* wild-type and mutant stipules and those induced by an auxin transport inhibitor demonstrate the entire diversity of laminated stipules observed in angiosperms. Protoplasma 250:223–234

Kuppusamy KT, Walcher CL, Nemhauser JL (2009) Cross-regulatory mechanisms in hormone signaling. Plant Mol Biol 69:375–381

Kutschera U, Bergfeld R, Schopfer P (1987) Cooperation of epidermis and inner tissues in auxin-mediated growth of maize coleoptiles. Planta 170:168–180

Liscum E, Reed JW (2002) Genetics of Aux/IAA and ARF action in plant growth and development. Plant Mol Biol 49:387–400

Liu JH, Reid DM (1992) Adventitious rooting in hypocotyls of sunflower (*Helianthus annuus*) seedlings. IV. The role of changes in endogenous free and conjugated indole-3-acetic acid. Physiol Plant 86:285–292

Lomax TL, Muday GK, Rubery PH (1995) Auxin transport. In: Davis PJ (ed) Plant Hormones. Kluwer Academic Publishers, Dordrecht, pp 509–530

Lyons RE, Widmer RE (1984) A gibberellin-auxin synergism associated with peduncle elongation and senescence in *Cyclamen persicum* Mill. 'Swan Lake'. Bot Gaz 145:170–175

Mano Y, Nemoto K (2012) The pathway of auxin biosynthesis in plants. J Exp Bot 63:2853–2872

Marchant A, Kargul J, May ST, Muller P, Delbarre A, Perrot-Rechenmann C, Bennett MJ (1999) AUX1 regulates root gravitropism in *Arabidopsis* by facilitating auxin uptake within root apical tissues. EMBO J 18:2066–2073

Marks TR, Ford Y-Y, Cameron RWF, Goodwin C, Myers PE, Judd HL (2002) A role for polar auxin transport in rhizogenesis. Plant Cell Tiss Org Cult 70:189–198

Marumo S, Abe H, Hattori H, Munakata K (1968a) Isolation of a novel auxin, methyl 4-chloroindoleacetate from immature seeds of *Pisum sativum*. Agric Biol Chem 32:117–118

Marumo S, Hattori H, Abe H, Munakata K (1968b) Isolation of 4-chloroindolyl-3-acetic acid from immature seeds of *Pisum sativum*. Nature 219:959–960

McKay MJ, Ross JJ, Lawrence NL, Cramp RE, Beveridge CA, Reid JB (1994) Control of internode length in *Pisum sativum*. Further evidence for the involvement of indole-3-acetic acid. Plant Physiol 106:1521–1526

McRae DH, Bonner J (1953) Chemical structure and antiauxin activity. Physiol Plant 6:485–510

Meicenheimer RD (1981) Changes in epilobium phyllotaxy induced by *N*-1-naphthylphthalamic acid and α-4-chlorophenoxyisobutyric acid. Am J Bot 68:1139–1154

Michalke W, Katekar GF, Geissler AE (1992) Phytotropin-binding sites and auxin transport in *Cucurbita pepo*: evidence for two recognition sites. Planta 187:254–260

Miyamoto K, Hoshino T, Hitotsubashi R, Yamashita M, Ueda J (2005a) Automorphosis-like growth in etiolated pea seedlings is induced by the application of chemicals affecting perception of gravistimulation and its signal transduction. Adv Space Res 36:1263–1268

Miyamoto K, Hoshino T, Yamashita M, Ueda J (2005b) Automorphosis of etiolated pea seedlings in space is simulated by a three-dimensional clinostat and the application of inhibitors of auxin polar transport. Physiol Plant 123:467–474

Miyamoto K, Uheda E, Oka M, Ueda J (2011) Auxin polar transport and automorphosis in plants. Biol Sci Space 25:57–68

Miyamoto K, Marasek-Ciołakowska A, Góraj J, Węgrzynowicz-Lesiak E, Ueda J, Saniewski M (2013) Morphactin stimulates stem elongation and thickening in decapitated shoots of *Bryophyllum calycinum*. Acta Agrobot 66:21–28

Morris DA, Kadir GO, Barry AJ (1973) Auxin transport in intact pea seedlings (*Pisum sativum* L.): the inhibition of transport by 2,3,5-triiodobenzoic acid. Planta 110:173–182

Mravec J, Kubeš M, Bielach A, Gaykova V, Petrášek J, Skůpa P, Chand S, Benková E, Zažímalová E, Friml J (2008) Interaction of PIN and PGP transport mechanisms in auxin distribution-dependent development. Development 135:3345–3354

Mravec J, Skůpa P, Bailly A, Hoyerová K, Křeček P, Bielach A, Petrášek J, Zhang J, Gaykova V, Stierhof Y-D, Dobrev PI, Schwarzerová K, Rolčik J, Seifertová D, Luschnig C, Benková E, Zažímalová E, Geisler M, Friml J (2009) Subcellular homeostasis of phytohormone auxin is mediated by the ER-localized PIN5 transporter. Nature 459:1136–1140

Muday GK, Haworth P (1994) Tomato root growth, gravitropism, and lateral development: correlation with auxin transport. Plant Physiol Biochem 32:193–203

Muday GK, Brunn SA, Haworth P, Subramanian M (1993) Evidence for a single naphthylphthalamic acid binding site on the zucchini plasma membrane. Plant Physiol 103:449–456

Muscolo A, Nardi S (1997) Auxin or auxin-like activity of humic matter. In: Drozd J, Gonet SS, Senesi N, Weber J (eds) The role of humic substances in the ecosystems and in enviromental protection. Proc the 8th Meeting of the International Humic Substances Society, Wrocław, Poland, September 9–14, 1996. PTSH-Polish Soc Humic Substances. Polish Chapter of the International Humic Substances Soc, pp 987–992

Nanda KK, Bhattacharya NC, Kaur NP (1973) Effect of morphactin on peroxidases and its relationship to rooting hypocotyl cuttings of *Impatiens balsamina*. Plant Cell Physiol 14:207–211

Naqvi S (1972) The effect of morphactin on the kinetics of indol-3yl-acetic acid –2-^{14}C transport in *Zea mays* L. coleoptile segments. J Exp Bot 23:763–767

Narayanaswami R, Veerraju V (1969) IAA synthesis in paddy soil as influenced by ammonium sulphate fertilization. Curr Sci 38:517–518

Nardi S, Concheri G, Dell'Agnola G (1996) Biological activity of humus. In: Piccolo A (ed) Humic substances in terrestrial ecosystems. Elsevier Science, Amsterdam, pp 361–406

Neumann PM, Doss RP, Sachs RM (1977) A new laboratory method used for investigating the uptake, translocation and metabolism of bark banded morphactin by trees. Physiol Plant 39:248–251

Niedergang-Kamien E, Leopold AC (1957) Inhibitors of polar auxin transport. Physiol Plant 10:29–38

Nishimura T, Matano N, Morishima T, Kakinuma C, Hayashi K, Komano T, Kubo M, Hasebe M, Kasahara H, Kamiya Y, Koshiba T (2012) Identification of IAA transport inhibitors including compounds affecting cellular PIN trafficking by two chemical screening approaches using maize coleoptile systems. Plant Cell Physiol 53:1671–1682

Noh B, Murphy AS, Spalding EP (2001) *Multidrug Resistance*-like genes of *Arabidopsis* required for auxin transport and auxin-mediated development. Plant Cell 13:2441–2454

Noh B, Bandyopadhyay A, Peer WA, Spalding EP, Murphy AS (2003) Enhanced gravi- and phototropism in plant *mdr* mutants mislocalizing the auxin efflux protein PIN1. Nature 423:999–1002

Oka M, Ueda J, Miyamoto K, Okada K (1998) Activities of auxin polar transport in inflorescence axes of flower mutants of *Arabidopsis thaliana*: Relevance to flower formation and growth. J Plant Res 111:407–410

Oka M, Miyamoto K, Okada K, Ueda J (1999) Auxin polar transport and flower formation in *Arabidopsis thaliana* transformed with indoleacetamide hydrolase (*iaaH*) gene. Plant Cell Physiol 40:231–237

Okada K, Shimura Y (1994) Genetic analyses of signaling in flower development using *Arabidopsis*. Plant Mol Biol 26:1357–1377

Okada K, Ueda J, Komaki MK, Bell CJ, Shimura Y (1991) Requirement of the auxin polar transport system in early stages of *Arabidopsis* floral bud formation. Plant Cell 3:677–684

Okubo H, Uemoto S (1985) Changes in endogenous gibberellin and auxin activities during first internode elongation in tulip flower stalk. Plant Cell Physiol 26:709–719

Op den Kelder P, Benschop M, De Hertogh AA (1971) Factors affecting floral stalk elongation of flowering tulips. J Am Soc Hortic Sci 96:603–605

Paál A (1919) Über phototropische Reizeitungen. Ber Deutschen Bot Gesellschaft 32:499–502

Parker-Rhodes AF (1940) Preliminary experiments on the estimation of traces of heteroauxin in soils. J Agric Sci (Camb) 30:654–671

Parry G, Delbarre A, Marchant A, Swarup R, Napier R, Perrot-Rechenmann C, Bennett MJ (2001) Novel auxin transport inhibitors phenocopy the auxin influx carrier mutation *aux1*. Plant J 25:399–406

Parups EV (1970) Effect of morphactin on the gravimorphism and the uptake, translocation and spatial distribution of indol-3yl-acetic acid in plant tissues in relation to light and gravity. Physiol Plant 23:1176–1186

Patel JD, Setia RC (1979) Effect of morphactin-EMD 7301 on cambial activity and xylogenesis in *Sterculia urens* Roxb. Indian J Exp Biol 17:350–353

Paterson KE (1983) Polarity of regeneration in excised leaves of *Crassula argentea*. II. Auxin transport inhibitors and auxin-cytokinin interaction. Can J Bot 61:1064–1071

Peer WA, Murphy AS (2007) Flavonoids and auxin transport: modulators or regulators? Trends Plant Sci 12:556–563

Petrášek J, Mravec J, Bouchard R, Blakeslee JJ, Abas M, Seifertová D, Wiśniewska J, Tadele Z, Kubeš M, Čovanová M, Dhonukshe P, Skůpa P, Benková E, Perry L, Křeček P, Lee OR, Fink GR, Geisler M, Murphy AS, Luschnig C, Zažímalová E, Friml J (2006) PIN proteins perform a rate-limiting function in cellular auxin efflux. Science 312:914–918

Phelps JE, Saniewski M, Smoliński M, Pieniążek J, McGinnes EA Jr (1974) A note on the structure of morphactin-induced wood in two coniferous species. Wood Fiber 6:13–17

Phelps JE, McGinnes EA Jr, Pieniążek J, Saniewski M, Smoliński M (1975) A scanning electron microscope study of the structures of normal, tension and morphactin IT 3456 induced wood in *Aesculus hippocastanum*. Bull Acad Pol Sci Ser Sci Biol 23:495–497

Phelps JE, McGinnes EA Jr, Smoliński M, Saniewski M, Pieniążek J (1977) A note on the formation of compression wood induced by morphactin IT 3456 in *Thuja* shoots. Wood Fiber 8:223–227

Power JF (1994) Understanding the nutrient cycling process. J Water Soil Conserv 49:16–23

Przemeck GKH, Mattsson J, Hardtke CS, Sung ZR, Berleth T (1996) Studies on the role of the *Arabidopsis* gene *MONOPTEROS* in vascular development and plant cell axialization. Planta 200:229–237

Purushothaman D, Marimuthu T, Venkataramanan CV, Kesavan R (1974) Role of actinomycetes in the biosynthesis of indole acetic acid in soil. Curr Sci 43:413–414

Rakusová H, Gallego-Bartolomé J, Vanstraelen M, Robert HS, Alabadi D, Blázquez MA, Benková E, Friml J (2011) Polarization of PIN3-dependent auxin transport for hypocotyl gravitropic response in *Arabidopsis thaliana*. Plant J 67:817–826

Rashotte AM, Brady SR, Reed RC, Ante SJ, Muday GK (2000) Basipetal auxin transport is required for gravitropism in roots of *Arabidopsis*. Plant Physiol 122:481–490

Raven JA (1975) Transport of indoleacetic acid in plant cells in relation to pH and electrical potential gradients, and its significance for polar IAA transport. New Phytol 74:163–172

Reed RC, Brady SR, Muday GK (1998) Inhibition of auxin movement from the shoot into the root inhibits lateral root development in *Arabidopsis*. Plant Physiol 118:1369–1378

Rincón A, Priha O, Sotta B, Bonnet M, Le Tacon F (2003) Comparative effects of auxin transport inhibitors on rhizogenesis and mycorrhizal establishment of spruce seedlings inoculated with *Laccaria bicolor*. Tree Physiol 23:785–791

Robert HS, Friml J (2009) Auxin and other signals on the move in plants. Nat Chem Biol 5:325–332

Roberts JL, Roberts E (1939) Auxin production by soil microorganisms. Soil Sci 48:135–139

Robinson RW, Cantliffe DJ, Shannon S (1971) Morphactin-induced parthenocarpy in the cucumber. Science 171:1251–1252

Rossi W, Grappelli A, Pietrosanti W (1984) Phytohormones in soil after atrazine application. Folia Microbiol 29:325–329

Rothert W (1894) Über Heliotropismus. Cohn's Beitr Biol Pfl 7:1–212

Rubery PH, Sheldrake AR (1974) Carrier-mediated auxin transport. Planta 118:101–121

Ruegger M, Dewey E, Hobbie L, Brown D, Bernasconi P, Turner J, Muday G, Estelle M (1997) Reduced naphthylphthalamic acid binding in the *tir*3 mutant of *Arabidopsis* is associated with a reduction in polar auxin transport and diverse morphological defects. Plant Cell 9:745–757

Sachs J (1882) Vorlesungen über Pflanzenphysiologie. Verlag von Wilhelm Engelmann, Leiptig

Sachs J (1887) Lectures on the physiology of plants. (trans: Ward HM). Clarendon Press, Oxford

Sachs T (2000) Integrating cellular and organismic aspects of vascular differentiation. Plant Cell Physiol 41:649–656

Saniewski M, De Munk WJ (1981) Hormonal control of shoot elongation in tulips. Sci Hortic Amsterdam 15:363–372

Saniewski M, Okubo H (1997) Auxin induces stem elongation in nonprecooled and precooled derooted and rooted tulip bulbs. J Fac Agric Kyushu Univ 42:53–61

Saniewski M, Okubo H (1998a) Effects of 2,3,5-triiodobenzoic acid (TIBA) on stem growth induced by indole-3-acetic acid (IAA) and naphthylacetic acid (NAA) in precooled rooted tulip bulbs. J Fac Agric Kyushu Univ 43:11–23
Saniewski M, Okubo H (1998b) Inhibitory effect of naphthylphthalamic acid (NPA) on stem growth induced by auxin in precooled tulip bulbs. J Fac Agric Kyushu Univ 43:59–66
Saniewski M, Okubo H, Puchalski J (1999) Effect of morphactin on stem growth in relation to auxin in precooled rooted tulip bulbs. Acta Physiol Plant 21:167–174
Saniewski M, Góraj J, Węgrzynowicz-Lesiak E, Miyamoto K, Ueda J (2014) Differential effects of auxin transport inhibitors on rooting in some Crassulaceae species. Acta Agrobot 67:85–92
Santelia D, Vincenzetti V, Azzarello E, Bovet L, Fukao Y, Düchtig P, Mancuso S, Martinoia E, Geisler M (2005) MDR-like ABC transporter AtPGP4 is involved in auxin-mediated lateral root and root hair development. FEBS Lett 579:5399–5406
Santelia D, Henrichs S, Vincenzetti V, Sauer M, Bigler L, Klein M, Bailly A, Lee Y, Friml J, Geisler M, Martinoia E (2008) Flavonoids redirect PIN-mediated polar auxin fluxes during root gravitropic responses. J Biol Chem 283:31218–31226
Sawhney S, Mahajan A (1995) Morphactin-induced malmorphogenetic effect and its transmission in *Kalanchoë*. Biol Plant 37:443–447
Scanlon MJ (2003) The polar auxin transport inhibitor *N*-1-naphthylphthalamic acid disrupts leaf initiation, KNOX protein regulation, and formation of leaf margins in maize. Plant Physiol 133:597–605
Scarpella E, Marcos D, Friml J, Berleth T (2006) Control of leaf vascular patterning by polar auxin transport. Gene Dev 20:1015–1027
Schiavone FM, Cooke TJ (1987) Unusual patterns of somatic embryogenesis in the domesticated carrot: developmental effects of exogenous auxins and auxin transport inhibitors. Cell Differ 21:53–62
Schneider G (1970) Morphactins: physiology and performance. Annu Rev Plant Physiol 21:499–536
Schneider EA, Kazakoff CW, Wightman F (1985) Gas chromatography–mass spectrometry evidence for several endogenous auxins in pea seedling organs. Planta 165:232–241
Sheldrake AR (1971) The occurrence and significance of auxin in the substrate of bryophytes. New Phytol 70:519–526
Smoliński M, Saniewski M, Pieniążek J (1972) The effect of morphactin IT 3456 on cambial activity and wood differentiation in *Picea excelsa*. Bull Acad Pol Sci Ser Sci Biol 20:431–435
Smoliński M, Pieniążek J, Saniewski M (1973) Induction of compression-like wood by morphactin in vertically growing shoots of *Pinus sylvestris* L. Proc Res Inst Pomol Skierniewice Poland Ser E 3:553–558
Smoliński M, Saniewski M, Pieniążek J (1974) The suppression of tension wood formation in bent shoots of *Aesculus hippocastanum* L. by morphactin IT 3456. Bull Acad Pol Sci Ser Sci Biol 22:809–812
Snyder WE (1949) Some responses of plants to 2,3,5-triiodobenzoic acid. Plant Physiol 24:195–206
Steinmann T, Geldner N, Grebe M, Mangold S, Jackson CL, Paris S, Gälweiler L, Palme K, Jürgens G (1999) Coordinated polar localization of auxin efflux carrier PIN1 by GNOM ARF GEF. Science 286:316–318
Stewart WS, Anderson MS (1942) Auxins in some American soils. Bot Gaz 103:570–575
Strzelczyk E, Pokojska-Burdziej A (1984) Production of auxins and gibberellin-like substances by mycorrhizal fungi, bacteria and actinomycetes isolated from soil and the mycorrhizosphere of pine (*Pinus sylvestris* L.). Plant Soil 81:185–194
Strzelczyk E, Kampert M, Dahm H (1973) Production and decomposition of indoleacetic acid (IAA) by microorganisms isolated from the root zone of two crop plants. Acta Microbiol Polonica Ser B 5:71–79
Strzelczyk E, Dahm H, Pachlewski R, Różycki R (1977) Production of indole compounds by the ectomycorrhizal fungus *Cantharellus cibarius* Fr. Pedobiologia 41:402–411

Sukumar P, Edwards KS, Rahman A, DeLong A, Muday GK (2009) PINOID kinase regulates root gravitropism through modulation of PIN2-dependent basipetal auxin transport in *Arabidopsis*. Plant Physiol 150:722–735

Sundberg B, Tuominen H, Little CHA (1994) Effects of the indole-3-acetic acid (IAA) transport inhibitors *N*-1-naphthylphthalamic acid and morphactin on endogenous IAA dynamics in relation to compression wood formation in 1-year-old *Pinus sylvestris* (L.) shoots. Plant Physiol 106:469–476

Sussman MR, Goldsmith MHM (1981) Auxin uptake and action of *N*-1-naphthylphthalamic acid in corn coleoptiles. Planta 150:15–25

Swarup R, Friml J, Marchant A, Ljung K, Sandberg G, Palme K, Bennett M (2001) Localization of the auxin permease AUX1 suggests two functionally distinct hormone transport pathways operate in the *Arabidopsis* root apex. Gene Dev 15:2648–2653

Swarup R, Kramer EM, Perry P, Knox K, Leyser HMO, Haseloff J, Beemster GTS, Bhalerao R, Bennett MJ (2005) Root gravitropism requires lateral root cap and epidermal cells for transport and response to a mobile auxin signal. Nat Cell Biol 7:1057–1065

Szajdak L, Maryganova V (2007) Occurrence of IAA auxin in some organic soils. Agron Res 5:175–187

Tamimi S, Firn RD (1985) The basipetal auxin transport system and the control of cell elongation in hypocotyls. J Exp Bot 36:955–962

Tatematsu K, Kumagai S, Muto H, Sato A, Watahiki MK, Harper RM, Liscum E, Yamamoto KT (2004) *MASSUGU2* encodes Aux/IAA19, an auxin-regulated protein that functions together with the transcriptional activator NPH4/ARF7 to regulate differential growth responses of hypocotyl and formation of lateral roots in *Arabidopsis thaliana*. Plant Cell 16:379–393

Thein M, Michalke W (1988) Bisulfite interacts with binding sites of the auxin-transport inhibitor *N*-1-naphthylphthalamic acid. Planta 176:343–350

Thimann KV (1935) On the plant growth hormone produced by *Rhizopus suinus*. J Biol Chem 109:279–291

Thimann KV (1977) Polarity and the transport of auxin. In: Hormone Action in the Whole Life of Plants. The University of Massachusetts Press, Amherst, pp 71–92

Thimann KV, Bonner WD Jr (1948) The action of triiodobenzoic acid on growth. Plant Physiol 23:158–161

Thomson K-S, Leopold AC (1974) *In vitro* binding of morphactins and 1-*N*-naphthylphthalamic acid in corn coleoptiles and their effects on auxin transport. Planta 115:259–270

Thomson K-S, Hertel R, Müller S, Tavares JE (1973) 1-*N*-naphthylphthalamic acid and 2,3,5-triiodobenzoic acid. In vitro binding to particulate cell fractions and action on auxin transport in corn coleoptiles. Planta 109:337–352

Tognoni F, Alpi A (1969) Morphactins, auxin transport and apical dominance in *Pisum sativum*. Ber Deutsch Bot Ges N.F. 3:53–60

Tukey HB, Went FW, Muir RM, van Overbeek J (1954) Nomenclature of chemical plant growth regulators. Report by a committee of the American Society of Plant Physiologists. Plant Physiol 29:307–308

Ueda J, Okada K, Komaki MK, Kato J, Shimura Y (1992) Auxin polar transport system and flower formation in *Arabidopsis*. Proc Nineteenth Annu Meet Plant Growth Regul Soc Am 19:175–180

Ueda J, Miyamoto K, Yuda T, Hoshino T, Fujii S, Mukai C, Kamigaichi S, Aizawa S, Yoshizaki I, Shimazu T, Fukui K (1999) Growth and development, and auxin polar transport in higher plants under microgravity conditions in space: BRIC-AUX on STS-95 space experiment. J Plant Res 112:487–492

Ueda J, Miyamoto K, Yuda T, Hoshino T, Sato K, Fujii S, Kamigaichi S, Izumi R, Ishioka N, Aizawa S, Yoshizaki I, Shimazu T, Fukui K (2000) STS-95 space experiment for plant growth and development, and auxin polar transport. Biol Sci Space 14:47–57

Ueda J, Miyamoto K, Uheda E, Oka M (2011) Auxin transport and a graviresponse in plants: relevance to ABC proteins. Biol Sci Space 25:69–75

Ueda J, Góraj J, Węgrzynowicz-Lesiak E, Miyamoto K, Saniewski M (2012) Morphactin substantially induced the fourth internode growth in decapitated tulips: relevance to endogenous levels of indole-3-acetic acid. J Fruit Ornamental Plant Res 20:161–171
Ueda J, Toda Y, Kato K, Kuroda Y, Arai T, Hasegawa T, Shigemori H, Hasegawa K, Kitagawa J, Miyamoto K, Uheda E (2013) Identification of dehydrocostus lactone and 4-hydroxy-ß-thujone as auxin polar transport inhibitors. Acta Physiol Plant 35:2251–2258
van der Weij HG (1932) Der Mechanismus des Wuchsstofftranportes. Recueil Trav Bot Néerl 29:379–496
van der Weij HG (1934) Der Mechanismus des Wuchsstofftransportes. II. Recueil Trav Bot Néerl 31:810–857
van Overbeek J, Blondeau R, Horne V (1951) Trans-cinnamic acid as an anti-auxin. Am J Bot 38:589–595
van Zeist W, Koevoets TCM (1951) *Kalanchoe blossfeldiana* treated with 2.3.5-triiodobenzoic acid (TIBA). Proc Kon Ned Akad Wetensch 54:126–132
Vieten A, Sauer M, Brewer PB, Friml J (2007) Molecular and cellular aspects of auxin-transport-mediated development. Trends Plant Sci 12:160–168
Wasternack C (2007) Jasmonates: An update on biosynthesis, signal transduction and action in plant stress response, growth and development. Ann Bot 100:681–697
Wasternack C, Hause B (2013) Jasmonates: biosynthesis, perception, signal transduction and action in plant stress response, growth and development. An update to the 2007 review in Annals of Botany. Ann Bot 111:1021–1058
Watanabe K, Ohno N, Yoshioka H, Gershenzon J, Mabry TJ (1982) Sesquiterpene lactones and diterpenoids from *Helianthus argophyllus*. Phytochemistry 21:709–713
Went FW (1928a) Wuchsstoff und Wachstum. Recueil Trav Bot Néerl 25:1–116
Went FW (1928b) Die Erklärung des phototropischen Krümmungsverlaufs. Recueil des Trav Bot Néerl 25A:483–489
Went FW, Thimann KV (1937) Phytohormones. The Macmillan Co, New York
Westing AH (1965) Formation and function of compression wood in gymnosperms. Bot Rev 31:381–480
Westing AH (1968) Formation and function of compression wood in gymnosperms. II. Bot Rev 34:51–78
Whippo CW, Hangarter RP (2006) Phototropism: bending towards enlightenment. Plant Cell 18:1110–1119
Whitehead DC (1963) Some aspects of the influence of organic matter on soil fertility. Soils Fertil 26:217–223
Wu G, Lewis DR, Spalding EP (2007) Mutations in *Arabidopsis multidrug resistance-like* ABC transporters separate the roles of acropetal and basipetal auxin transport in lateral root development. Plant Cell 19:1826–1837
Wu G, Cameron JN, Ljung K, Spalding EP (2010) A role for ABCB19-midiated polar auxin transport in seedling photomorphogenesis mediated by cryptochrome 1 and pytochrome B. Plant J 62:179–191
Yamada K, Nakano H, Yokotani-Tomita K, Bruinsma J, Yamamura S, Hasegawa K (2000) Repetition of the classical Boysen-Jensen and Nielsen's experiment on phototropism of oat coleoptiles. J Plant Physiol 156:323–329
Yamaguchi K, Itoh T, Shimaji K (1980) Compression wood induced by 1-*N*-naphthylphthalamic acid (NPA), an IAA transport inhibitor. Wood Sci Technol 14:181–185
Yamaguchi K, Shimaji K, Itoh T (1983) Simultaneous inhibition and induction of compression wood formation by morphactin in artificially inclined stems of Japanese larch (*Larix leptolepis* Gordon). Wood Sci Technol 17:81–89
Yang T, Law DM, Davies PJ (1993) Magnitude and kinetics of stem elongation induced by exogenous indole-3-acetic acid in intact light-grown pea seedlings. Plant Physiol 102:717–724
Young LM, Evans ML (1996) Patterns of auxin and abscisic acid movement in the tips of gravistimulated primary roots of maize. Plant Growth Regul 20:253–258
Zazimalova E, Napier RM (2003) Points of regulation for auxin action. Plant Cell Rep 21:625–634

Chapter 9
Transformations of Organic Matter in Soils Under Shelterbelts of Different Ages in Agricultural Landscape

Lech Wojciech Szajdak, Victoria Maryganova, Eugene Skakovskii, and Ludmila Tychinskaya

Abstract In this review we have summarized the results of long-term investigations on the transformations of soil organic matter (SOM) in soils under shelterbelts of different ages located in the Agroecological Landscape Park in Turew (40 km south of Poznań, West Polish Lowland). The first shelterbelts were planted in Turew ~ 200 years ago. The youngest shelterbelt was created in 1993. Here we present the data on the effect of the age of these shelterbelts on some biologically active substances in soil, such as phytohormone indole-3-acidic acid (IAA), free-extractable lipids, and especially humic acids (HA), their chemical structure, and hydrophobic–hydrophilic properties. The conversion from arable cropping to shelterbelts not only influenced the accumulation of SOM and biologically active substances in soils but also changed the composition, structure, and stability of free-extractable lipids and HA with the age of shelterbelts being the principal factor. The SOM under the old shelterbelt during ~ 200 years has undergone the most significant biochemical and chemical transformations (oxidation, hydrolysis, polymerization) and advanced stages of humification with the accumulation of resistant compounds in humic substances (HS) and lipids and destruction of some anthropogenic contaminants.

L.W. Szajdak (✉)
Institute for Agricultural and Forest Environment, Polish Academy of Sciences, Str. Bukowska 19, 60-809 Poznań, Poland
e-mail: lech.szajdak@isrl.poznan.pl

V. Maryganova
Institute for Nature Management, National Academy of Sciences of Belarus, Str. F. Skorina 10, 220114 Minsk, Belarus
e-mail: vmaryganova@mail.ru

E. Skakovskii • L. Tychinskaya
Institute of Physico-Organic Chemistry, National Academy of Sciences of Belarus, Str. Surganova 13, 220072 Minsk, Belarus
e-mail: spektr@ifoch.bas-net.by

L.W. Szajdak (ed.), *Bioactive Compounds in Agricultural Soils*,
DOI 10.1007/978-3-319-43107-9_9

Keywords Shelterbelts • Organic matter • Molecular weight • ^{13}C NMR spectroscopy • IR and Vis spectroscopy • Hydrophobic interaction chromatography • Thermal analyses • Auxin • Lipids

9.1 Introduction

The simplification of the structure of agroecosystems connected with the intensification of agricultural production increases the hazards of leaching, wind and water erosion, and volatilization of chemical substances from soil (Ryszkowski et al. 1997; Szajdak and Ryszkowski 2002). One of the methods of controlling circulation of substances in agricultural landscape with a high level of fertilization in cultivated fields is the creation of biogeochemical barriers in the form of shelterbelts. Shelterbelts or midfield afforestations planted on agricultural land belong to the stable elements in the landscape, which:

- Restrain soil erosion.
- Separate agricultural fields from the watercourses.
- Improve microclimate for agricultural production.
- Regulate water regime in soils.
- Decrease the concentration of many chemical compounds migrating with groundwater outflow from adjoining cultivated fields (Peterjohn and Correll 1984; Ryszkowski and Bartoszewicz 1989; Cooper 1990; Bartoszewicz and Ryszkowski 1996; Prusinkiewicz et al. 1996; Ryszkowski et al. 1997, 2002; Bartoszewicz 2000; Szajdak and Matuszewska 2000; Szajdak and Życzyńska-Bałoniak 2002; Szajdak 2011).

Shelterbelts are introduced to the global carbon cycle because of their capacity to mitigate the greenhouse effect by sequestration of carbon (C) and nitrogen (N) (Post and Kwon 2000; Szajdak 2002; Szajdak and Ryszkowski 2002; Paul et al. 2002; Poulton et al. 2003; Zhang Bin and Peng Xin-Hua 2006; Jandl et al. 2007).

SOM is a critical component of soil–plant ecosystem, and it changes with land use or agricultural management practices. The response of soils to their removal from agricultural use in most cases is an increase in the content of organic matter (OM), which is primarily related to the cessation of the removal of organic material with the crop yields and its accumulation within the soil profile (Post and Kwon 2000; Poulton et al. 2003; Zhang Bin and Peng Xin-Hua 2006; Jandl et al. 2007). The afforestation of former agricultural land in most cases increases the content of SOM because input of OM with crop residues for arable soils is usually lower compared with litter input for forest soils (Kögel-Knabner 2002; Lützow et al. 2006). Accumulation of soil C pool occurs until the soil reaches a new equilibrium between C input (litterfall, rhizodeposition) and C output (respiration, leaching).

SOM is composed of many organic substances of various chemical compositions, structures, and stabilities. Dissolved organic matter (DOM) is included in the labile pool of SOM; HA and lipids represent the most recalcitrant and stable reservoir of organic C (Lützow et al. 2006). The chemistry of SOM is influenced by litter quality, the intensity of litter decomposition, and the related production and accumulation of microbially derived substances. Changes in land use not only influence the total soil organic carbon (SOC) and N stocks in soils but can also alter the structure and stability of SOM fractions.

Although DOM represents a small part of SOM, it plays an important role in the transport, mineralization, and stabilization of organic matter in soils and influences nutrient availability, the mobility, and toxicity of metals and organic pollutants (Kalbitz et al. 2000; Chantigny 2003). The primary sources of DOM are leaching of substances from fresh litter and the products of plant residue decomposition. DOM is an important substrate for microbial growth. Its production and consumption are dependent mainly on microbial activities and the equilibrium with the solid phase of SOM (Zsolnay 1996; Chantigny 2003).

Land use may significantly influence the amount and the composition of DOM in soil (Chantigny 2003). Afforestation of agricultural soils leads to a significant increase in dissolved organic carbon (DOC) concentration (Zsolnay 1996; Chantigny 2003) and induces a gradual qualitative change in DOM, with significantly increasing C/N ratios in soil solution over time (Rosenqvist et al. 2010). DOC concentrations and fluxes with throughfall were strongly positively correlated with tree height and stand age, while dissolved organic nitrogen (DON) showed no such trends, suggesting different origins of DOC and DON in throughfall (Rosenqvist et al. 2010).

Plant hormones called phytohormones or plant growth regulators represent a wide group of organic substances (auxins, gibberellins, cytokinins, abscisic acid, and ethylene) which are biosynthesized in soils, plants, and sediments by microorganisms (fungi, bacteria, actinomycetes, algae) and translocated to another part of the plant or environment to impact on a wide range of physiological and development processes at low quantities.

One of the most important phytohormones is auxin IAA. IAA is formed in soils from tryptophan by enzymatic conversion (Sarwar et al. 1992; Martens and Frankenberger 1993). A principal feature of IAA is its ability to affect growth, development, and health of plants (Dahm et al. 1997) by activating root morphology and metabolic changes in the host plant (Bandurski and Schulze 1977). The physiological impact of this substance is involved in cell elongation, apical dominance, root initiation, parthenocarpy, abscission, callus formation, and respiration (Tena et al. 1986; Strzelczyk et al. 1997).

The major and the most important part of SOM consists of HS which comprise up to 70–80 % of SOM in mineral soil. HS are considered to be a complex heterogeneous and polydisperse system of various macromolecules of amphiphilic nature, containing different functional groups and aromatic and aliphatic structural units, and produced by biological and chemical degradation of organic residues in the natural environment. The properties of HS in the environment involve their structural characteristics and composition, which are controlled by the process of humification of organic matter and are dependent on soil type, vegetation, and climatic conditions (Orlov 1990; Kögel-Knabner 1993; Stevenson 1994; Zech et al. 1997; Quideau et al. 2001).

Amphiphilic character and complex chemical composition of HS are responsible for their solubility, conformation, and susceptibility to biodegradation and enable them to interact with a wide variety of organic and inorganic compounds through chemical bonding and hydrophobic interactions. Because of that, HS have a significant influence on the biogeochemical cycles of both natural and anthropogenic substances (Weber 1988; Stevenson 1994; Salloum et al. 2002). Hydrophilic fractions are the most reactive and mobile components of HS. Hydrophobic components of soil humus are stable and responsible for the formation of hydrophobic surfaces, reactions of hydrophobic interactions, the accumulative type of humus profile, and stability of soil aggregates (Piccolo and Mbagwu 1999; Milanowskii and Shein 2002).

A change in land use from agriculture to forestry affects processes of humification and mineralization in soil and thereby the content and chemical composition of HS (Alriksson and Olsson 1995). At present, investigations of the transformations of HS in soils under shelterbelts are scarce. Szajdak et al. (2002) investigated transformations of inorganic and organic forms of nitrogen and chemical structure of HA in two soils (mineral and mineral–organic) under a shelterbelt located between cultivated field and a pond in agricultural landscape. HA from mineral–organic soil were characterized by higher degree of humification than HA from mineral soil. For both kinds of soils, an increase in the distance from the edge of the shelterbelt was accompanied by a decrease in the degree of humification of HA, though the content of inorganic and organic forms of nitrogen in mineral soil decreased with the distance from the edge of the shelterbelt while those in mineral–organic soil increased in this direction.

The study of the changes in HS in a Norway spruce chronosequence (18–91 years) on former arable land showed that the HA and fulvic acid (FA) contents in the organic horizons increased significantly with afforestation age up to 73 years and then remained constant or even decreased (Cerli et al. 2008). The accumulation of alkyl C moieties in the HA, as shown by the elemental composition and the ^{13}C NMR spectra, increased with stand age from the beginning of afforestation due to selective preservation of stable compounds from resins, waxes, and other lipids and resynthesis of aliphatic microbial products. The aromatic C content in the HA increased up to 64 years and then decreased, while lignin-derived phenolic structures in the HA increased up to 29 years and then decreased.

One of the important constituents of SOM is the lipid fraction which includes organic compounds insoluble in water but soluble in common organic solvents such as hexane, benzene, chloroform, ethanol, etc. (Bergman 1963; Dinel et al. 1990; Stevenson 1994). Solvent-extractable lipids constitute a small portion of the total SOM (Fridland 1982; Stevenson 1994). Nevertheless, they play an important role in transformations of SOM and formation of stable SOM fractions (Fridland 1982; Stevenson 1994; Lorenz et al. 2007). The content and composition of lipids are most sensitive to changes in biochemical processes in soil. The lipid fraction of SOM can serve as a diagnostic index of soil biological activity (Fridland 1982; Bull et al. 2000). Also lipids are very important for such soil features as aggregate stability and water retention (Dinel et al. 1990; Amblès et al. 1993; de Blas et al. 2010). Lipids naturally present in soils were found to compete for or block sorption sites on the organic matter and thereby reduce the sorption affinity of the soils for aromatic sorbates (Chilom et al. 2005; Wang and Xing 2007; Ahangar et al. 2009). Besides, data regarding the content and composition of lipids can be useful for determination of some organic pollutants in soils (Fridland 1982; Kiang and Grob 1986).

Lipids of plant and microbial residues contribute mainly to soil lipids with plant lipids being predominant (Fridland 1982; Dinel et al. 1990; van Bergen et al. 1997). The content and composition of soil lipids depend mainly on the amount and composition of plant residues, as well as on the environmental conditions of their formation (Fridland 1982; Miller and Donahue 1995). Conversion from arable cropping to a forested soil leads to changes in the amount and composition of plant residues and therefore to changes in the content and composition of soil lipids. No literature exists on the concentration and composition of the lipid fraction in the soils under shelterbelts and the effect of the age of shelterbelts on these parameters.

In this review we have summarized the results of long-term investigations on the transformations of OM in soils under shelterbelts of different ages located in the Agroecological Landscape Park in Turew (40 km south of Poznań, West Polish Lowland). The first shelterbelts were planted in Turew due to the initiative of General Dezydery Chłapowski in the beginning of the nineteenth century. The youngest shelterbelt was created in 1993. Here we present the data on the effect of the age of these shelterbelts on some biologically active substances in soil, such as phytohormone IAA, free-extractable lipids, and especially HA, their chemical structure, and hydrophobic–hydrophilic properties, for better understanding of their role in functioning shelterbelts as biogeochemical barriers in agricultural landscape.

9.2 Materials and Methods

9.2.1 Soil Samples

Investigations were carried out on soils under two shelterbelts (the first was 14 years old (young) and the second was ~200 years old (old)) and adjoining cultivated fields located in the Agroecological Landscape Park in Turew (40 km south of Poznań, West Polish Lowland). Intensive agriculture is observed in this region. Cultivated fields are represented by 70 %, 12 % meadows, and 14 % shelterbelts. The shelterbelts and cultivated fields were introduced on Hapludalfs soils (Soil Taxonomy 1968). The old shelterbelt consists mainly of *Robinia pseudoacacia* with admixture of *Quercus pentraea* and *Quercus robur*. It is 30 m wide and 1200 m long. The young shelterbelt is composed of *Quercus pentraea*, *Quercus robur*, *Larix deciduas*,

Pinus silvestris, *Sorbus aucuparia*, *Sorbus intermedia*, *Tilia cordata*, and some others, totaling 24 tree species. This one is 30 m wide and 400 m long. Soils of both adjoining fields were under continuous rye cropping. Soil samples were taken from the upper 20 cm of soils (humus horizon) in the middle of the shelterbelts areas (No. 1) and from adjoining cultivated fields 100 m from the shelterbelts (No. 2) during the whole vegetative growth season of 2007. For additional extraction of HS, soil samples were taken from soils under shelterbelts of different ages and adjoining arable fields also in 2007, when the young shelterbelt was 14 years old.

Total organic carbon (TOC) was analyzed on a carbon analyzer TOC 5050A with Solid Sample Module (SSM-5000A) produced by Shimadzu, Japan. For the estimation of DOC, soil samples in redistilled water were heated at 100 °C for 2 h under reflux condenser. Extracts were separated by a mean filter paper (Whatman qualitative filter paper, pore size 11 μ) and analyzed on a carbon analyzer TOC 5050A (Shimadzu, Japan) (Smolander and Kitunen 2002). Total nitrogen was estimated by Kjeldahl method and ammonium and nitrate ions by Spurwaya method (Ryszkowski et al. 2002). Soil pH was assayed by potentiometric titration in 1 N KCl (1:2.5, v/v).

9.2.2 Determination of IAA in Soils

The method of the determination of IAA is described in Chap. 10.

All experiments were done in triplicate and the results were averaged. All chemicals used in this study were of analytical grade. The precision based on replicate analyses was ±0.01 for pH measurements, ±3 % for TOC, ±3 % for DOC, ±4 % for IAA, ±3 % for total nitrogen, and ±4.5 % for ammonium and nitrates.

9.2.3 Extraction and Purification of Humic Substances

Dilute aqueous alkali is an efficient extractant removing most of organic matter from mineral soils (Kononova 1966; Schnitzer and Khan 1978; Stevenson 1994). Because of that, HA extracted with 0.1 M NaOH are suggested to be the most representative part of soil humus. Neutral solution of sodium pyrophosphate is considered as a milder and more selective extractant compared to sodium hydroxide, though the amount of SOM recovered is considerably less than with dilute alkali, but less alteration occurs. HS extracted with 0.1 M $Na_4P_2O_7$ at pH 7 are thought to be a more labile part of soil organic matter, representing the young newly formed organic matter in a greater degree than alkaline solution (Piccolo and Mirabella 1987; Hayes 1991; Stevenson 1994).

HS from air-dried soil samples were extracted separately with 0.1 M NaOH and 0.1 M $Na_4P_2O_7$ at pH 7 using an extractant/soil ratio of 5:1 under an N_2 atmosphere at room temperature. The system was shaken for 4 h and after that stored for 20 h. The dark-colored supernatant solutions after separation from the residual soils by centrifugation (4000 g for 30 min) were adjusted to pH 1.3 with 6 M HCl and allowed to stand for 24 h at room temperature for the coagulation of the HA fractions which were separated by centrifugation. The purification of the HA was performed using the following method. The HA fractions were dissolved in 400–500 ml of distilled water and adjusted to pH 7. The solutions were centrifuged at 6.000 g at 24 °C for 1 h to separate the clay, adjusted to pH 1.35, and centrifuged after 24 h of storage at room temperature. The procedure was repeated three times. The finally precipitated HA were freeze-dried and stored in a vacuum desiccator over P_2O_5 (Szajdak and Maryganova 2003).

HA from the young shelterbelt and adjoining cultivated field were referred to as HA-1 young and HA-2 young; HA from the 200-year-old shelterbelt and adjoining field were referred to as HA-1 old and HA-2 old, respectively.

9.2.4 Characterization of Humic Substances

A detailed characterization of HS and their environmental processes requires the application of different instrumental techniques such as size exclusion chromatography (SEC), thermal analysis (TA), ultraviolet–visible (UV–Vis), infrared (IR), and ^{13}C nuclear magnetic resonance (^{13}C NMR) spectroscopy.

9.2.4.1 Determination of Molecular Weight

SEC (often referred to in the past as gel permeation chromatography) is the most convenient and widely used technique for the determination of molecular weight distributions in HS. SEC separates dissolved macromolecules according to their size or, more specifically, their hydrodynamic volume. Molecular weights are typically estimated with the use of some standards, assuming that the relationship between size and molecular weight is the same for the standards and samples being analyzed (Swift and Posner 1971; Kumada and Miyara 1973; Goh and Williams 1979; Mori et al. 1987; De Nobile and Chen 1999).

The molecular weight distribution of HA was estimated by using SEC on Sephadex G-100, superfine (Pharmacia, Sweden). Sephadex was swollen in 0.1 M Tris-HCl buffer at pH 9.0 during 3 days, packed in a 1.6×36 cm column to give an effective gel bed, and eluted upward with 0.1 M Tris-HCl buffer at pH 9.0 under gravity. A 2 mL of 0.2 % humic compound solution was pumped onto the gel at the column flow-rate of 0.8 mL/min via a K-PA-digital (Ika Labortechnik, Germany). Eluted material was collected in 3 mL fractions and the absorbance of each fraction was monitored at $\lambda = 340$ nm using a Specoll 11 colorimeter (Carl Zeiss Jena, Germany) in 1 cm cells. The void (exclusion) volume of the column was determined using blue dextran 2000 (Pharmacia, Sweden). The total effective column volume of the gel material was determined using potassium dichromate solution. For the estimation of weight-average apparent molecular weights of fractions, a calibration of the column with globular proteins was used. The term "apparent" is used here because the MW measured for HS macromolecules using this technique may differ variously from their actual MW due to the differences in configuration and chemistry between HS and globular proteins. The relative content of each fraction of humic compounds was estimated from the areas under corresponding peaks on the curves of their molecular weight distribution (Szajdak and Maryganova 2003).

9.2.4.2 Thermal Analyses

TA techniques are powerful analytical tools in the investigation of the thermal properties of such complex materials as HS, their chemical structure, and the humification process of SOM. These rapid and accurate methods require little sample preparation and are widely used for the study of HS (Shurygina 1971; Schnitzer et al. 1974; Dziadowiec 1979; Leiweber an Schulten 1992; Gonet and Wegner 1993; Francioso et al. 2003).

Thermal properties of HA extracted with 0.1 M NaOH were investigated using an OD-103-derivatograph (MOM-Paulink-Paulink-Erdey, Hungary). The HA samples were placed in platinum crucibles and heated from 20 to 700 °C at a heating rate of 10°/min with Al_2O_3 as a referent substance. The curves of differential thermal analysis (DTA), thermogravimetry (TG), and differential thermogravimetry (DTG) were recorded simultaneously. Weight losses at different steps of thermal decomposition were calculated from the TG curves.

9.2.4.3 Spectroscopic Methods

Spectroscopic methods are very informative and popular in the study of HS. Using UV–Vis spectroscopy, some classical parameters such as the absorbance at 465 nm (E_4) and the E_4/E_6 ratio can be determined. They are useful to estimate the development of aromatic condensation and polyconjugation in humic macromolecules and, hence, the degree of humification of HS of various origins (Kononova 1966; Chen et al. 1977; Orlov 1990).

The E_4/E_6 ratios were determined by dissolving 3 mg of HA in 10 mL of 0.05 M $NaHCO_3$ (pH ± 9.0) and measuring optical densities at $\lambda = 465$ nm (E_4) and $\lambda = 665$ nm (E_6) on BECKMAN DU®-68 spectrophotometer with 1 cm thickness of layer, USA (Chen et al. 1977).

IR spectroscopy can provide considerable information on the nature, reactivity, and structural arrangements of different functional groups in HS (Bloom and Leenheer 1989; Orlov 1990; Stevenson 1994). IR spectra of HA were run in pellets with KBr on a Specord-80 spectrometer (Germany) within 4000 and 400 cm^{-1}.

^{13}C NMR spectroscopy is one of the most powerful techniques for studying the chemical structure of HS. Using this technique it is possible to obtain the direct information on the carbon skeleton of HS (Bloom and Leenheer 1989; Preston

1996; Kögel-Knabner 1997; Mao et al. 2000), as well as to calculate the degrees of their hydrophobicity and hydrophilicity (Piccolo et al. 1999; Spaccini et al. 2006).

Solution ^{13}C NMR spectra were recorded in 0.5 M NaOD on a modified Tesla BS587A spectrometer (Czech Republic) at the frequency of 20.182 MHz in a "quantitative" regime without nuclear Overhauser enhancement with interpulse delay >5 T_1, where T_1 is the time of spin–lattice relaxation. Relative content of different types of C was calculated with an integrator.

^{13}C NMR spectra of HA were divided into five chemical shift regions. In the alkyl C region (0–45 ppm), almost all saturated hydrocarbons, as well as alkyl groups of amino acids and alkyl-substituted aromatics, resonate. O,N-Alkyl carbons (45–65 ppm) are associated with carbohydrates, methoxyls of lignins, and amino acids. Carbons resonating at 65–108 ppm (O-alkyl) are representative mainly of the contribution of carbohydrates and carbohydrate-like components, even though carbons belonging to different structural groups resonate in this region. The region between 108 and 165 ppm is characterized by aromatic C. Signals in the region 165–200 ppm are assigned with carboxyl and other carbonyl-related C (165–200 ppm) (Preston 1996). Areas of alkyl and aromatic carbons were attributed to hydrophobic carbons, whereas those of O- and N-alkyl carbons and carbonyl-related carbons represented hydrophilic carbons (Piccolo et al. 1999). The percentages of hydrophobic carbons (HB) and hydrophilic carbons (HI) were used to calculate the degrees of hydrophobicity (HB/HI) for all HA (Spaccini et al. 2006).

9.2.4.4 Hydrophobic Interaction Chromatography

Hydrophobic interaction chromatography (HIC) is a perspective method for the study of amphiphilic properties of HS (Blondeau and Kalinowski 1986; Milanowskii 2000; Milanowskii and Shein 2002). This method includes the use of hydrophobic interactions between hydrophobic centers of the investigating substance and hydrophobic ligands attached to the uncharged gel matrix. The most hydrophilic components of the sample do not adsorb on the gel and are eluted with a starting buffer. The adsorbed components are fractionated on the basis of successively reducing the strength of their hydrophobic interactions with hydrophobic matrix.

HIC of the HA samples was carried out on a column (13 × 1.6 cm) packed with Octyl-Sepharose 4 Fast Flow (Farmacia, Sweden) in 0.05 M Tris-HCl buffer at pH 8.0 containing 3 M NaCl. The fractionation was performed using step-by-step elution with the following sequence of eluents: (i) 0.05 M Tris-HCl buffer at pH 8.0 containing 3 M NaCl, (ii) 0.05 M Tris-HCl buffer at pH = 8.0, (iii) distilled water, and (iv) 0.1 M NaOH solution. The relative contents of the chromatographic fractions of the HA samples were estimated from the areas under corresponding peaks on the chromatograms (Maryganova et al. 2004).

9.2.5 *Extraction and Characterization of Soil Lipids*

Lipids were extracted from soil samples with *n*-hexane at a ratio of soil/hexane 1:6 in a Soxhlet apparatus for 10 h. After removal of *n*-hexane under a nitrogen stream, residues were weighted, and the total lipid contents were calculated as weight fractions of extractable lipids (g) per kg soil samples.

For NMR analyses lipids were dissolved in $CDCl_3$. ^1H and ^{13}C NMR spectra were recorded on an AVANCE 500 spectrometer (Bruker, Germany) at the frequencies 500 and 126 MHz for ^1H and ^{13}C, respectively, in standard 5 mm tubes. The proton spectra were acquired using a pulse angle of 30°, a relaxation delay of 5 s and 128 scans. The ^{13}C spectra were acquired with a pulse angle of 45°, a relaxation delay of 3 s and 12,000 scans. Chemical shifts of lines in the proton spectra were referenced to $CHCl_3$ (a touch of solvent) signal ($\delta = 7.27$ ppm), and for ^{13}C NMR spectra the signal of $CDCl_3$ ($\delta = 77.7$ ppm) was used. For the signal identification, the method of distortionless enhancement by polarization transfer (DEPT) was used (Derome 1987). Also spectra of individual compounds (hydrocarbons, fatty acids, aldehydes and alcohols, resin acids, steroids, ethers, and esters including mono-, di-, and triglycerides of fatty acids) were recorded in identical conditions.

9.3 Conclusions

9.3.1 *Chemical Properties of Soils*

Among the most important factors affecting the accumulation of C in soils under afforestations are their age, the productivity and composition of plant communities, and physical and biological conditions in soils. A soil under afforestation of a higher age usually contains higher amounts of SOM and organic C and N (Paul et al. 2002; Szajdak 2002; Dell'Abate et al. 2002; Poulton et al. 2003; Jandl et al. 2007; De Marco et al. 2013). Old-growth forests have the highest C density, whereas younger stands have a larger C sink capacity. Stand productivity declines significantly in mature forest stands. However, even very old forests can sequester large amounts of C (Knohl et al. 2003). Tree species also affect the C storage of the ecosystem. Trees with a high wood density (deciduous tree species) accumulate more C than trees with light wood (coniferous species) (Jandl et al. 2007). The increase in soil C was greater under plantations of N_2-fixing species than in plantations of non-N_2-fixing species (Paul et al. 2002; Resh et al. 2002). The biomass production of a mixed stand was higher than that for pure stands (Resh et al. 2002; Jandl et al. 2007).

As shown in Table 9.1, all the soils investigated were acidic throughout the vegetative growth season. The pH values of the soil under the young shelterbelt had minor differences with those of adjoining cultivated field (3.73–5.25 and 3.72–4.15, respectively), whereas the soil under the old shelterbelt was much more acidic (pH 3.19–3.51) compared to the corresponding cultivated soil which was slightly acidic

Table 9.1 Chemical properties of soils under shelterbelts of different ages and adjoining arable fields (Szajdak and Maryganova 2009)

	Fields adjoining shelterbelts		Shelterbelts	
Soil chemical property	14 years old	200 years old	14 years old	200 years old
pH (1 N KCl)	3.9 (3.72–4.15)	5.3 (5.03–5.65)	4.5 (3.73–5.25)	3.4 (3.19–3,51
Organic carbon (gC kg^{-1})	4.6 (4.0–5.6)	4.5 (3.8–5.9)	8.3 (5.1–11.6)	55.2 (38.7–76.3)
Dissolved organic carbon (gC kg^{-1})	0.22 (0.16–0.29)	0.22 (0.18–0.29)	0.40 (0.22–0.76)	2.47 (1.5–3.31)
N-total (g kg^{-1})	0.7 (0.6–1.1)	1.1 (0.5–1.9)	0.8 (0.5–1.2)	3.3 (1.6–4.9)
N-NO_3^- (mg kg^{-1}) of soil	27.8 (11.7–43.3)	13.9 (5.4–20.3)	13.6 (6.66–24.2)	310.1 (140.0–701.3)
N-NH_4^+ (mg kg^{-1}) of soil	9.2 (3.6–19.7)	15.0 (5.5–33.5)	17.8 (6.6–39.8)	33.9 (15.5–67.0)

Data in parentheses indicate ranges

(pH 5.03–5.65). The enhanced acidification of soil is a well-known effect of afforestation of former arable soil (Alriksson and Olsson 1995; Ritter et al. 2003; Rosenqvist et al. 2010). Another effect of afforestation is an age-related decline in base cations, especially Ca, from the soil/solution system (Ritter et al. 2003; Rosenqvist et al. 2010), which can be another reason for pH decreasing.

The contents of TOC in the soils of cultivated fields adjoining the shelterbelts are similar and equal on average to 4.5–4.6 gC kg^{-1} (Table 9.1). The average content of TOC in the soil of the young shelterbelt increased to 8.3 gC kg^{-1} during 14 years. For the old shelterbelt, the accumulation of organic matter during ~ 200 years has led to the substantial amount of TOC equal on average to 55.2 gC kg^{-1}. The accumulation of TOC in the soils under shelterbelts means that the humification processes with accumulation of stable SOM dominate over processes of the OM mineralization.

According to Smolander and Kitunen (2002), the microbial biomass and activities are correlated with the total concentrations of DOC. The DOC contents in the soils of both adjoining cultivated fields were similar (on average 0.22 gC kg^{-1}). In the soil under the young shelterbelt, the DOC content was 2 times higher. The highest DOC contents available for microbial and biochemical activities were observed in the soil under the old shelterbelt. This mean value was 10 times higher than those for the soils of the cultivated fields (Table 9.1).

The concentrations of total nitrogen and its mineral forms (nitrate and ammonium ions) in the soil under the old shelterbelt revealed the highest values during the whole vegetative growth season among all the soils under study (Table 9.1). Especially high values were determined for nitrate in the soil under the old shelterbelt with the mean content being 22–23 times higher than in the soil under the young shelterbelt and the soils of adjoining arable field. These results are in line with earlier studies (Szajdak 2002; Liao et al. 2006), which reported that a soil under an afforestation of a higher age contained higher amounts of total and

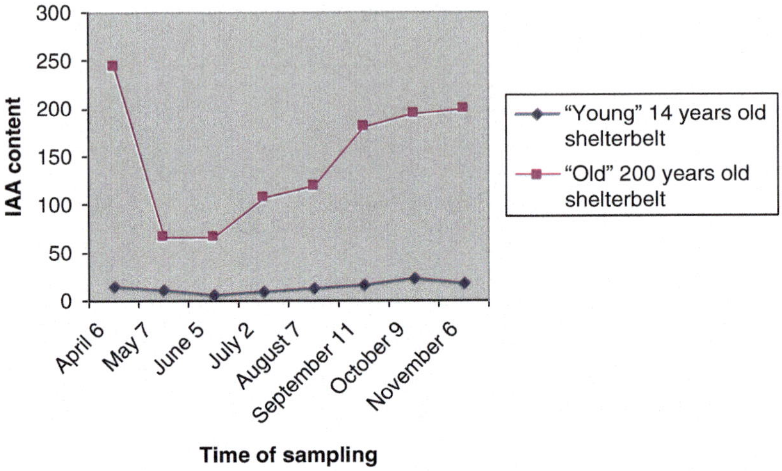

Fig. 9.1 The concentrations of IAA in soils (μg·kg^{-1} of soil) under the young and the old shelterbelts during entire vegetation season (Szajdak and Maryganova 2009)

inorganic N. As well these differences may be due to the different composition of shelterbelt plants. The main plant of the old shelterbelt *Robinia pseudoacacia* is a nitrogen-fixing tree; hence, it significantly increases the total N content in soil (De Marco et al. 2013). This is not the case for plants of the young shelterbelt.

9.3.2 Indole-3-acetic Acid

The IAA concentrations in the soil under the young shelterbelt and both adjoining cultivated fields ranged from 5.8 to 21.8 μg·kg^{-1}, and there were no significant differences among their mean values (Fig. 9.1).

The IAA contents in the soil under the old shelterbelt were found to be much higher during the all period of study (64.4–241.5 μg·kg^{-1}). The mean IAA amount in the soil under the old shelterbelt (145 μg·kg^{-1}) was 11–13 times higher than in the other soils under study (11.1–13.1 μg·kg^{-1}).

The transformations of the IAA content in the soil under the old shelterbelt observed in different phenological periods of plants proceeded in a similar way to the transformations of the IAA content in the other investigated soils. For all the soils, the highest amounts of IAA were measured in the beginning of April after wintering (Fig. 9.1). This value in the soil under the old shelterbelt (241.5 μg·kg^{-1}) was 16–17 times higher than in the other soils under study. At the beginning of May, a decrease in the IAA amount was observed in all the soils with especially dramatic lowering of IAA value in the soil under the old shelterbelt (Fig. 9.1). In June, the IAA contents were the lowest in the all investigated soils. From July to November, the IAA concentrations progressively increased, especially in the soil under the old shelterbelt. A similar trend was found in changes of free amino acid concentrations (Ryszkowski et al. 1998; Szajdak 1996) and inorganic forms of nitrogen (NH_4^+ and NO_3^-) (Futomo et al. 1985) in soils during the whole vegetative season.

The highest IAA content in the soil under the old shelterbelt corresponded to the highest concentrations of total N and its inorganic forms, total organic C and DOC (Table 9.1, Fig. 9.1).

Up to now, there is a lack of data concerning the transformation of auxins in soil. Our results are in agreement with earlier studies of IAA in arable soils, which reported a wide range of the IAA contents (0.20–45 $\mu g \cdot kg^{-1}$). We have also found that the concentrations of IAA in organic soils were much higher than in mineral soils. The IAA contents in the different kinds of peat ranged from 57.9 to 210.2 $\mu g \cdot kg^{-1}$; in the secondary transformed peat–moorsh soils, they were in the range of 69.3–186.1 $\mu g \cdot kg^{-1}$. There was a positive correlation between the concentrations of IAA and the contents of HS in peats (Szajdak and Maryganova 2007).

Stimulatory effects of HS on plant growth have been observed and extensively documented (Gorovaja et al. 1995; Muscolo and Nardi 1997; Muscolo et al. 1998; Chen et al. 2004). Muscolo and Nardi (1997) and Muscolo et al. (1998) suggested that IAA was actually present in the HS preparations. The biological activity of HS is similar to IAA, but it is very complex. Beneficial effects of HS on plant growth (leaf chlorophyll concentration, shoot and root fresh and dry weight, the number of root initials, and the number of flower buds) are well known.

Up to 80 % of all the rhizosphere bacteria produce IAA (Kampert and Strzelczyk 1975; Patten and Glick 1996), but also the IAA decomposition takes place in the soil mainly due to the microbial activity (Raczkowska-Błach et al. 1995). The soil flora also produces appreciable amounts of auxins under natural conditions, particularly when organic material is present to support microbial growth. An additional source of auxins in soil is organic manures (Stevenson 1967).

Consequently, we can conclude that numerous factors influence the IAA concentrations in soil. Taking them into account, we can suggest that a very high IAA content in the soil under the old shelterbelt comparable with the largest values for peats may be due to the high contents of SOM and DOM, large amounts of N and its inorganic forms, and high microbial activity. The main plant of the old shelterbelt *Robinia pseudoacacia* as a nitrogen-fixing tree may also introduce to the high IAA content in this soil.

9.3.3 Humic Substances

Land-use changes such as those which result from afforestation of former agricultural soils affect processes of humification and mineralization in soil and thereby the content, chemical composition, and properties of HS.

9.3.3.1 Molecular Weight Distribution

HS can be considered as a mixture of weak-acid polyelectrolytes having various molecular weight moieties. The molecular weights of HS are reported to range from several hundred to several hundred thousand Daltons, and the chemical structures show no repetitive pattern (Stevenson 1994).

Using SEC the HA extracted from soils under shelterbelts of different ages (14-year-old and 200-year-old) and adjoining arable fields were fractionated on two fractions with different molecular weights, i.e., exhibiting a bimodal molecular weight distribution. In accordance with the boundaries of exclusion on Sephadex G-100 (Lindqvist 1967; Swift and Posner 1971), the fraction representing peak I revealed a molecular weight in the range of 100,000 or higher. The material representing peak II corresponded to a lower-molecular-weight fraction (Table 9.2).

For HA extracted with 0.1 M NaOH, the percentage of a higher-molecular-weight fraction (≥100,000) was found to vary from 29 to 40 %. A lower-molecular-weight fraction with an average molecular weight of 7,763–10,116 contributed from 60 to 71 % of the HA. The HA from the soils under the young shelterbelt and adjoining cultivated field (HA-1 young and HA-2 young) had practically similar molecular weight distribution. By contrast, the HA from the soil under the old shelterbelt (HA-1 old) exhibited a larger percentage of the higher-molecular-weight

Table 9.2 Molecular weight distribution of HA extracted with 0.1 M NaOH and 0.1 M $Na_4P_2O_7$ at pH 7 from soils under shelterbelts of different ages and adjoining arable fields (Szajdak and Maryganova 2003)

HA samples	Fraction 1 Average MW	Fraction 1 % percentage of fraction	Fraction 2 Average MW	Fraction 2 % percentage of fraction
HA extracted with 0.1 M NaOH				
HA-1 young (14-year-old shelterbelt)	>100, 000	34	7,763	66
HA-2 young (field adjoining 14-year-old shelterbelt)	> 100, 000	36	7,763	64
HA-1 old (200-year-old shelterbelt)	> 100, 000	40	7,763	60
HA-2 old (field adjoining 200-year-old shelterbelt)	> 100, 000	29	10,116	71
HA extracted with 0.1 M $Na_4P_2O_7$ at pH 7				
HA-1 young (14-year-old shelterbelt)	> 100, 000	18	10,116	82
HA-2 young (field adjoining 14-year-old shelterbelt)	> 100, 000	25	7,763	75
HA-1 old (200-year-old shelterbelt)	> 100, 000	27	13,366	73
HA-2 old (field adjoining 200-year-old shelterbelt)	> 100, 000	32	17,660	68

fraction and a lower proportion of the lower-molecular-weight fraction than that from the soil of the adjoining field (HA-2 old).

It was found that 0.1 M $Na_4P_2O_7$ at pH 7 extracted from soils the lower-molecular-weight organic matter than 0.1 M NaOH. This finding is in accordance with the results of Cameron et al. (1972) and Piccolo and Mirabella (1987). The HA extracted with 0.1 M $Na_4P_2O_7$ at pH 7 from the soils under both shelterbelts contained higher amounts of the lower-molecular-weight fraction compared to those from the soils of adjoining cultivated fields. It can be noted that the HA from the young shelterbelt (HA-1 young) is characterized by a higher content of the lower-molecular-weight fraction than that from the soil under the old shelterbelt (HA-1 old).

9.3.3.2 Vis Spectroscopy

The best method for measuring the degree of humification is still being discussed because there is not a well-defined model of the HS structure. However, the humification process has been studied with regard to the changes in chemistry and structure of plant residues during the decomposition process and the effects of land use and soil management on soil organic matter characteristics (Zech et al. 1997). Therefore, the humification degree is usually evaluated through indirect measurements reflecting the structural changes that occur during the humification process.

Absorbance at 465 nm (E_4), as well as E_4/E_6 ratio has been considered to characterize the degree of aromatic condensation and polyconjugation in the humic molecules. The E_4 value reflects the development of polyconjugation systems in humic macromolecules and increases with increasing in degree of polyconjugation (Orlov 1990). The E_4/E_6 ratio supposedly decreases with progressive humification and increases in the aromatic condensation (Kononova 1966). Chen et al. (1977) regarded the influence of aromaticity on the E_4/E_6 ratio values to be a secondary factor only, but Niger et al. (2002) confirmed the reliability of the E_4/E_6 ratio as an indication of the aromatic condensation level. Campbell et al. (1967) found a relationship between this ratio and the stability of soil humus: the humic material with the lowest mean residence time had the highest E_4/E_6 ratio.

As shown in Table 9.3, all the HA extracted with 0.1 M NaOH are characterized by higher E_4 values and lower E_4/E_6 ratios than those isolated with 0.1 M $Na_4P_2O_7$ at pH 7. This reflects the higher degrees of aromatic condensation and polyconjugation in the macromolecules of the formers. The HA extracted with 0.1 M NaOH from the soils under the young 14-year-old shelterbelt and adjoining cultivated field were found to have practically similar E_4 and E_4/E_6 ratios. For the HA-1 old, E_4/E_6 was lower compared to the HA-2 old indicating more developed systems of polyconjugation and a greater degree of humification for the HA from the soil under the old shelterbelt compared with the reference HA.

As already noticed, HA extracted with 0.1 M $Na_4P_2O_7$ at pH 7 represent a more labile part of soil humus than those isolated with 0.1 M NaOH. For the formers, differences in the parameters of the Vis spectra are more evident. HA-1 old reveals the higher E_4 and the lower E_4/E_6 ratio than HA-2 old reflecting more developed

Table 9.3 Parameters of Vis spectra of HA extracted with 0.1 M NaOH and 0.1 M $Na_4P_2O_7$ at pH 7 from soils under shelterbelts of different ages and adjoining arable fields (Szajdak and Maryganova 2003)

HA samples	Absorbance at 465 nm (E_4)	Absorbance at 665 nm (E_6)	E_4/E_6
HA extracted with 0.1 M NaOH			
HA-1 young (14-year-old shelterbelt)	0.963	0.218	4.42
HA-2 young (field adjoining 14-year-old shelterbelt)	0.951	0.212	4.49
HA-1 old (200-year-old shelterbelt)	0.918	0.167	5.50
HA-2 old (field adjoining 200-year-old shelterbelt)	0.931	0.162	5.75
HA extracted with 0.1 M $Na_4P_2O_7$ at pH 7			
HA-1 young (14-year-old shelterbelt)	0.737	0.069	10.68
HA-2 young (field adjoining 14-year-old shelterbelt)	0.808	0.106	7.62
HA-1 old (200-year-old shelterbelt)	0.886	0.107	8.28
HA-2 old (field adjoining 200-year-old shelterbelt)	0.812	0096	8.46

systems of polyconjugation in the molecules of the HA from the soil under the old shelterbelt compared to the reference HA. By contrast, HA-1 young is characterized by the lower E_4 and the higher E_4/E_6 ratio than HA-2 young. It indicates that the HA from the young shelterbelt is characterized by less developed systems of aromatic polyconjugation compared to the HA from the soil of the adjoining field.

9.3.3.3 Thermal Analyses

The main thermal parameters of HA extracted with 0.1 N NaOH from soils under shelterbelts of different ages (14-year-old and 200-year-old) and adjoining arable fields are summarized in Table 9.4. For all HA, the shapes of DTA and DTG curves are compatible. Therefore, each thermal effect on the DTA curve corresponds to a weight loss at a definite step of thermal decomposition of HA. The DTA curves of all HA are characterized by an endothermic peak at 100 °C attributed to reactions of dehydration and three exotherms, within the 240–320 °C, 390–480 °C, and 590–670 °C ranges, indicating the thermal reactions of organic components of increasing thermal stability. Below 350 °C predominantly degradation of carbohydrates, dehydration and decarboxylation of oxygen containing functional groups, and destruction of other thermally labile fragments of HA are considered to occur, while in the temperature region of 350–500 °C, the reactions of more thermally stabile part of HA, particularly aromatic structures, take place. In the temperate region from 500 to 700 °C, the destruction of especially thermally stable fragments of the HA molecules, organomineral components, and the secondary reactions of the coke formation are involved (Rakovskiy and Filimonov 1967; Schnitzer et al. 1974; Sheppard and Forgeron 1987; Dell'Abbate et al. 2002).

Table 9.4 Data of thermal analysis of HA extracted with 0.1 M NaOH from soils under shelterbelts of different ages and adjoining arable fields (Szajdak and Maryganova 2003)

	Weight losses (%) and peak temperatures (°C) of endo- and exo-effects in the temperature regions (°C)										
	Up to 150 endo		150–350 exo-1		350–500 exo-2		500–700 exo-3		Up to 700	residue at 700	
HA samples	Weight loss	Peak temp.	Weight loss	Peak temp.	Weight loss	Peak temp.	Weight loss	Peak temp.	Weight loss	Weight loss	Z = 150–350 / 350–500
HA-1 young (14-year-old shelterbelt)	5.09	100	21.38	240	34.11	390	4.07	590	64.65	35.35	0.63
HA-2 young (field adjoining 14-year-old shelterbelt)	4.79	100	22.34	240	36.17	390	7.45	590	70.75	29.25	0.62
HA-1 old (200-year-old shelterbelt)	1.53	100	26.46	250	47.83	400	4.58	600	80.40	19.60	0.55
HA-2 old (field adjoining 200-year-old shelterbelt)	4.76	100	28.58	320	26.47	490	27.00	670	86.81	13.19	1.08

Parameter Z represents the ratio of the weight losses in the temperature regions 150–350 °C and 350–500 °C. It reflects the ratio between thermally labile (predominantly aliphatic) and thermally stable (predominantly aromatic) parts of the humic molecules. HA from the soils under shelterbelts of different ages and adjoining cultivated fields reveal different thermal behavior. It is interesting to notice different thermal stability of the HA from the soils of the fields adjoining the young and the old shelterbelts (HA-2 young and HA-2 old). For the HA-2 old, the weight loss in the temperature region up to 350 °C is higher and that in the interval 350–500 °C lower than those for the HA-2 young. Accordingly, the Z value for the HA-2 old is much higher. These data show that the HA-2 old contains more thermally labile structures in its molecules than the HA-2 young. Also the HA-2 old is characterized by a significantly higher weight loss in the third exothermic region of 500–700 °C, due to the combustion of organomineral, probably clay–humus complexes (Lishtvan et al. 1984).

For the HA from the soils under both shelterbelts, weight losses taking place up to 700 °C are lower, and the residues at 700 °C larger than those calculated for the HA from the soils of adjoining cultivated fields. There are not much differences in the weight losses up to 350 °C and also in parameter Z between the HA from the soil under the young shelterbelt and that from the reference soil (HA-1 young and HA-2 young, respectively). By opposite, the HA-1 old is characterized by a lower weight loss in the first exothermic region up to 350 °C and a significantly higher weight loss in the second exothermic region of 350–500 °C compared to those observed for the HA-2 old. As a result, Z parameter for the HA-1 old is two times lower. These data reflect much more intensive accumulation of more thermally stable and, hence, more humified HA in the soil under the old shelterbelt compared to the reference HA, as well as to the HA from the soil under the young shelterbelt.

9.3.3.4 IR Spectroscopy

Figure 9.2 shows the IR spectra of HA extracted with 0.1 M Na OH from soils under 14-year-old and 200-year-old shelterbelts and adjoining arable fields. They have a number of typical absorption bands at the following regions: 3500–3300 cm^{-1} (hydrogen-bonded OH), 2920 and 2860 cm^{-1} (aliphatic C–H stretch), 1725 cm^{-1} (C=O stretch in undissociated COOH groups), 1630–1600 cm^{-1} (conjugated C=C in aromatic rings and olefins and/or COO^- groups), 1660 and 1540 cm^{-1} (C=O stretching of amide I and NH deformation in amide II, respectively, indicating the presence of polypeptide groups), 1450 cm^{-1} (CH deformation of $C–CH_3$ and $C–CH_2$), 1260 cm^{-1} (phenolic C–O stretch), 1230 cm^{-1} (C–O stretch of COOH groups), and 1100–1000 cm^{-1} (C–O stretch of polysaccharides) (Orlov and Osipova 1988; Bloom and Leenheer 1989; Stevenson 1994). Strong absorption both at 1050 cm^{-1} and in

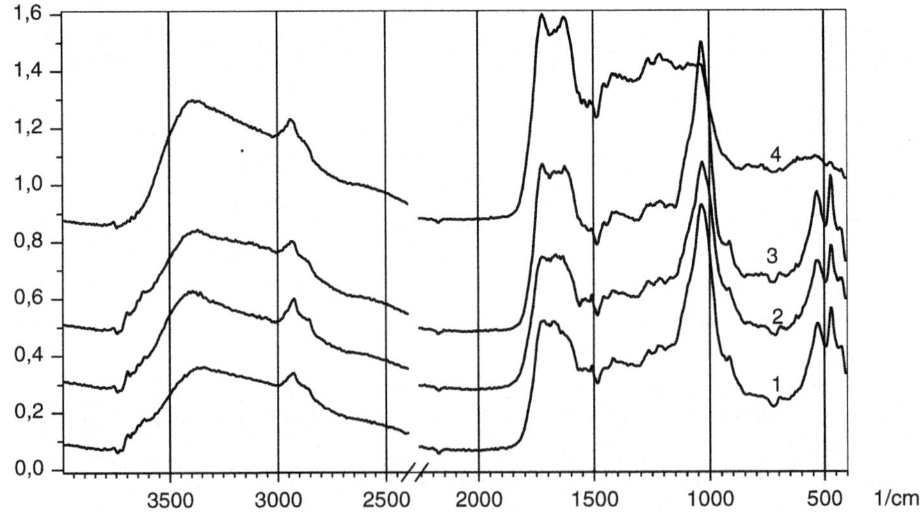

Fig. 9.2 IR spectra of HA-2 young (1), HA-1 young (2), HA-2 old (3), and HA-1 old (4) extracted with 0.1 M NaOH from soils under shelterbelts of different ages and adjoining arable fields (Maryganova and Szajdak 2011)

the 600–400 cm^{-1} region in the IR spectra of all the HA with the exception of that from the soil under the old shelterbelt is associated with silicate and other mineral impurities (Orlov and Osipova 1988; Francioso et al. 2000). Since the extraction and purification procedures were the same for all the HA samples, it can be deduced that the HA from the soil under the old shelterbelt was not tightly bound to the soil inorganic matter, opposite to the other HA samples.

Inspection of the IR spectra reveals some structural differences between the HA samples. The HA from the soil under the young shelterbelt shows relatively more intense bands of aliphatic structures both at 2920–2860 cm^{-1} region and 1450 cm^{-1}, as well as absorption of polypeptides at 1660 and 1540 cm^{-1} as compared to the HA from the soil of adjoining cultivated field.

The main difference between the IR spectrum of the HA from the soil under the old shelterbelt and that of the HA from the soil of adjoining arable field is that, together with an increased intensity of aliphatic bands at 2920–2860 and 1450 cm^{-1}, the former is characterized by an increase in absorption of carboxylic groups at 1720 and 1230 cm^{-1} and aromatic C=C at 1625 cm^{-1} as well as a decrease in absorption of the amide I and amide II bands at 1660 and 1540 cm^{-1} compared with the HA from the soil of the arable field. It can be noticed also a rather significant absorption of polysaccharides in the 1100–1000 cm^{-1} region of the IR spectrum of the HA from the soil under the old shelterbelt. Absorption of polysaccharides in the other HA samples cannot be evaluated because of a strong band of Si–O in this region of the IR spectra.

Consequently, IR spectroscopy evidence suggests that the transition from the cultivated soil to the newly forested one (the young shelterbelt) leads to the increase in CH_2 and CH_3 groups and polypeptides in the structure of the HA. Comparison of the HA from the soil under the old shelterbelt with that from the corresponding cultivated soil shows that the former is relatively richer in aromatics, CH_2, CH_3, and carboxylic groups, but contains lower amounts of polypeptides.

9.3.3.5 ^{13}C NMR Spectroscopy

The relative contents of the different carbon types, as determined from the integration of the chemical shift regions in solution ^{13}C NMR spectra, as well as percentages of hydrophobic and hydrophilic carbon atoms, and the degrees of hydrophobicity HB/HI for the HA extracted with 0.1 M $Na_4P_2O_7$ at pH 7 HA from soils under 14-year-old and 200-year-old shelterbelts and adjoining arable fields, are reported in Table 9.5.

The relative contents of different structural groups in the HA of both cultivated soils (HA-2 young and HA-2 old) are similar. They contain 17–18 % alkyl C, 22 % O,N-alkyl C, 25 % O-alkyl C, 23 % aromatic C, and 12–13 % carboxyl C. Compared with the HA of the adjoining cultivated soil, the HA from the soil under the young shelterbelt (HA-1 young) is characterized by lower contents of aromatic C and carboxyl C, as well as a higher percentage of hetero-alkyl C, in particular O,N-alkyl C, associated mainly with methoxyls of lignins and amino acids (Table 9.5). By con-

Table 9.5 Carbon distribution (%) over chemical shift regions (ppm) in solution ^{13}C NMR spectra of HA extracted with 0.1 M $Na_4P_2O_7$ at pH 7 from soils under shelterbelts of different ages and adjoining arable fields (Maryganova et al. 2010)

HA samples	0–45 alkyl	45–65 O–N-- alkyl	65–108 O-alkyl	108–165 aromatic	165–200 carboxyl	HB[a]	HI[b]	HB/HI[c]
HA-1 young (14-year-old shelterbelt)	18	25	26	21	10	39	61	0.64
HA-2 young (field adjoining 14-year-old shelterbelt)	18	22	25	23	12	40	60	0.67
HA-1 old (200-year-old shelterbelt)	20	18	21	26	15	46	54	0.85
HA-2 old (field adjoining 200-year-old shelterbelt)	17	22	25	23	13	40	60	0.67

HB[a] percent of hydrophobic carbon atoms, *HI[b]* percent of hydrophilic carbon atoms, *HB/HI[c]* degree of hydrophobicity

trast, the HA from the soil under the old shelterbelt (HA-1 old) is richer in aromatic C, alkyl C, and carboxyl C, but contains lower amounts of hetero-alkyl C compared with those of adjoining cultivated soil.

The amount and composition of plant litter were shown to be essential controlling factors for the formation of SOM and humification processes in terrestrial ecosystems (Swift et al. 1979; Kögel-Knabner 2002). The chemical composition of the primary resources controls decomposition and humification by changing the turnover rates. Resources rich in phenols, waxes, and lignins are known to decompose more slowly and contribute markedly to the stable SOM pool in comparison with proteins and sugars. The residues of young plants decompose more rapidly than those of older plants, the latter often containing more stable cell wall compounds (Swift et al. 1979). In soils with high total N and low C/N ratio, aromaticity of HA tended to be higher (Zech et al. 1992).

The input of OM with crop residues for arable soils is usually lower than litter input for forest soils (Kögel-Knabner 2002; Lützow et al. 2006). Conversion from arable cropping to a newly forested soil (under a young shelterbelt) leads to an increase in the fresh litter input followed by increasing microbiological activity. The organic compounds released in the soil during mineralization of the fresh residues of young plants are involved in the processes of humification with the formation of the young immature humic molecules enriched with carbohydrate and peptide structural units (Spaccini et al. 2000). Because of that, the HA from the soil under the young shelterbelt has a lower percentage of aromatic C and higher proportions of O,N-substituted aliphatic C, i.e., has a lower degree of humification and is younger than the HA of the corresponding cultivated field.

Advanced stages of humification in the soil under the old shelterbelt are characterized by the higher aromatic C content (Zech et al. 1992). As shown in studies of forest soil profiles and litter bag experiments (Kögel-Knabner 1993), the relative amount of alkyl C increases during biodegradation and humification, whereas the amount of O-alkyl carbon shows a relative decrease. This is associated with loss of the most easily metabolizable carbohydrates and amino acids and an accumulation of alkyl C in such recalcitrant biopolymers as cutin and suberin. Strong microbial utilization of HA from different soils, including a forest soil, has already been established (Filip et al. 1999; Filip and Tesařová 2004). Aliphatic structural units in HA, mainly carbohydrates and peptides, are preferentially utilized by microbes, and remaining HA contain more condensed aromatic structures (Filip and Tesařová 2004). The DOC content and thereby the microbial activity in the soil under the old shelterbelt are much higher than in all other soils (Table 9.1), which as well may be the reason for the highest aromaticity of the corresponding HA. Old trees of *Robinia pseudoacacia* and a significant total N content in the soil may contribute to the aromaticity of HA-1 old. The increase in carboxyl C may be related to the side chain oxidation of plant-derived lignin–phenolic compounds and/or incorporation of carbonyl-rich material from fresh vegetal tissues (Zech et al. 1992; Zech and Guggenberger 1996). Therefore, the HA from the soil under the old shelterbelt can be regarded as the most humified and mature HA among all the HA under study.

Hydrophobic–hydrophilic properties of HA depend on their chemical composition, structure, and conformation (Stepanov 2005; Maryganova et al. 2006; Dębska and Gonet 2007; Dębska et al. 2007; Stepanov 2008). The percentage of hydrophobic carbons HB and the degree of hydrophobicity HB/HI calculated from the ^{13}C NMR data are highest for the HA from the old shelterbelt and lowest for the HA from the soil under the young shelterbelt. For the HA from both cultivated soils, the degrees of hydrophobicity were found to be similar and a little higher than the value for the young shelterbelt (Table 9.5).

9.3.3.6 Hydrophobic Interaction Chromatography

Hydrophobic interaction chromatograms of the HA extracted with 0.1 M $Na_4P_2O_7$ at pH 7 from soils under 14-year-old and 200-year-old shelterbelts and adjoining arable fields are shown in Fig. 9.3, and the relative contents of their chromatographic fractions with different hydrophobic and hydrophilic properties in Table 9.6.

All the HA samples were fractionated on five fractions differing by their capacity to react with hydrophobic gel, as it was shown for HA from peat, sapropel, and brown coal (Maryganova et al. 2006). Each eluent eluted one fraction except for the 0.1 M NaOH solution which eluted two fractions. The applied sequence of eluents reduces the strength of hydrophobic interactions between the hydrophobic sites of the HA and the hydrophobic ligands on the gel matrix. This leads to a gradual increase in the hydrophobicity of the HA fractions from fraction 1 to fraction 5.

The most hydrophilic fraction 1 of the HA samples did not adsorb on the hydrophobic gel and was eluted with a starting buffer. It comprised from 34 to 48 % of the

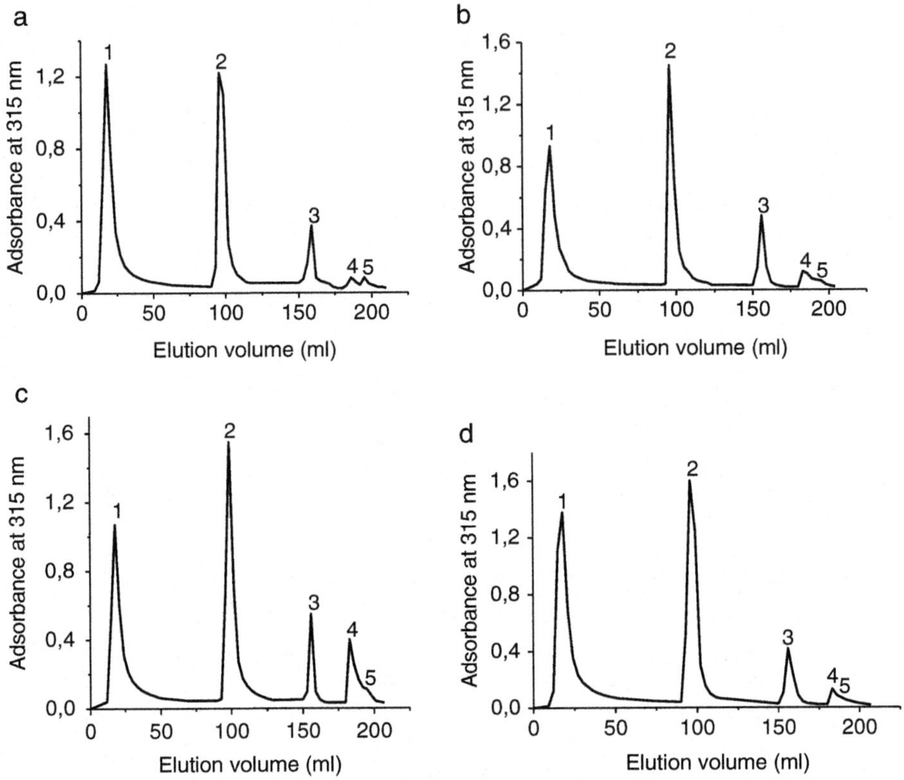

Fig. 9.3 Chromatography of hydrophobic interaction of HA-1 young (**a**), HA-2 young (**b**), HA-1 old (**c**), and HA-2 old (**d**) on Octyl-Sepharose 4 Fast Flow column (13 × 1.6 cm) using step-by-step elution with 0.05 M Tris-HCl buffer at pH = 8.0 containing 3 M NaCl (peak 1), 0.05 M Tris-HCl buffer at pH = 8.0 (peak 2), distilled water (peak 3), and 0.1 M NaOH (peaks 4 and 5) (Maryganova et al. 2010)

total areas of the HA chromatograms. The proportion of more hydrophobic fraction 2 ranged from 39 to 44 %. The relative contents of the most hydrophobic components (fractions 3+4+5) ranged from 11 to 23 %. According to the previous study (Maryganova et al. 2006), the hydrophilic components of peat HA obtained by the HIC fractionation were characterized by the highest content of carboxylic groups and the lowest proportion of saturated hydrocarbon chains. By contrast, the most hydrophobic fraction of peat HA contained the largest amount of saturated aliphatic structures and the least percentage of carboxylic groups.

The HA-1 young from the soil under the young shelterbelt is characterized by the largest proportion of the most hydrophilic fraction 1 equal to 48 % and the lowest content of the most hydrophobic fractions 3+4+5 (11 %). For the HA-2 young from the soil of adjoining arable field, the content of the most hydrophilic fraction is 8.3 % lower, and the percentage of the most hydrophobic components 45.5 % higher compared with the former (Table 9.6). By contrast, the HA-1 old from the

Table 9.6 Relative contents of chromatographic fractions (%) obtained by HIC of HA extracted with 0.1 M $Na_4P_2O_7$ at pH 7 from soils under shelterbelts of different ages and adjoining cultivated fields on Octyl-Sepharose 4 Fast Flow column (13×1.6 cm) using step-by-step elution with 0.05 M Tris-HCl buffer at pH = 8.0 containing 3 M NaCl (fraction 1), 0.05 M Tris-HCl buffer at pH = 8.0 (fraction 2), distilled water (fraction 3), and 0.1 M NaOH (fractions 4 and 5) (% of the total area of the HA chromatograms) (Maryganova et al. 2010)

	Fractions				
HA samples	1	2	3	4	5
HA-1 young (14-year-old shelterbelt)	48	41	8	2	1
HA-2 young (field adjoining 14-year-old shelterbelt)	44	39	11	3	2
HA-1 old (200-year-old shelterbelt)	34	43	11	9	3
HA-2 old (field adjoining 200-year-old shelterbelt)	44	44	9	2	1

soil under the old shelterbelt is characterized by the lowest content of the most hydrophilic fraction 1 equal to 34 % and twice the proportion of the most hydrophobic components compared to that of the corresponding cultivated soil (Table 9.6).

Thus, fractionation of HA under study by HIC allowed to separate polydisperse and heterogeneous systems of HA on humic components with different hydrophobic–hydrophilic properties. Our data showed that HA with different chemical composition and structure had different amphiphilic properties. HA-1 young from the soil under the young shelterbelt revealed the lowest content of hydrophobic aromatic C and the highest percentage of hydrophilic hetero-alkyl C, especially O,N-alkyl C (Table 9.5). In accordance with this, the content of the most hydrophilic fraction was the highest, while the proportion of hydrophobic components was the lowest (Table 9.6). The increased percentage of the hydrophilic fraction in the HA from the soil under the young shelterbelt is accounted for by incorporation of the organic compounds released in this soil during mineralization of additional fresh plant residues (Spaccini et al. 2000).

For HA-1 old from the soil under the old shelterbelt, the opposite was observed: this HA was richest in aromatic C and alkyl C, which can be regarded as hydrophobic carbon atoms, but contained the least amounts of hydrophilic hetero-alkyl C atoms (Table 9.5). So, despite the highest proportion of hydrophilic carboxyl carbons, this HA was characterized by the lowest content of the most hydrophilic fraction 1 and the highest percentage of the most hydrophobic components (Table 9.6). This result is understandable, taking into account the high microbiological activity in this soil (the highest DOC content) which results in strong microbial utilization of mainly hydrophilic fractions in HA-1 old and its enrichment with hydrophobic compounds.

Consequently, the hydrophobic and hydrophilic properties of the HA under study, determined using the HIC method, are in line with the ^{13}C NMR spectroscopy data. In light of these results, HA extracted from the soil under the young shelterbelt contains the highest percentage of the most hydrophilic fraction and is more hydrophilic than HA from the soil of adjoining cultivated field. By contrast, HA isolated from the soil under the old shelterbelt is characterized by the lowest proportion of hydrophilic compounds and is considerably more hydrophobic (therefore, more stable) that HA from the cultivated field and from the soil under the young shelterbelt.

Summarizing all the results in the study of HA from the soils under shelterbelts of different ages and adjoining arable fields, we can conclude the following. The age of shelterbelts is the principal factor affecting the composition, molecular structure, and amphiphilic properties of HA. Neutral solution of sodium pyrophosphate was found to be a more sensitive reagent for the HA extraction than sodium hydroxide because HA extracted with the former from the soils under the young shelterbelt and adjoining cultivated field revealed differences in their chemical structure opposite to the HA extracted with the latter.

The HA from the soil under the old shelterbelt was found to be richer in aromatic, alkyl, and carboxyl carbon atoms, but contained lower amounts of heteroalkyl C, exhibited the higher degrees of aromatic condensation and polyconjugation than that of adjoining cultivated soil, and, hence, can be regarded as more humified and chemically mature HA. The HA extracted from the soil under the young shelterbelt was characterized by a lower amount of aromatic C, and a higher percentage of O,N-alkyl C, predominantly in carbohydrate and polypeptide structures, than HA from the adjoining arable field, i.e., had a lower degree of humification and chemical maturity and was younger. HA from the soil under the old shelterbelt was characterized by a significantly lower content of hydrophilic fraction and twice the proportion of the hydrophobic components, compared to the HA of the adjoining cultivated field, i.e., more stable. By contrast, for HA from the soil under the young shelterbelt, the amount of the hydrophilic components was higher, and the percentage of the hydrophobic fractions lowers, compared with HA from the soil of the corresponding cultivated field.

9.3.4 Lipids

Total lipid contents in soils under study ranged from 0.20 to 2.40 g kg^{-1} soil. Concentrations of the hexane-extractable lipid fractions were equal for the soils of both cultivated fields and amounted for 0.20 g kg^{-1}. The content of lipids in the hexane extract of the soil under the young shelterbelt was 2.5 times higher (0.50 g kg^{-1}), whereas the highest content of lipids (2.40 g kg^{-1}) was found in the extract of the soil under the old shelterbelt. Comparison of these results with the TOC data of the soil samples (Table 9.1) shows correlation between the lipid fraction content and TOC. These data are in line with the conclusion that the maximum of the total lipid content corresponds to soils rich in OM and the minimum to soils poor in OM (Fridland 1982).

Most of the studies have been performed on the lipids extracted from whole soils and their particle size fractions with the application of gas chromatography/mass spectrometry or gas chromatography/pyrolysis-mass spectrometry (van Bergen et al. 1997; Bull et al. 2000; Naafs et al. 2004; Nierop et al. 2005; Quénéa et al. 2006; Jeannotte et al. 2011). On the contrary, only very limited work was related to their analysis with nondestructive techniques such as NMR spectroscopy

(Almendros et al. 2001; Lodygin and Beznosikov 2005; Drori et al. 2006; Ahangar et al. 2009), though NMR spectroscopy allows us to characterize lipid structures present in nonvolatile complex material.

^{1}H and ^{13}C NMR spectra of hexane-extractable lipids from the soils of both shelterbelts and adjoining cultivated fields are qualitatively similar and differ by intensities of lines showing different contents of individual compounds. Figure. 9.4 shows ^{1}H NMR spectrum of lipids extracted from the soil of arable field adjoining the young shelterbelt as example spectrum.

This spectrum indicates that hexane-extractable lipids under study consist of a complex mixture of organic compounds. The most intense lines (in Fig. 9.4 they are represented with amplification) are characteristic for saturated hydrocarbons (alkans): the chemical shifts $\delta_{CH2} = 1.26$ ppm and $\delta_{CH3} = 0.89$ ppm. In this region also absorb hydrogen atoms of methylene and methyl groups of aliphatic chains of fatty acids, aldehydes, alcohols, ethers, and esters including mono-, di-, and triglycerides of fatty acids.

In addition, in ^{1}H NMR spectra of lipids under study, there are lines characteristic for the following classes of compounds: fatty acids, $\delta_{CH2COOH} = 2.36$ ppm; aldehydes, $\delta_{CH2CO} = 2.43$ ppm and $\delta_{COH} = 9.77$ ppm; alcohols, $\delta_{CH2O} = 3.65$ ppm; esters, $\delta_{CH2OCO} = 4.05$ ppm; triglycerides of fatty acids, $\delta_{CHO} = 5.28$ ppm, $\delta_{CH2O} = 4.26$, and 4.15 ppm; mono- and diglycerides of fatty acids, $\delta_{CHO} = 5{,}03$ м.д. and $\delta_{CH2O} = 4{,}22 \div 3{,}66$ ppm; and sitosterol, $\delta_{CHO} = 3.59$ ppm. The content of the latter is very small.

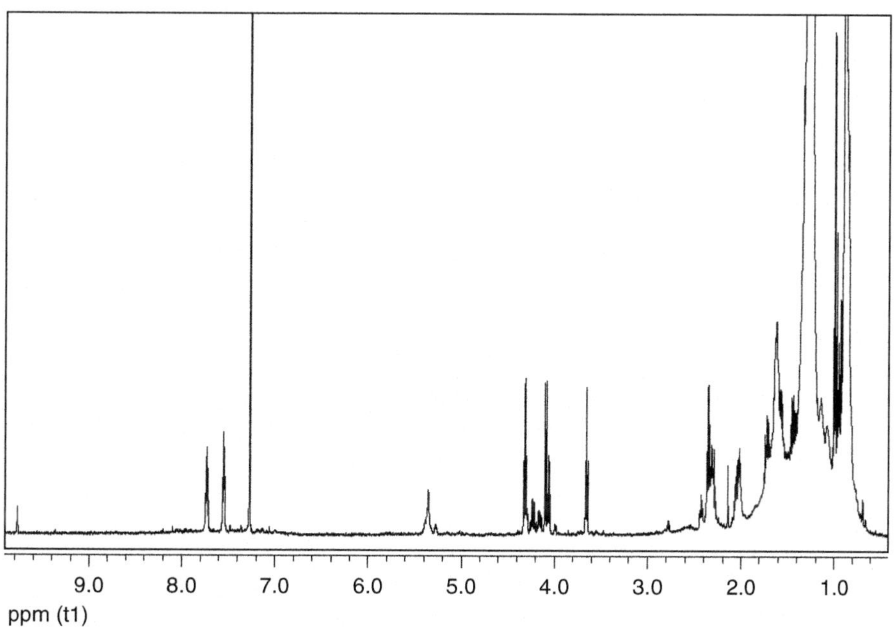

Fig. 9.4 ^{1}H NMR spectrum ($CDCl_3$) of hexane-extractable lipids from the soil of arable field adjoining 14-year-old shelterbelt (Szajdak et al. 2015)

Intense signal in the region of double bonds with maximum at 5.36 ppm indicates the presence of unsaturated fatty acids both as free compounds and combined as esters. Mainly this is oleic acid which is confirmed by ^{13}C NMR spectra of the samples under study. The characteristic signal of methylene group located between two double bonds $\delta_{CH2}=2.78$ ppm testifies to the presence of unsaturated linoleic acid. The relative content of unsaturated fatty acids in the total amount of fatty acids can be estimated by integral intensities of corresponding characteristic lines. These values are approximately 50 % for all the samples with the exception of lipids extracted from the soil under the old shelterbelt. For the latter this value is not more than 10 %.

In the ^{1}H NMR spectrum of lipids extracted from the soil under the young shelterbelt, there are also lines characteristic for ethers: $\delta_{CH2OCH2}=3.37$ ppm, as well as slight signals of dehydroabietic acid connected probably with the presence of coniferous trees in this shelterbelt (*Pinus silvestris)*.

Additionally, in the ^{1}H NMR spectra of all the samples, there are signals corresponding to esters of *o*-phthalic acid: $\delta_{CH}=7.72$ and 7.54 ppm of normal ($\delta_{CH2O}=4.32$ ppm) and β-branched ($\delta_{CH2O}=4.10$ and 4.22 ppm) structures. Ecological monitoring carried out by Kiang and Grob (1986) showed that esters of *o*-phthalic acid which are widely used as plasticizers could get to the soils from air and water. In the region of 6.8 ÷ 8.2 ppm, there are also slight signals of other aromatic compounds; their contents are negligible compared to those of esters of *o*-phthalic acid.

^{13}C NMR spectra of hexane-extractable lipids are dominated by signals of long-chain alkyl carbon (Fig. 9.5). The most intense lines are characteristic of saturated

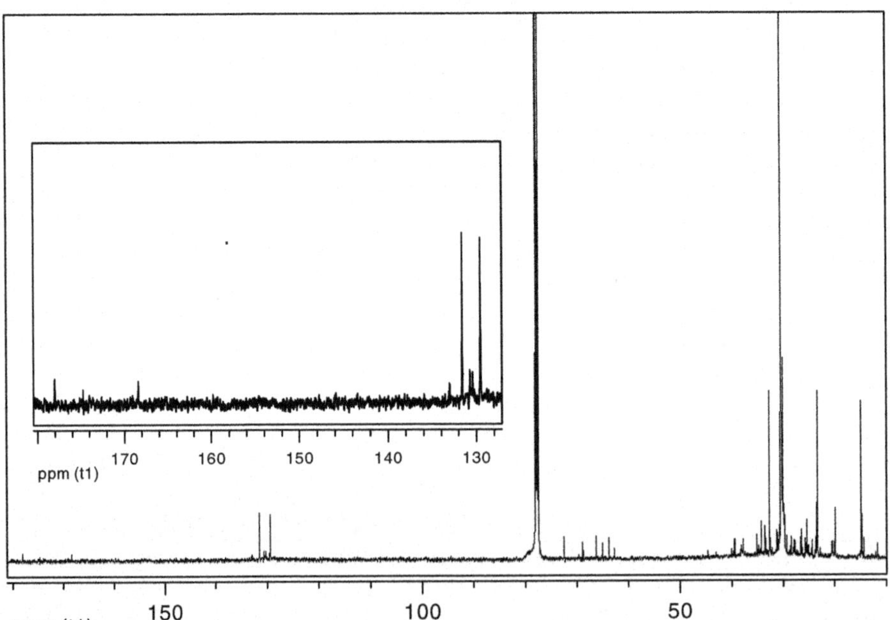

Fig. 9.5 ^{13}C NMR spectrum ($CDCl_3$) of hexane-extractable lipids from the soil of arable field adjoining 14-year-old shelterbelt (Szajdak et al. 2015)

hydrocarbons: δ_{CH2}=32.64, 30.41, 30.07, and 23.40 ppm and δ_{CH3}=14.81 ppm. Additionally, there are signals of fatty alcohols, δ_{CH2OH}=63.82 ppm and δ_{CH2}=33.46, 32.64, 30.39, 30.36, 30.31, 30.14, 26.45, and 23.40 ppm and δ_{CH3}=14.78 ppm; fatty acids, δ_{COOH}=178.05 ppm and $\delta_{CH=CH}$=130.71 and 130.40 ppm (C of double bonds of oleic acid) and δ_{CH2}=34.79 …, 27.91, 27.84, and 25.35 ppm; fatty aldehydes, δ_{COH}=203.67 ppm and δ_{CH2COH}=37.80 ppm; esters, δ_{COO}=174.73 ppm and δ_{CH2O}=65.11 ppm and δ_{CH2}=35.14, 32.64, … 26.64, 25.44, and 23.40 ppm, as well as weak signals of glycerides of fatty acids. These spectra confirm the presence of esters of *o*-phthalic acid in the lipid fractions: δ_{COO}=168.40 ppm, δ_{CH2O}=72.49 and 66.26 ppm, δ_{C}=133.07 and 133.02 ppm, and δ_{CH}=131.59 and 129.54 ppm.

Figure 9.6 presents identical regions (3.3–5.5 ppm) of ^{1}H NMR spectra of lipids extracted from soils under the young and the old shelterbelts. Judging from these spectra, the relative content of unsaturated fatty acids (signal at 5.36 ppm) in the lipid fraction from the soil under the old shelterbelt is much lower than that in the soil under the young shelterbelt, in spite of the fact that the former contains the highest relative amount of total fatty acids (Table 9.7). Besides, in ^{1}H NMR spectrum of lipids from the soil under the old shelterbelt, triplet of protons of CH_2O group of ethers (3.37 ppm) is absent, but intensities of the signals of esters (4.05 ppm) and alcohols (3.65 ppm) rise, indicating increase in their relative content (Table 9.7). It is interesting to note that lines of esters of *o*-phthalic acid in the proton spectrum of lipids from the soil under the old shelterbelt (multiplet signals at 4.32, 4.22, and 4.10 ppm) reveal very low intensities showing their minimal content.

Consequently, ^{1}H and ^{13}C NMR spectra of hexane-extractable lipids indicate that saturated hydrocarbons are the most abundant components. Their amount in the lipid fraction of the soil of the arable field adjoining the young shelterbelt equal ~ 60 %, in the case of that adjoining the old shelterbelt— ~ 70 %. Lipids from the soil under the young shelterbelt contain ~75 % of paraffinic hydrocarbons, and their highest amount was found in the lipid fraction of the soil under the old shelterbelt (more than 80 %). It may be connected with the fact that woody species, more abundant in the soil under the old shelterbelt, tend to have higher proportions of alkyl C compared to agricultural species (Lorenz and Lal 2005). Moreover, saturated hydrocarbons in the lipid fraction of the soil under the old shelterbelt are predominantly higher-molecular-weight compounds compared to the other lipid samples under study. It indicates more developed processes of polymerization occurring in the soil under the old shelterbelt.

The other significant class of organic compounds detected in hexane extracts of soils under study are fatty acids with relative content amounts 21–38 molar percent in the range of compounds presented in Table 9.7. In spite of the similar lipid content in soils of both arable fields, the relative content of fatty acids in lipids of the arable field adjoining the young shelterbelt is 1.5 times higher compared to those of the arable field adjoining the old shelterbelt (Table 9.7). The relative amount of fatty acids in the lipid fraction of the soil under the young shelterbelt is slightly higher than in that of the adjoining field, whereas the lipid content of the soil under the old shelterbelt is 1.8 higher compared to that of the adjoining arable field.

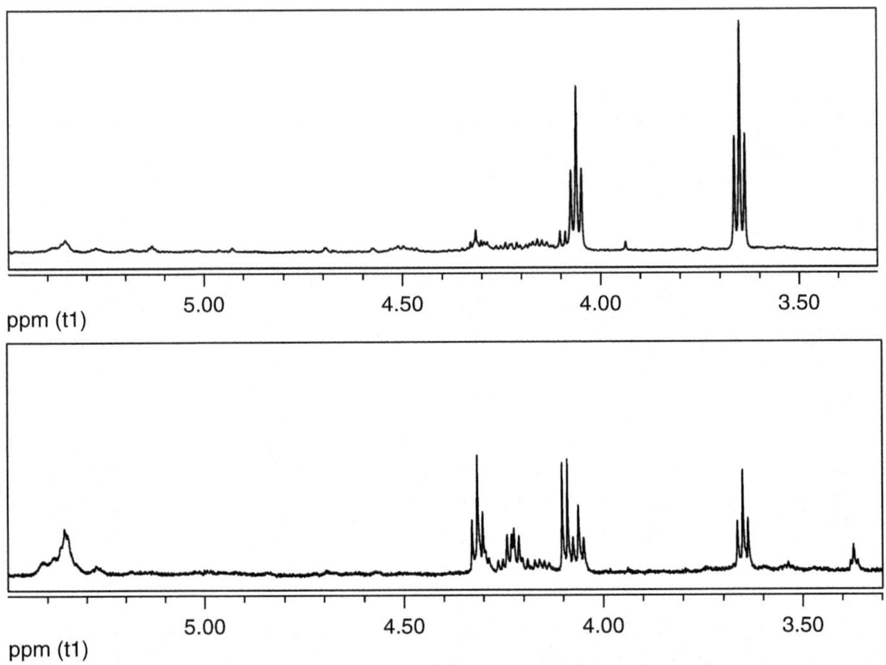

Fig. 9.6 ^{1}H NMR spectra ($CDCl_3$, region 3.3–5.5 ppm) of hexane-extractable lipids from soils under 14-year-old shelterbelt (*below*) and 200-year-old shelterbelt (*above*) (Szajdak et al. 2015)

Table 9.7 Relative contents (molar percent) of some classes of compounds in hexane-extractable lipids from soils under shelterbelts of different ages and adjoining cultivated fields (the sum of these compounds is equal to 100 %) (Szajdak et al. 2015)

Soil samples	Fatty acids	Esters of fatty acids	Fatty aldehydes	Fatty alcohols	Triglycerides of fatty acids	Esters of *o*-phthalic acid
14-year-old shelterbelt	35	11	15	14	6	19
Field adjoining 14-year-old shelterbelt	31	11	12	16	7	23
200-year-old shelterbelt	38	19	14	23	3	3
Field adjoining 200-year-old shelterbelt	21	10	11	16	5	37

The relative content of esters of fatty acids in the lipid fraction of the soil under the old shelterbelt is two times higher than in other lipid samples (Table 9.7) which indicates more developed processes of polycondensation. In this lipid fraction, also the highest relative amount of fatty alcohols and the lowest relative quantity of triglycerides of fatty acids have been detected showing a high degree of fat destruction.

Esters of *o*-phthalic acid were detected in all the lipid fractions, but in different quantities. Their largest relative content was found in the lipids of the soil of the arable field adjoining the old shelterbelt, and the lowest amount in the soil under the old shelterbelt. The relative content of esters of *o*-phthalic acid in lipids of the soil under the young shelterbelt is somewhat lower compared to the adjoining arable field.

Accordingly, the lipid fraction of the soil under the old shelterbelt contains the highest amounts of saturated hydrocarbons, fatty alcohols, and fatty acids and their esters and the lowest proportions of easily oxidative unsaturated fatty acids (oleic and linoleic acids), ethers, and triglycerides of fatty acids. Apparently, in the soil under the old shelterbelt, during 200 years of organic matter accumulation, the processes of oxidation, hydrolysis, and polymerization are developed in the highest degree. It is known that alkyl C as in polymethylenic compounds is among the biologically most stable forms of soil organic carbon in aerobic soils (Lützow et al. 2006). Resistant aliphatic polymers may be formed in soils upon oxidative polymerization of some lipids (presumably unsaturated) (de Leeuw 2007). Consequently, the highest content of saturated hydrocarbons and the lowest amount of unsaturated fatty acids in the lipid fraction of the soil under the old shelterbelt can indicate the advanced stages of the lipid transformation with the accumulation of stable compounds. The lowest amount of esters of *o*-phthalic acid in the lipid fraction of this soil is due probably to their destruction in the soil with high biological activity.

The hexane-extractable lipids of the soil under the young shelterbelt contain somewhat larger amounts of saturated hydrocarbons, fatty acids, and aldehydes and smaller proportions of *o*-phthalates compared to those of adjoining arable field. To all appearance, conversion from arable cropping to forested soil 14 years ago caused some accumulation of abovementioned components in the lipid fraction and a modest destruction of *o*-phthalats resulting from increased biological activity of this soil.

The summary in this review result shows that the conversion from arable cropping to shelterbelts under study not only influenced the accumulation of SOM and some biologically active substances in the soils under shelterbelts but also changed the composition, structure, and stability of free-extractable lipids and HA with the age of shelterbelts being the principal factor.

Conversion from arable cropping to a newly forested soil (under a young shelterbelt) leads to an increase in the fresh litter input followed by increasing microbiological activity. The organic compounds released in the soil during mineralization of the fresh residues of young plants are involved in the processes of humification with the formation of the young immature humic molecules enriched with carbohydrate and peptide structural units. Because of that, the HA from the soil under the young shelterbelt has a lower degree of humification and is more hydrophilic and young than the HA of the corresponding cultivated field. The hexane-extractable lipids of this soil contain larger amounts of saturated hydrocarbons, fatty acids, and aldehydes and a lower content of anthropogenic contaminates *o*-phthalats compared to those of adjoining arable field.

The OM in the soil under the old shelterbelt during ~ 200 years has undergone the most significant biochemical and chemical transformations (oxidation, hydrolysis, polymerization) and advanced stages of humification with the accumulation of resistant compounds. The HA from the soil under the old shelterbelt is characterized by the highest amounts of alkyl and aromatic C and can be regarded as the most humified, hydrophobic, and stable HA among all the HA under study. The lipid fraction of this soil also accumulated the largest proportions of recalcitrant compounds, especially saturated hydrocarbons, and had the lowest relative proportions of easily oxidative unsaturated fatty acids (oleic and linoleic), as well as esters of *o*-phthalic acid among all the samples under study.

References

Ahangar AG, Smernik RJ, Kookana RS, Chittleborough DJ (2009) The effect of lipids on the sorption of diuron and phenanthrene in soils. Chemosphere 74:1062–1068

Almendros G, Tinoco P, Gonzalez-Vila FJ, Ludeman Y-D, Sanz J, Velasco F (2001) ^{13}C-NMR of forest soil lipids. Soil Sci 166:186–196

Alriksson A, Olsson M (1995) Soil changes in different age classes of Norway spruce (*Picea abies (L.) Karst.*) on afforested farmland. Plant Soil 168–169:103–110

Amblès A, Jambu P, Jacquesy I-C, Parlanti E, Secouet B (1993) Changes in the ketonic portion of lipidic components during the decomposition of plant debris in a hydromorphic forest podzol. Soil Sci 156:49–56

Bandurski RS, Schulze A (1977) Concentration of indole-3-acetic acid and its derivatives in plants. Plant Physiol 60:211–213

Bartoszewicz A (2000) Effect of the change of soil utilization on the concentration of nitrogen mineral forms in soils and ground water. Pol J Soil Sci 33(2):13–20

Bartoszewicz A, Ryszkowski L (1996) Influence of shelterbelts and meadows on the chemistry of ground water. In: Ryszkowski L, French N, Kędziora A (eds) Dynamics of an agricultural landscape. Państwowe Wydawnictwo Rolnicze i Leśne, Poznań, pp 98–100

Bergman W (1963) Geochemistry of lipids. In: Breger IA (ed) Organic geochemistry. Pergamon, New-York, pp 503–542

Blondeau R, Kalinowski E (1986) Fractionation of humic substances by hydrophobic interaction chromatography. J Chromatogr 35:585–589

Bloom PR, Leenheer JA (1989) Vibrational, electronic, and highenergy spectroscopic methods for characterizing humic substances. In: Hayes MHB, MacCarthy P, Malcolm RL, Swift RS (eds) Humic substances II-in search of structure. Wiley, Chichester, pp 409–446

Bull ID, van Bergen PF, Nott CJ, Poulton PR (2000) Organic geochemical studies of soils from the Rothamsted classical experiments – V. The fate of lipids in different long-term experiments. Org Geochem 31:389–408

Cameron RS, Thornton BK, Swift RS, Pozner AM (1972) Molecular weight and shape of humic acid from sedimentation and diffusion measurements on fractional extracts. J Soil Sci 23:394–408

Campbell CA, Paul EA, Rennie DA, McCallum KJ (1967) Applicability of the carbon dating method of analysis to soil humus studies. Soil Sci 104:217–224

Cerli C, Celi L, Kaiser K, Guggenberger G, Johansson M-B, Cignetti A, Zanini E (2008) Changes in humic substances along an age sequence of Norway spruce stands planted on former agricultural land. Org Geochem 39:1269–1280
Chantigny MH (2003) Dissolved and water-extractable organic matter in soils: a review on the influence of land use and management practices. Geoderma 113:357–380
Chen Y, Senesi N, Schnitzer M (1977) Information provided on humic substances by E_4/E_6 ratios. Soil Sci Soc Am J 41:352–358
Chen Y, De Nobili M, Aviad T (2004) Stimulatory effects of humic substances on plant growth. In: Magdoff F, Weil RR (eds) Soil organic matter in sustainable agriculture. CRC Press, Boca Raton, pp 103–129
Chilom G, Kohl SD, Rice JA (2005) The influence of lipids on the energetics of uptake of polycyclic aromatic hydrocarbons by natural organic matter. J Environ Qual 34:1055–1062
Cooper AB (1990) Nitrate depletion in the riparian zone and stream channel of small headwater catchments. Hydrobiology 202:13–26
Dahm H, Sitek JM, Strzelczyk E (1997) Synthesis of auxins by bacteria isolated from roots pine seedlings inoculated with rusty forest soil. Pol J Soil Sci 10:131–137
de Blas E, Rodrígues-Alleres M, Almendros G (2010) Speciation of lipid and humic fractions in soils under pine and eucalyptus forest in northwest Spain and its effect on water repellency. Geoderma 155:242–248
de Leeuw JW (2007) On the origin of sedimentary aliphatic macromolecules: a comment on recent publications by Gupta et al. Org Geochem 38:1585–1587
De Marco A, Esposito F, Berg B, Giordano M, De Santo AV (2013) Soil C and N sequestration in organic and mineral layers of two coeval forest stands implanted on pyroclastic material (Mount Vesuvius, South Italy). Geoderma 209–210:128–135
De Nobile M, Chen Y (1999) Size exclusion chromatography of humic substances: limits, and prospectives. Soil Sci 164:825–833
Dębska B, Gonet I (2007) Shape of hydrophobic and hydrophilic fractions in humic acids formed as a result of post-harvest residue decomposition. Pol J Soil Sci 40(1):57–65
Dębska B, Drag M, Banach-Szott M (2007) Molecular size distribution and hydrophilic and hydrophobic properties of humic acids isolated from forest soils. Soil Water Resour 2:45–53
Dell'Abate MT, Benedetti A, Trinchera A, Dazzi C (2002) Humic substances along the profile of two Typic Haploxerert. Geoderma 107:281–296
Derome AE (1987) Modern NMR techniques for chemistry research. Pergamon Press, Oxford
Dinel H, Schnitzer M, Mehuys GR (1990) Soil lipids: origin, nature, content, decomposition and effect on soil physical properties. In: Bollag JM, Stozky G (eds) Soil biochemistry, vol 6. Marcel Dekker, New York, pp 397–429
Drori Y, Lam B, Simpson AJ, Aizenshtat Z, Chefetz B (2006) The role of lipids on sorption characteristics of freshwater- and wastewater-irrigated soils. J Environ Qual 35:2154–2161
Dziadowiec H (1979) Zmiany energetyczne towarzyszące humifikacji ściółek leśnych. Studia Society Science. Toruniensis. Sectio D 11:104 (in Polish)
Filip Z, Tesařová M (2004) Microbial degradation and transformation of humic acids from permanent meadows and forest soils. Int Biodeterior Biodegrad 54:225–231
Filip Z, Pecher W, Berthelin J (1999) Microbial utilization and transformation of humic acids extracted from different soils. J Plant Nutr Soil Sci 162:215–222
Francioso O, Ciavatta C, Sánchez-Cortes S, Tugnoli V, Sitti L, Gessa C (2000) Spectroscopic characterization of soil organic matter in long-term amendment trials. Soil Sci 165:495–504
Francioso O, Ciavatta C, Montecchio D, Tugnoli V, Sánchez-Cortés S, Gessa C (2003) Quantitative estimation of peat, brown coal and lignite humic acids using chemical parameters, ^{1}H NMR and DTA analyses. Bioresour Technol 88:189–195
Fridland EV (1982) Nekotorye osobennosti pochvennykh lipidov v svyazi s ehkologicheskimi usloviyami (Some properties of soil lipids as related to ecological conditions). Pochvovedenie (2):38–46 (in Russian)

Futomo M, Fry WW, Blevis RL (1985) Function of legume cover crops in no-till and conventional till corn production. In: Hargrove WL, Boswell FC, Langdale GW (eds) The rising hope of our land. Proceedings of the 1985 Southern region no-till conference, Griffin, Georgia, 16–17 July 1985, pp 63–68

Goh KM, Williams MR (1979) Changes in molecular weight distribution of soil organic matter during soil development. J Soil Sci 30:747–755

Gonet SS, Wegner K (1993) Impact of long-term organic and mineral fertilization on the properties of humic acids. Zeszyty Naukowe Akademii Rolniczej. Kraków 37(277):51–63

Gorovaja AI, Orlov DS, Sherbenko OV (1995) Guminovye veshchestva (Humic substances). Naukova Dumka, Kiev (in Russian)

Hayes MHB (1991) Concepts of the origins, composition and structure of humic substances. In: Wilson WS et al (eds) Advances in soil organic matter research: the impact on agriculture and the environment. Royal Society of Chemistry, London, pp 3–22

Jandl R, Lindner M, Vesterdal L, Bauwens B, Baritz R, Hagedorn F, Johnson DW, Minkkinen K, Byrne KA (2007) How strongly can forest management influence soil carbon sequestration? Geoderma 137:253–268

Jeannotte R, Hamel C, Jabaji S, Whalen JK (2011) Pyrolysis-mass-spectrometry and gas chromatography-flame ionization detection as complementary tools for soil lipid characterization. J Anal Appl Pyrolysis 90:232–237

Kalbitz K, Solinger S, Park J-H, Michalzik B, Matzner E (2000) Controls on the dynamics of dissolved organic matter in soils: a review. Soil Sci 165:277–304

Kampert M, Strzelczyk E (1975) Synthesis of auxin by fungi isolated from the roots of pine seedlings (*Pinus silvestris* L.) and from soil. Acta Microbiol Pol 7:223–230

Kiang PH, Grob RL (1986) Development of screening method for the determination of 49 priority pollutants in soil. J Environ Sci Health A 21(1):15–53

Knohl A, Schulze E-D, Kolle O, Buchmann N (2003) Large carbon uptake by an unmanaged 250-year-old deciduous forest in Central Germany. Agric For Meteorol 118:151–167

Kögel-Knabner I (1993) Biodegradation and humification processes in forest soils. In: Bollag J-M, Stotzky G (eds) Soil biochemistry. Marcel Dekker, New-York, pp 101–137

Kögel-Knabner I (1997) ^{13}C and ^{15}N spectroscopy as a tool in soil organic matter studies. Geoderma 80:243–270

Kögel-Knabner I (2002) The macromolecular organic composition of plant and microbial residues as inputs to soil organic matter. Soil Biochem 34:139–162

Kononova MM (1966) Soil organic matter. Pergamon Press, Oxford

Kumada K, Miyara E (1973) Sephadex gel fractionation of humic acids. Soil Sci Plant Nutr 19:255–263

Leiweber L, Schulten HR (1992) Differential thermal analysis, thermogravimetry and in source pyrolysis-mass spectrometry studies on the formation of soils organic matter. Thermochim Acta 200:151–167

Liao JD, Boutton TW, Jastrow JD (2006) Storage and dynamics of carbon and nitrogen in soil physical fractions following woody plant invasion of grassland. Soil Biol Biochem 38:3184–3196

Lindqvist I (1967) Adsorption effect in gel filtration of humic acid. Acta Chem Scand 21:2564–2566

Lishtvan II, Dudarchik VM, Bambalov NN (1984) Izuchenie vzaimodeystviya guminovych veshchestv s glinistymi materialami (Study of the interaction of humic substances with clayey materials) Vestsi AN Belarusi. Chem Serious 1:87–90 (in Russian)

Lodygin ED, Beznosikov VA (2005) Influence of soil moisture on concentrations and ^{13}C NMR profiles of lipids in three Albeluvisols. Geoderma 127:253–262

Lorenz K, Lal R (2005) The depth distribution of soil organic carbon in relation to land use and management and the potential of carbon sequestration in subsoil horizons. Adv Agron 88:35–66

Lorenz K, Lal R, Preston CM, Nierop KGJ (2007) Strengthening the soil organic carbon pool by increasing contribution from recalcitrant aliphatic bio(macro)molecules. Geoderma 142:1–10
Lützow MV, Kögel-Knabner I, Ekschmitt K, Matzner E, Guggenberger G, Marschner B, Flessa H (2006) Stabilization of organic matter in temperate soils: mechanisms and their relevance under different soil conditions: a review. Eur J Soil Sci 57:426–445
Mao J-D, Hu W-G, Schmidt-Rohr K, Davies G, Ghabbour EA, Xing B (2000) Quantitative characterization of humic substances by solid-state carbon-13 nuclear magnetic resonance. Soil Sci Soc Am J 64:873–884
Martens DA, Frankenberger WT Jr (1993) Metabolism of tryptophane in soil. Soil Biol Biochem 25:1679–1687
Maryganova V, Szajdak LW (2011) Composition and molecular structure of humic substances from soils under shelterbelts of different age and adjoining arable fields. In: Szajdak LW (ed) Shelterbelts: efficient elements of the landscape. LAP (Lambert Academic Publishing), Saarbrücken, pp 67–86
Maryganova V, Bambalov N, Szajdak L (2004) Hydrophobic and hydrophilic properties of peat humic substances. In: Paivanen J (ed) Proceedings of the 12th international peat congress, Tampere, Finland, 6–11 June 2004, vol 1, pp 315–320
Maryganova VV, Szajdak L, Tychinskaya L Yu (2006) Sravnitelnaya otsenka amfifilnykh svoystv guminovykh veshchestv razlichnogo genezisa po dannym khromatografii gidrofobnogo vzaimodeystviya (Comparative estimation of amphiphilic properties of humic substances of various genesis as determined by hydrophobic interaction chromatography). In: Topf v reshenii problem ehnergetiki, selskogo khozyaystva i ehkologii (Peat in solution of energy, agriculture and ecology problems). Proceedings of the international conference, Minsk, Belarus, May 29–June 2, 2006, pp 225–229 (in Russian)
Maryganova V, Szajdak LW, Tychinskaya LYu (2010) Hydrophobic and hydrophilic properties of humic acids from soils under shelterbelts of different age. Chem Ecol 26(Suppl 2):25–33
Milanowskii EYu (2000) Amfifilnye komponenty gumusovykh veshchestv pochv (Amphiphilic components of soil humus substances). Pochvovedenie 6:706–715 (in Russian)
Milanowskii EYu, Shein EB (2002) Funktsionalnaya rol amfifilnykh komponentov gumusovykh veshchestv v protsessakh gumuso-strukturoobrazovaniya i v genezise pochv (Functional role of amphiphilic humus components in humus structure formation and soil genesis). Pochvovedenie 10:1201–1213 (in Russian)
Miller RW, Donahue RL (1995) Soils and our environment. Prentice Hall, New York
Mori S, Hiraide M, Mizuike A (1987) Aqueous size exclusion chromatography of humic acids on a Sephadex gel column with diluted phosphate buffers as eluents. Anal Chem Acta 193:231–238
Muscolo A, Nardi S (1997). Auxin or auxinlike activity of humic matter. In: Drozd J, Gonet SS, Senesi N, Weber J (eds) The role of humic substances in the ecosystems and in environmental protection. Proceedings of the 8th meeting of the international humic substances society, Wrocław, Poland, 9–14 September, 1996, PTSH – Polish Society of Humic Substances. Polish Chapter, International Humic Substances Society, pp 997–992
Muscolo A, Cutrupi S, Nardi S (1998) IAA detection in humic substances. Soil Biol Biochem 30:1199–1201
Naafs DFW, van Bergen PF, Boogert SJ, de Leeuw JW (2004) Solvent extractable lipids in an acidic andic forest soil: variations with depth and season. Soil Biol Biochem 36:297–308
Negre M, Vindrola D, Spera S, Ferraris L, Gennari M (2002) Effect of the chemical composition of soil humic acids on their viscosity, surface pressure, and morphology. Soil Sci 167:636–651
Nierop KGJ, Naafs DFW, van Bergen PF (2005) Origin, occurrence and fate of extractable lipids in Dutch coastal dune soils along pH gradient. Org Geochem 36:555–566
Orlov DS (1990) Pochvennye gumusovye kisloty i obshchaya teoriya gumifikatsii (Soil humus acids and the general theory of humification). MGU, Moskva (in Russian)

Orlov DS, Osipova NI (1988) Infrakrasnye spektry pochv i pochvennykh komponentov (Infrared spectra of soils and soil components). MGU, Moskva (in Russian)
Patten CL, Glick BR (1996) Bacterial biosynthesis of indole-3-acetic acid. Can J Microbiol 42(3):207–220
Paul KI, Polglase PJ, Nyakuengama JG, Khanna PK (2002) Change in soil carbon following afforestation. For Ecol Manag 168:241–257
Peterjohn WT, Correll DL (1984) Nutrient dynamics in agricultural watershed: observation on the role of a riparian forest. Ecology 65:1466–1475
Piccolo A, Mbagwu JSC (1999) Role of hydrophobic components of soil organic matter in soil aggregate stability. Soil Sci Soc Am J 63:1801–1810
Piccolo A, Mirabella A (1987) Molecular weight distribution of peat humic substances extracted with different inorganic and organic solutions. Sci Total Environ 62:39–46
Piccolo A, Conte P, Cozzolino A (1999) Effects of mineral and monocarboxylic acids on the molecular associations of dissolved humic substances. Eur J Soil Sci 50:687–694
Post WM, Kwon KC (2000) Soil carbon sequestration and land-use change: processes and potential. Glob Change Biol 6(3):317–327
Poulton PR, Pye E, Hargreaves PR, Jenkinson D (2003) Accumulation of carbon and nitrogen by old arable land reverting to woodland. Glob Change Biol 9:942–955
Preston CM (1996) Application of NMR to soil organic matter analysis: history and prospects. Soil Sci 161:144–166
Prusinkiewicz Z, Pokojska U, Józefowicz-Kotlarz J, Kwiatkowska A (1996) Studies on the functioning of biogeochemicals barriers. In: Ryszkowski L, French N, Kędziora A (eds) Dynamics of an agricultural landscape. Państwowe Wydawnictwo Rolnicze i Leśne, Poznań, pp 110–119
Quénéa K, Largeau C, Derenne S, Spaccini R, Bardoux G, Mariotti A (2006) Molecular and isotopic study of lipids in particle size fractions of a sandy cultivated soil (Cestas cultivation sequence, southwest France): sources, degradation, and comparison with Cestas forest soil. Org Geochem 37:20–44
Quideau SA, Chadwick OA, Benesi A, Graham RC, Anderson MA (2001) A direct link between forest vegetation type and soil organic matter composition. Geoderma 104:41–60
Raczkowska-Błach E, Różycki H, Strzelczyk E, Pokojska A (1995) Decomposition of indoleacetic acid (IAA) in soil and by bacterial strains isolated from soil and from the root zone of Scots pine (*Pinus sylvestris L.*). Microbiol Res 150(3):265–270
Rakovskiy V, Filimonov V (1967) Mechanizm peaktsii termicheskogo razlozheniya kislorodsoderzhashchikh komponentov topliv (Mechanism of the reaction of thermal destruction of oxygen containing components of fuels). Khimiya i khimicheskaya tekhnologiya, vol 3, Nedra, Moskva, pp 302–311 (in Russian)
Resh SC, Binkley D, Parrotta JA (2002) Greater soil carbon sequestration under nitrogen-fixing trees compared with eucalyptus species. Ecosystems 5:217–231
Ritter E, Vesterdal L, Gundersen P (2003) Changes in soil properties after afforestation of former intensively managed soils with oak and Norway spruce. Plant Soil 249:319–330
Rosenqvist L, Kleja DV, Johansson M-B (2010) Concentrations and fluxes of dissolved organic carbon and nitrogen in a *Picea abies* chronosequence on former arable land in Sweden. For Ecol Manag 259:275–285
Ryszkowski L, Bartoszewicz A (1989) Impact of agricultural landscape structure on cycling of inorganic nutrients. In: Clarholm M, Bergström L (eds) Ecology of arable land. Kluwer Publ, Dordrecht, pp 241–246
Ryszkowski L, Bartoszewicz A, Kędziora A (1997) The potential role of mid-field forest as buffer zones. In: Haycock NE, Burt TP, Gouldingf KW, Pinay G (eds) Buffer zones: their processes and potential in water protection. Quest Environmental, Harpenden, pp 171–191
Ryszkowski L, Szajdak L, Karg J (1998) Effects of continuous cropping of rye on soils biota and biochemistry. Crit Rev Plant Sci 38:225–244

Ryszkowski L, Szajdak L, Bartoszewicz A, Życzyńska-Bałoniak I (2002) Control of diffuse pollution by mid-field shelterbelts and meadow strips. In: Ryszkowski L (ed) Landscape ecology in agroecosystems management. CRC Press, Boca Raton/London/New York/Washington, DC, pp 111–143

Salloum MJ, Chefetz B, Hatcher PG (2002) Phenanthrene sorption by aliphatic-rich natural organic matter. Environ Sci Technol 36:1953–1985

Sarwar M, Arshad M, Frankenberger WTJ (1992) Tryptophan-dependent biosynthesis of auxins in soil. Plant Soil 147:207–215

Schnitzer M, Khan SU (1978) Soil organic matter. Elsevier, Amsterdam

Schnitzer M, Turner RC, Hofman JA (1974) A thermogravimetric study of organic matter of representative Canadian soils. Can J Soil Sci 44:7–13

Sheppard JD, Forgeron DW (1987) Differential thermogravimetry of peat fractions. Fuel 66:232–236

Shurygina EA (1971) Differential thermal analysis (DTA) and thermogravimetry (TG) of soils humus substances. Geoderma 6:169–177

Smolander A, Kitunen V (2002) Soil microbial activities and characteristics of dissolved organic C and N in relation to tree species. Soil Biol Biochem 34:651–660

Soil Taxonomy (1968) A basic system of soil classification for making and interpreting soil surveys. U.S. Department of Agriculture Soil Conservation Service. Robert E Krieger Publishing Company, Malabar, pp 362–364

Spaccini R, Piccolo A, Haberhauer G, Gerzarek MN (2000) Transformation of organic matter from maize residues into labile and humic fractions of three European soils as revealed by ^{13}C distribution and CPMAS-NMR spectra. Eur J Soil Sci 51:583–594

Spaccini R, Mbagwu JSC, Conte P, Piccolo A (2006) Changes of humic substances characteristics from forested to cultivated soils in Ethiopia. Geoderma 132:9–19

Stepanov AA (2005) Structural features of amphiphilic humic acid fractions from a southern chernozem. Eurasian Soil Sci 38:843–847

Stepanov AA (2008) Separation and characterization of amphiphilic humus acid fractions. Mosc Univ Soil Sci Bull 63:125–129

Stevenson FJ (1967) Organic acids in soil. In: McLaren AD, Peterson GH (eds) Soil biochemistry. Marcel Dekker INC, New York, pp 119–146

Stevenson FJ (1994) Humus chemistry: genesis, composition, reactions, 2nd edn. Wiley, New York

Strzelczyk E, Dahm H, Pachlewski R, Różycki R (1997) Production of indole compounds by the ectomycorrhizal fungus *Cantharellus cibarius* Fr. Pedobiologia 41:402–411

Swift RS, Posner AM (1971) Gel chromatography of humic acids. J Soil Sci 22:237–249

Swift MJ, Heal OW, Anderson JM (1979) Decomposition in terrestrial ecosystems. Blackwell, Oxford

Szajdak L (1996) Impact of crop rotation and phenological periods on rhodanese activity and sulphur amino-acids concentrations in brown soils under continuous rye cropping and crop rotation. Environ Int 22:563–569

Szajdak L (2002) The estimation of the evolution of nitrous oxide and carbon dioxide from soils under shelterbelt. In: Ilnicki P (ed) Restoration of carbon capacity and biodiversity in abandoned grassland on peatland in Poland. Wydawnictwo Akademii Rolniczej im. A. Cieszkowskiego, Poznań, pp 150–162

Szajdak LW (2011) Shelterbelts: efficient element of the landsacape. LAP (Lambert Academic Publishing), Saarbrücken

Szajdak L, Maryganova VV (2003) Structural characterization of humic substances from soils under shelterbelts of different age and adjoining arable fields. Humic Subst Environ 3:37–44

Szajdak L, Maryganova V (2007) Occurrence of IAA auxin in some organic soils. Agron Res 5(2):175–187

Szajdak L, Maryganova V (2009) Impact of age and composition of shelterbelts plants on IAA content as allelochemical in soils. Allelopathy J 23:461–468

Szajdak L, Matuszewska T (2000) Reaction of woods in changes of nitrogen in two kinds of soil. Pol J Soil Sci 33(1):9–17

Szajdak L, Ryszkowski L (2002) The estimation of the evolution of nitrous oxide from soils under shelterbelts. Działalność Naukowa PAN 12:133–134 (in Polish)

Szajdak L, Życzyńska-Bałoniak I (2002) Influence of mid-field afforestation on the changes of organic nitrogen compounds in ground water and soil. Pol J Environ Stud 11(1):91–95

Szajdak L, Maryganova V, Meysner T, Tychinskaja L (2002) Effect of shelterbelt on two kinds of soils on the transformation of organic matter. Environ Int 28:383–392

Szajdak LW, Maryganova V, Skakovskiy E, Tychinskaya L (2015) ^{1}H and ^{13}C NMR spectroscopic studies of hexan-extractable lipids from soils under shelterbelts of different age and composition of plants. Chemosphere 119:1422–1427

Tena M, Pinilla JA, Magallanes M (1986) L-phenylalanine deaminating activity in soil. Soil Biol Biochem 18(3):321–325

van Bergen PF, Bull ID, Poulton PR, Evershed RP (1997) Organic geochemical studies of soils from Rothamsted Classical Experiments. – total lipid extracts, solvent insoluble residues and humic acids from Broadbalk Wilderness. Org Geochem 20(1–2):117–135

Wang X, Xing B (2007) Roles of acetone-conditioning and lipid in sorption of organic contaminants. Environ Sci Technol 41:5731–5737

Weber JH (1988) Binding and transport of metals by humic materials. In: Frimmel FH, Christman RF (eds) Humic substances and their role in environment. Willey, Chichester, pp 165–178

Zech W, Guggenberger G (1996) Organic matter dynamics in forest soils of temperate and tropical ecosystems. In: Piccolo A (ed) Humic substances in terrestrial ecosystems. Elsevier, Amsterdam, pp 101–170

Zech W, Ziegler F, Kögel-Knabner I, Haumaier L (1992) Humic substances distribution and transformation in forest soils. Sci Total Environ 117–118:155–174

Zech W, Senesi N, Guggenberger G, Kaiser K, Lehmann J, Miano TM, Miltner A, Schrith G (1997) Factors controlling humification and mineralization of soil organic matter in the tropics. Geoderma 79:117–161

Zhang Bin, Peng Xin-Hua (2006) Organic matter enrichment and aggregate stabilization in a severely degraded ultisol after reforestation. Pedosphere 16(6):699–706

Zsolnay A (1996) Dissolved humus in soil waters. In: Piccolo A (ed) Humic substances in terrestrial ecosystems. Elsevier, Amsterdam, pp 171–223

Chapter 10
Phytohormone in Peats, Sapropels, and Peat Substrates

Lech Wojciech Szajdak

Abstract The concentration of indole-3-acetic acid was measured in a low-moor peat, a high-moor peat, moorshes with a different degree of decomposition, and commercial substrates used for horticulture and pomology. In addition, the amount of IAA was determined in sapropels and in soils treated with different types of fertilizers.

Keywords Auxin • Peats • Sapropels • Bounded amino acids • Total nitrogen • Different kinds of fertilizers

10.1 Introduction

Plant rhizosphere in soil harbors many bacteria, viruses, fungi, protozoa, and algae. This community is supported nutritionally by a high input of organic compounds delivering from the degradation of plants, microbes, plant roots, and root exudates in aerobic and anaerobic conditions. These substances are necessary for microbial growth (Lynch 1990; Smolander and Kitunen 2002). The quantitative and qualitative composition of root exudates varies depending on a plant species (Smith 1976), a type of soil, and the physical properties of the environment such as land use, the degree of decomposition of plants, pH, ionic strength, humidity, and temperature (Martin and Kemp 1980; Aeshad and Frankenburger 1991; Szajdak and Maryganova 2007, 2008; Shahab et al. 2009).

Actinomycetes are Gram-positive bacteria. They are the most widely distributed group of microorganisms in nature and are also well known as saprophytic soil inhabitants (Takizawa et al. 1993). Most *Actinomycetes* belong to the genus *Streptomyces* (Goodfellow and Simpson 1987; Suzuki et al. 2000). They are a source of biologically active compounds such as amino acids, carbohydrates,

L.W. Szajdak (✉)
Institute for Agricultural and Forest Environment, Polish Academy of Sciences, Str. Bukowska 19, 60-809 Poznań, Poland
e-mail: lech.szajdak@isrl.poznan.pl

L.W. Szajdak (ed.), *Bioactive Compounds in Agricultural Soils*,
DOI 10.1007/978-3-319-43107-9_10

enzymes, vitamins, antifungal, and antibacterial compounds. Plant growth-promoting substances have been developed for agricultural originate from this genus (Bandurski and Schulze 1977; Ilic et al. 2007).

Streptomyces populate the rhizosphere of plants and can enhance plant growth by producing plant growth-promoting substances, for instance, auxin or gibberellin (Tresner et al. 1961; Kaunat 1969; Brown 1972; Merckx et al. 1987). Plant growth-regulating compounds called phytohormones represent a wide group of organic substances produced by plants and microorganisms in the rhizosphere. Their role is to enhance seed germination and plant growth. These substances are synthesized in one part of the plant and then transported to another part to influence a whole range of physiological and developmental processes at low concentrations. These compounds stimulate plant growth and biofertilization (the fixation of atmospheric N_2 and solubilization of nutrients) and can protect against plant pathogens. Numerous plant processes are known to be controlled by plant growth regulators; however, the exact mechanisms of control have remained elusive (Chalvignac and Mayaudon 1971; Goldsmith 1977; Cohen and Bandurski 1982; Lebuhn et al. 1995; Dahm et al. 1997; Beyeler et al. 1999; Chen et al. 2004; Kennedy et al. 2004).

Many naturally occurring hormones contain heterocyclic aromatic structures and carboxylic and amine groups. In lichens, some organic compounds [auxins, such as indole-3-acetic acid (IAA)] (Fig. 10.1) are produced by fungi and algae (Kampert and Strzelczyk 1975).

Auxins represent a broad group of compounds which are derivatives of indole. Their chemical structure explains the relationship between quantity and activity of their members. Since the discovery of auxins, efforts have been made to manipulate plant growth and development. Mycorrhizal fungi produce auxins (IAA is preferred) which affect the growth of susceptible roots by modifying the apices so that they assume the structure of the host tissues of mycorrhizas (Harley and Smith 1983; Strzelczyk and Pokojska-Burdziej 1984; Strzelczyk et al. 1997).

Auxins have the ability to improve plant growth by stimulating cell elongation, root initiation, seed germination, and seedling growth (El-Tarabily 2008). IAA is a common natural auxin and is a product of L-tryptophan metabolism in microorganisms. Approximately 80 % of rhizosphere bacteria can produce IAA (Bhavdish et al. 2003). *Streptomyces spp.*, inhabiting the rhizosphere of various plants, also serve as a good source of IAA. A rich supply of substrates available in root exudates gives the *Streptomycetes* the opportunity to synthesize and release IAA (Wheeler et al. 1984; Kravchenko et al. 1991; Martens and Frankenberger 1994). Several *Streptomyces* species such as *S. olivaceoviridis, S. rimosus, S. rochei, and Streptomyces spp.* populating the tomato rhizosphere have the ability to produce IAA and improve plant growth by increasing seed germination and dry weight and causing root elongation (Sanderson et al. 1987; Aldesuquy et al. 1998;

Fig. 10.1 Indole-3-acetic acid

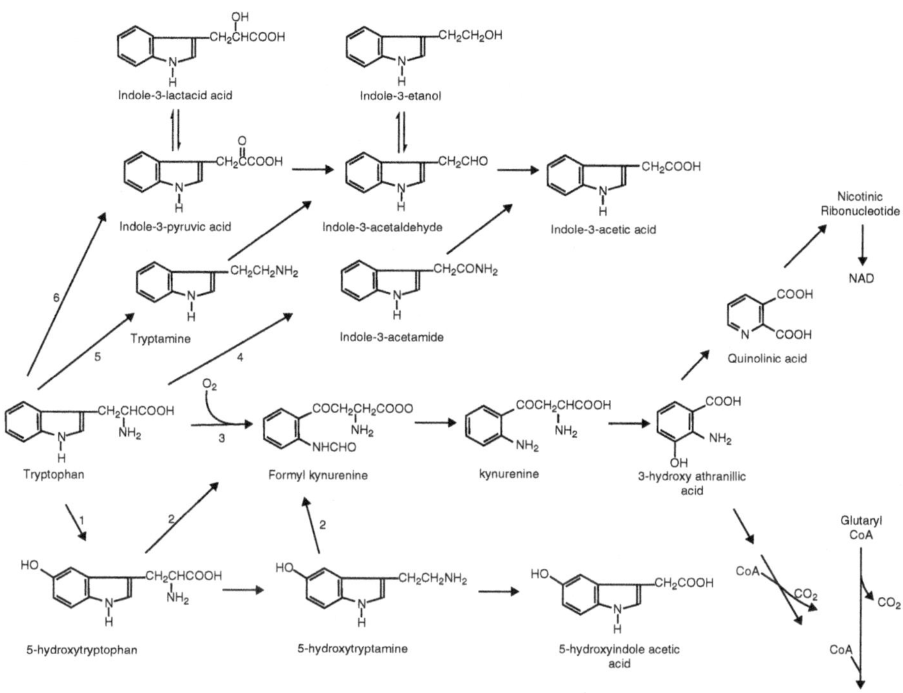

Fig. 10.2 L-Tryptophan pathways in bacteria. Enzymes in pathways are as follows: 1, tryptophan-5-hydroxylase; 2, indole 2,3-dioxygenase; 3, tryptophan 2,3-dioxygenase; 4, tryptophan monooxygenase; 5, tryptophan decarboxylase; 6, tryptophan aminotransferase (Martens and Frankenberger 1993)

Tokala et al. 2002; El-Tarabily 2008). Therefore, *Streptomyces* and *Actinomycetes* have a practical usage. Mycostop, based on the strain K61 of *S. griseoviridis* and *S. lydicus* WYEC108, produces IAA to promote plant growth (Mahadevan and Crawford 1997; Tsavkelova et al. 2006).

IAA seems to play an important function in nature due to its ability to regulate plant growth and development. This compound activates root morphology and metabolic changes in the host plant (Brown 1972; Bhavdish et al. 2003; Chen et al. 2004).

Therefore, the formation of chemical compounds with a biological activity such as auxins is extremely important in the productivity of soils and the growth and development of plants (Rovira 1965; Frankenberger and Arshad 1991a, b).

The conversion of L-tryptophan in soils, plants, animals, and the human body undergoes several pathways (Bender 1989; Tena et al. 1986; Müller et al. 1989; Sarwar et al. 1992; Szajdak et al. 2004a, 2007; Szajdak and Maryganova 2008). L-tryptophan can be converted to kynurenine [L-2-amino-4-(2-aminophenyl)-4-oxobutanoic acid] by tryptophan 2,3-dioxygenase (EC 1.13.11.11) (Fig. 10.2).

L-tryptophan is a substrate for the synthesis of vitamin, niacin. It is a metabolic energy substrate (Martens and Frankenberger 1993). Moreover, L-tryptophan is converted by hydroxylase to 5-hydroxytryptophan, 5-hydroxytryptamine (serotonin), and 5-hydroxy-indole-3-acetic acid. L-tryptophan is incorporated into the structure of proteins or peptides. In addition, 5-hydroxytryptophan and 5-hydroxytryptamine can be included in the reaction into the kynurenine route by indole-2,3- dioxygenase (EC 1.13.11.17), (Bender 1989). Furthermore, L-tryptophan may be converted by soil microorganisms into IAA.

The formation of IAA in soils is regulated by L-aminotransferase which is specific for amino acids and uses L-tryptophan as a substrate. However, the activity of this enzyme (L-aminotransferase tryptophan) requires the presence of the following amino acids: L-asparagine, L-lysine, and L-alanine. The stimulation of L-tryptophan aminotransferase by different amino acids may explain why rhizosphere in soils exhibits a higher concentration of auxin in comparison with a root-free soil (Martens and Frankenberger 1993). Furthermore, Müller et al. (1989) postulated that the ability of soil microbiota to form and release IAA in their environment was dependent on the catabolic conversion of L-tryptophan. The enzymes involved in the catabolism of L-tryptophan may regulate the formation of auxin in soils.

In addition, Kutacek and Kefeli (1970), Law (1987), McQueen-Mason and Hamilton (1989), and Tsurusuki et al. (1990) postulated that IAA was formed from D-tryptophan in plants. Furthermore, Kuraishi and Sakurai (1988) showed that in plants, D-tryptophan activated the IAA formation to a higher degree than L-tryptophan. Therefore, D-stereospecificity of aminotransferase is required for the reaction. Müller et al. (1989) postulated that the ability of microorganism in soils to the formation and utilization of IAA depended on the balance between anabolism and catabolism of both L and D isomer of tryptophan. While L-tryptophan was identified as the precursor of IAA formation, D-tryptophan may also serve as a substrate in IAA formation in plants (Law 1987; McQueen-Mason and Hamilton 1989; Tsurusuki et al. 1990; Martens and Frankenberger 1993).

In addition, mineral and organic soils contain enzyme peroxidase. It participates in the oxidation of IAA (Gelinas 1973). The process of the oxidation of IAA by peroxidases is inhibited by natural inhibitors of this pathway—phenolic compounds of low molecular weight (caffeic acid, gallic acid, hydroquinone, catechol, 3,4-dimethylphenol, 4-methoxyphenol, 2-aminophenol, 6-chloro-4-nitro-2--aminophenol) (Krylov et al. 1994). The authors showed that only a minimum concentration of these inhibitory compounds was required for the cessation of IAA (1 mM) oxidation catalyzed by peroxidase (Table 10.1).

The lowest threshold inhibitory concentration was reported for caffeic acid and the highest one for 3,4-dimethylphenol. Stonier and Yoneda (1967, 1970) reported that several phenolic inhibitors from *Pharbitis* had a high molecular weight ranging from 5000 to 10,000 (Krylov et al. 1994). Phenolic compounds are common in peat-forming plants and, therefore, they can be also found in peat. Even though some part of these substances is bound and insoluble, others can be extracted with organic solvents. Due to the decomposition of lignin, new phenolic compounds in peat are created. Lignin, however, cannot be the only origin of phenolic compounds as the most important peat-forming moss—*Sphagnum*—contains no lignin, but it does contain phenolic compounds. The most common phenolic compound is

Table 10.1 Minimal inhibitor concentration required for the cessation of IAA (1 mM) oxidation catalyzed by peroxidase (0.1 μM)

Inhibitor	Threshold inhibitor concentration (μM)	Structures
Caffeic acid	0.24–0.26	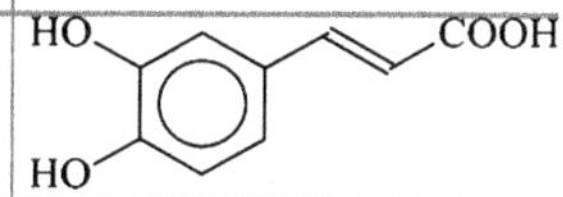
Hydroquinone	0.52–0.64	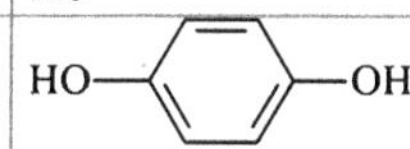
Catechol	0.64–0.80	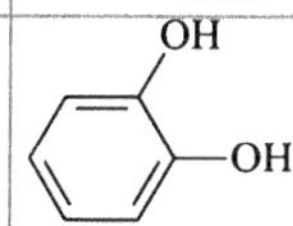
4-Methoxyphenol	0.96–1.04	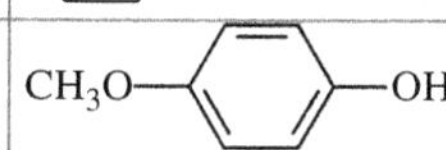
2-Aminophenol	0.96–1.04	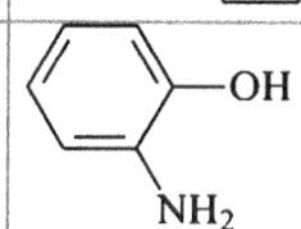
6-Chloro-4-nitro-2-aminophenol	1.20–1.28	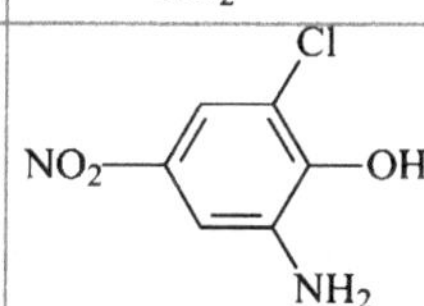
Gallic acid	1.20–1.32	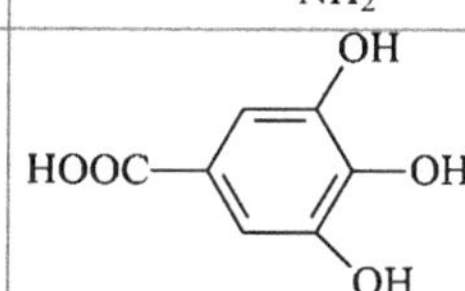
3,4-Dimethylphenol	18.4–20.0	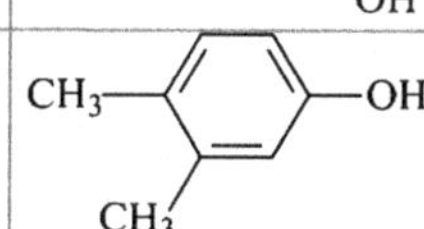

Krylov et al. (1994); Szajdak—structures

p-hydroxy-β-(carboxymethyl)-cinnamic acid (sphagnol, sphagnum acid), which decomposes fast to several substances such as p-hydroxybenzoic acid, p-hydroxy-acetophenone, and hydroxybutenolide. Some phenolic compounds in peat are water soluble and hence their contents decreases during leaching (Uosukainen and Pihlaja 2006).

Moreover, recent studies have shown that atrazine (herbicide, derivate of S-triazine) inhibits IAA formation both in the field and in vitro (Grapelli and Rossi 1979; Rossi et al. 1984; Maryganova and Szajdak 2007; Szajdak et al. 2007; Szajdak and Maryganova 2008). Atrazine was found to have an adverse impact on the root development of vegetation cuttings as well as on the height of several species of greenhouse-grown flowers and nursery-produced ornamental plants.

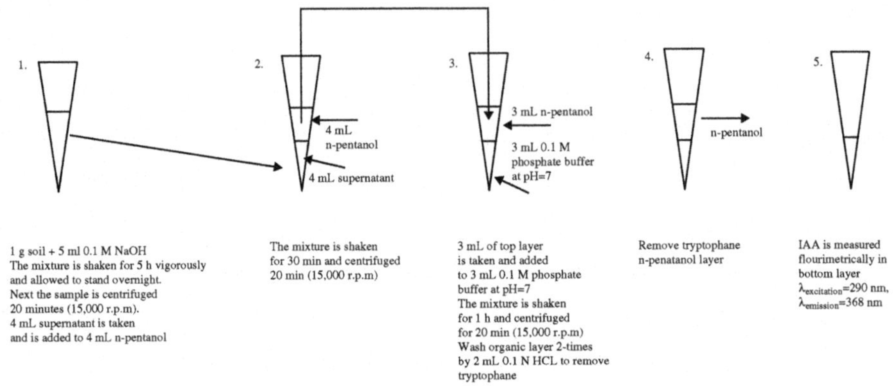

Fig. 10.3 Scheme of the extraction and determination of IAA in soils

IAA is commonly formed in mineral and organic soils (peat, moorsh, and sapropels) (Szajdak et al. 2004a; Maryganova and Szajdak 2007; Szajdak and Maryganova 2007, 2008). It is also created under natural conditions in compost as well as in inorganic and organic soils under different cultivated plants (Musculo and Nardi 1997; Muscolo et al. 1998).

Molecular components of humic substances regulate plant growth phytohormones. Both humic and fulvic acids inhibit the activity of indole acetic acid oxidase (IAA-oxidase) by hindering IAA destruction. When IAA is protected from its degrading enzymes, it continues to stimulate growth processes. In addition, the results revealed that unfractionated humic acid regulates plant growth hormones most effectively. While the activity of growth regulators is optimized within plant tissues, plant metabolism remains functional and normal growth processes continue to occur.

Some kind of peats and sapropels is commonly use as substrates, growing media for agriculture, horticulture, floriculture, and pomology. Therefore, the knowledge of the content of IAA and other organic chemical compounds of a well-known structure in organic soils—peat soils and sapropels—is needed.

10.2 The Determination of Indole-3-Acetic Acid (IAA) in Soils by Fluorimetric Method

Reagents IAA standard, 0.1 N NaOH, n-pentanol, 0.1 M phosphate buffer at pH = 7, and 0.1 n HCl

Instruments pH meter, shaker, centrifuge, and spectrofluorometer

Method (Fig. 10.3)

(a) Add 2-g soil to 10-mL 0.1 N NaOH. The mixture is shaken for 5.0 h vigorously and is allowed to stand overnight. Next day the sample is centrifuged for 20 min (15,000 r.p.m.).

(b) 4 mL of supernatant is taken and added to 4 mL of n-penthanol.
(c) The mixture is shaken for 30 min.
(d) The mixture is again centrifuged for 20 min (15,000 r.p.m.).
(e) 3 mL of top layer is taken and added to 3 mL of 0.1 M phosphate buffer at pH=7.0. The mixture is shaken for 1.0 h and again centrifuged for 20 min (15,000 r.p.m).
(f) Wash organic layer two times by 2 mL 0.1 N HCl to remove tryptophane.
(g) IAA in the resulting soil extraction is measured fluorometrically in the bottom layer at $\lambda_{excitation}=290$ nm and $\lambda_{emission}=368$ nm.
(h) The IAA concentration is calculated from the analytical curve. The IAA concentration ranged from 50 to 300 ng mL^{-1} and was prepared similarly to investigated soils samples.

The highest fluorescence of IAA is measured at pH 7 (Figs. 10.4 and 10.5). Therefore, the measurement of IAA should be performed at this pH value. The analytical curve expresses the relationship between the fluorescence and the concentration of IAA with high correlation coefficient ($r=0.998$). The slopes of IAA (i) before extraction, (ii) without any washing with 0.1 N HCl, and (iii) after washing with 0.1 N HCl are used to calculate IAA in soil samples (Fig. 10.5).

10.3 Peat

Peat represents a raw material. The transformation of vegetable matter into peat is a process whose continuation leads to the formation of lignite, coal, and anthracite. During peat formation, the humification of organic substances takes place.

Fig. 10.4 Fluorescence of IAA in different pH

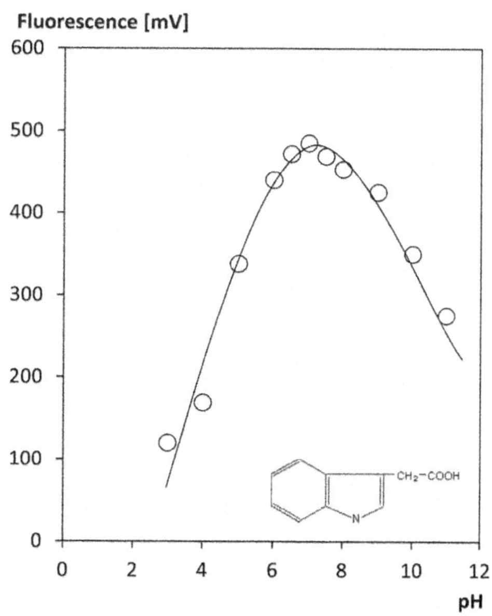

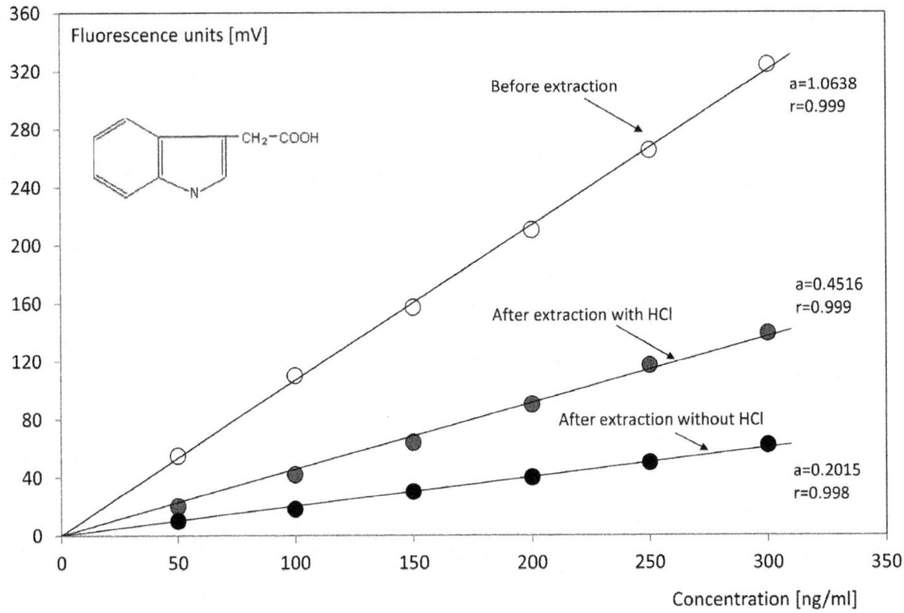

Fig. 10.5 Analytical curve of IAA

Subsequent steps of diagenesis include dewatering and compaction. Throughout humification process, the concentration of hydrogen and oxygen continually decreases, while the concentration of organic matter increases. Peat is a heterogeneous material due to a variety of plants whose residues contribute to peat formation as well as due to varied environmental conditions in which humification takes place. Peatland formation depends not only on climate but also on land geomorphology (except for the coastal or highly mountainous areas). Peat is comprised of relatively unstable substances whose reactivity contributes to its usefulness. Peat is characterized by a colloidal behavior and an irreversible loss of wettability. The physiological activity of peat is observed in the stimulation and promotion of plant growth (Van Dijk 1971; Maciak et al. 1977; Lüttig 1986; Sokolov and Bambalov 2000; Szajdak et al. 2011).

Peat and the peat processing products (peat substrate, growing media, fertilizer, sorbents, the ion exchanger) play an important role in the environment protection. A degree of decomposition, botanical composition, ash content, acidity, moisture content, heat capacity, and heat diffusivity determines the structure of peat deposits, their role in the formation of biospheric processes at a regional and an interregional level, as well as their influence on aqueous territories. Due to hydraulic engineering and agrochemical amelioration, peat deposits can be transformed into fertile peat grounds whose economic use differs because of the necessity to observe certain rules of agro-techniques and hydro-amelioration (Lishtvan 2004).

In 2001 world resources of peat were estimated at 1.9 trillion tons, out of which the countries from the former Soviet Union (FSU) possessed ca. 770 billion tons and Canada, ca. 510 billion tons. Domestic deposits of peat occurred in all 50 states of the USA, with estimated resources of about 310 billion tons or about 16 % of the world total (US Geological Survey, Mineral Commodity Summaries, January 1998).

Approximately 95 % of peat in the USA is being sold for horticultural/agricultural usage, including general soil improvement, potting soils, earthworm culture, nursery business, and golf course maintenance and construction. Other applications included seed inoculants, vegetable cultivation and mushroom culture, mixed fertilizers, and packing for flowers and plants. In the industrial sector, peat found widespread use as an oil absorbent, an efficient filtration medium for the removal of waterborne contaminants in mine waste streams, and municipal storm drainage.

The concentrations of IAA ranged from 124.5 to 386.05 $\mu g \cdot kg^{-1}$ d.m. in high-moor peats, 57.90–273.20 $\mu g \cdot kg^{-1}$ d.m. in low-moor peats, and 87.10–296.94 $\mu g \cdot kg^{-1}$ d.m. in moorshes (Tables 10.2 and 10.3).

However, the content of IAA in commercial substrates ranged from 22.9 to 373.9 $\mu g \cdot kg^{-1}$ d.m. (Table 10.4).

10.4 Sapropel

The term sapropel defines bottom sediments occurring in freshwater basins. The following terms are used for sapropel in the scientific literature:

Sapropel, gytia (Polish)

- Sapropel, sapropelite, clay sapropel, calcareous gyttja—sapropel (English)
- Sapropel, sapropelit, dy, gyttja, lergyttja, kalkgyttja (Swedish)
- пресноводный органогенный, сапропелит м; сапропель м, сапропель, глинистый м, сапропель, известковый м. (Russian)
- Sapropel n., Sapropelit m., Sapropelgestein n., Faulschlammkohle f, Sapropel n., Faulschlamm m., Gyttja f., Mudde f, Tonmudde f, Kalkgyttja, Kalkmudde f., kalkhaltiger Faulschlamm m., Seekreide f., Alm m. (German)
- muta, mätälieju, sapropeliitti; sapropeli, mätälieju, savisapropeli, savimätälieju, kalkkilieju (Finish)

Sapropel is located near the peat. It is formed during different long-term chemical, biochemical, biological, and physical processes. The following processes dominate: the degradation of aquatic vegetation, the decomposition of life forms remains and/or organic and inorganic soil pieces drifted by water, and the conversions of organic compounds. Sapropel includes three main components: water (from 60 to 97 %), ash (sand, loam, carbonates, phosphates, silica, calcium, magnesium and iron compounds, less than 30 %, etc.), and organic compounds (at least 15 %) of a well-known and unknown structure (Belkevitch 1962; Ilnicki 2002; Kurzo 2010). Sapropel contains inorganic compounds (less than 30 % in dry mass: silica, calcium, iron, magnesium, potassium, aluminum, sulfur, phosphorous (Co, Mn, Cu, B, Zn, I, Br, Mo, Cr, Be, Ni, Ag, Sn, Pb, Sr, Ti) (Kireicheva and Khokhlov 2000).

Table 10.2 Peats, degree of the decomposition, total organic carbon, dissolved organic carbon, and the concentrations of IAA in different kind of peats and IAA in (μg kg^{-1} d. m.)

No of place	Place of sampling / Depth in cm	Type of soil	Degree of decomposition in von Post scale	pH in 1 N KCl	TOC g kg^{-1}	DOC g kg^{-1}	IAA μg kg^{-1} d.m.
High-moor peatlands							
1	Kusowskie Bagno 0–25 cm	Cotton *Sphagnum*	H2	2.56	609.35	18.36	191.81 ± 8.6
2	Kusowo 25–50 cm	Cotton *Sphagnum*	H3	2.58	639.15	14.77	167.17 ± 7.5
3	Kusowo 50–75 cm	Cotton *Sphagnum*	H4	2.57	643.05	13.02	168.05 ± 7.8
4	Kusowo 75–100 cm	Cotton *Sphagnum*	H2	2.62	650.45	13.17	156.21 ± 6.5
5	Orekhovskiy Mokh 0–25 cm	*Augustifolium*	H2	3.3	511.1	15.4	124.5 ± 4.7
6	Orekhovskiy Mokh 30–50 cm	*Magellanicum*	H2	3.5	536.7	10.9	144.2 ± 4.6
7	Ducora 0–25 cm	Pine *Sphagnum*	H4	2.3	564.2	18.1	203.4 ± 7.7
8	Ducora 30–50 cm	*Eriophorum Sphagnum*	H4	2.4	567.9	15.4	210.2 ± 7.6
9	Ducora 75–100 cm	*Eriophorum Sphagnum*	H4	2.6	579.9	11.8	185.2 ± 7.0
10	High ryam 0–25 cm	Pine-Cotton grass, *Sphagnum*, raised bog	H4, H5	4.0	443.6 ± 15.9	14.36 ± 0.52	291.02 ± 10.2
11	High ryam 50–75 cm	Wood-Cotton grass, transitional mire	H6	4.5	447.1 ± 15.2	9.74 ± 0.34	314.78 ± 14.2
12	Low ryam 0–50 cm	Fuscum peat	H1	3.5	394.5 ± 14.3	394.5 ± 14.3	386.05 ± 16.2
	Low ryam 50–75 cm	Fuscum peat	H1	4.0	394.5 ± 14.3	394.5 ± 14.3	279.14 ± 10.9
1	Low moor peatlands						
1	Stążka 0–25 cm	Sedge–moss	H3	6.53	580.60	18.26	244.09 ± 10.8
2	Stążka 25–50 cm	Sedge–moss	H3	5.82	589.95	9.28	269.89 ± 12.1
3	Stążka 50–75 cm	Sedge–moss	H3	5.61	598.50	6.53	219.09 ± 9.4

(continued)

Table 10.2 (continued)

No of place	Place of sampling / Depth in cm	Type of soil	Degree of decomposition in von Post scale	pH in 1 N KCl	TOC g kg^{-1}	DOC g kg^{-1}	IAA μg kg^{-1} d.m.
4	Stążka 75–100 cm	Sedge–moss	H5	5.60	625.65	6.23	188.77±8.5
5	Tagan 0–25 cm	Grasses	H4	5.79	417.90	10.34	267.29±12.0
6	Tagan 50–75	Grasses	H4	5.67	433.40	7.18	273.20± 12.2
7	Peatland near village Ducora 0–25 cm	Sedge	H3	5.8	532.4	16.6	131.2±4.6
8	Peatland near village Ducora 30–50 cm	Wood-sedge	H4	5.9	559.5	16.1	128.4±4.4
9	Peatland near village Ducora 75–100 cm	Sedge	H3	5.9	579.8	13.1	134.3±4.7
10	Peatland near village Rusakovichi 0–25 cm	Wood–reed	H6	5.3	495.2	14.9	133.3±5.1
11	Peatland near village Rusakovichi 30–50 cm	Reed	H4	5.8	564.3	12.4	108.6±3.8
12	Peatland near village Rusakovichi 75–100 cm	Reed	H4	5.7	597.1	10.9	57.9±1.9
13	Ptich 0–30	Hypnum	H2	5.4	422.7	16.3	74.1±2.7

Maryganova and Szajdak (2007); Szajdak unpublished data

In addition, there are organic chemicals of a well-known structure in sapropel (biologically active substances: amino acids, nucleotides, nucleosides, peptides, carbohydrates, bitumes, fatty acids, terpenoids, phenols and phenolic acids, nucleic acids, flavones, derivatives of purine and pyrimidine bases, carotenoids, enzymes as catalase, peroxidase, reductase urease, xanthine oxidase, protease); alkaloids (derivatives of different structures); amines; vitamins (B complex vitamins including B_1, B_{12}, B_3, B_6) E, C, D, K, etc.; and unknown structures (humic and fulvic acids).

Sapropel contains also phytohormones: gibberelic acid, cytokinin, ethylene, abscisic acid, brassinosteroids, IAA, and its derivatives. The concentration of IAA

Table 10.3 Moorshes, degree decomposition, pH, and the concentrations of TOC, DOC, and IAA (μg kg^{-1} d.m)

No of place	Place of sampling Depth in cm	Kind of peat	Degree of decomposition in von Post scale	pH in 1 N KCl	TOC [%]	DOC [%]	IAA μg kg^{-1} d.m.
1	Czarna Wieś 5–10 cm	Moorsh	–	5.19	37.19	12.81	186.10 ± 8.37
2	Czarna Wieś 50–70 cm	Sedge–moorsh peat	H_1	5.16	44.02	5.80	110.86 ± 4.85
3	Otoczne 5–10 cm	Moorsh	–	5.46	38.10	10.80	87.10 ± 3.96
4	Otoczne 45–50 cm	Sedge–reed peat	H_5	5.63	45.58	7.55	122.70 ± 5.58
5	Kwatera 17; 5–10 cm	Moorsh	–	4.70	38.20	19.39	128.70 ± 5.11
6	Kwatera 17; 70–80 cm	Alder peat	H_6	5.39	40.23	5.34	69.30 ± 3.16
7	Shelter 0–25 cm	Moorsh soil	–	6.28	343.70	9.74	296.94 ± 13.3
8	Shelterbelt 25–50 cm	Sedge peat with wooden	R3	6.33	475.95	8.65	235.02 ± 10.6
9	Shelterbelt 50–75 cm	Sedge peat	R3	6.30	533.00	7.29	208.35 ± 9.4
10	Shelterbelt 75–100 cm	Sedge peat	R2	6.55	543.45	6.32	216.29 ± 9.7
11	Hirudo 0–25 cm	Moorsh soil	–	4.98	442.80	14.50	255.39 ± 11.5
12	Hirudo 25–50 cm	Alder swamp peat	R3	5.93	492.70	9.78	198.43 ± 8.9
13	Hirudo 50–75 cm	Sedge peat with wooden	R3	6.09	553.70	5.99	166.75 ± 7.5
14	Hirudo 75–100 cm	Sedge peat	R3	5.85	566.80	4.22	129.18 ± 5.8
11	Zbęchy 0–25 cm	Moorsh soil with pea	–	5.73	254.80	6.16	211.34 ± 9.2
12	Zbęchy 25–50 cm	Alder swamp peat	R3	6.16	466.30	6.98	212.84 ± 9.5
13	Zbęchy 50–75 cm	Sedge with wooden peat	R3	5.91	540.20	5.90	156.90 ± 7.8
14	Zbęchy 75–100 cm	Sedge	R3	6.07	544.30	4.49	139.53 ± 6.2
15	Mostek 0–25 cm	Moorsh soil with peat	–	5.83	396.50	11.03	229.65 ± 10.3

(continued)

Table 10.3 (continued)

No of place Place of sampling Depth in cm		Kind of peat	Degree of decomposition in von Post scale	pH in 1 N KCl	TOC [%]	DOC [%]	IAA μg kg^{-1} d.m.
16	Mostek 25–50 cm	Sedge peat	R3	5.80	486.30	9.97	210.84 ± 9.2
17	Mostek 50–75 cm	Sedge peat	R3	5.82	489.40	8.45	190.65 ± 7.6
18	Mostek 75–100 cm	Sedge peat	R3	5.71	528.50	6.46	171.05 ± 7.7
19	Tagan 0–25 cm	Wooden	H3	6.21	395.7	10.32	273.20 ± 12.2
20	Tagan 50–75 cm	Wooden grasses	H3	5.84	425.1	6.65	249.45 ± 11.6

Szajdak and Maryganova (2008)
TOC, total organic carbon; DOC, dissolved organic carbon; IAA, concentration of indole-3-acetic acid; μg kg^{-1} of IAA ±95 % confidence interval

Table 10.4 Quantities of IAA (μg · kg^{-1} s.m.) in commercial substrates obtained from the Research Institute of Pomology and Floriculture, Skierniewice, Poland

Substrates	IAA μg · kg^{-1} d.m.
1. "A-19," 10 % liquid fertilizer includes extract from vermicompost	22.9 ± 0.8
2. "A-1," solid substrate prepared on the basis of vermicompost, cow manure, and poultry litter	342.8 ± 11.9
3. "Eko 1," loose substrate prepared on the basis of waste products	373.9 ± 14.5
4. "S 1," loose substrate prepared on the basis of sewage sludge	323.9 ± 11.3
5. "Brykiet," dark, high-moor peat with high degree of decomposition	273.9 ± 9.8
6. "KL RHP 15," specialist substrate includes clay and perlite	181.7 ± 9.0
7. "Old kokos" prepared from fiber of kokos after 3 years of cultivation	179.1 ± 7.1
8. Hollas, high-moor peat with medium degree of decomposition	171.2 ± 7.7
9. "Alonet" 3, high-moor peat with low degree of decomposition	158.0 ± 7.9
10. "KL 4," substrate with high content of high black frozen peat	113.2 ± 5.8
11. "KL Pelargonia," specialist substrate includes different kinds of peats	102.7 ± 4.6
12. "New kokos"	101.8 ± 5.3

Szajdak (2004)
μgkg^{-1} of IAA ±95 % confidence interval

ranged from 161.01 to 373.9 μg · kg^{-1} d.m. in the organic sapropel and from 14.52 to 65.33 μg · kg^{-1} d.m. in the mineral one (Table 10.5) (Szajdak and Sokolov 1997; Szajdak and Maryganova 2008).

Sapropel as a nonhazardous fertilizer activates many biochemical and chemical conversions and pathways in soils and plants leading to an increase

Table 10.5 The content of IAA (μgkg^{-1} d.m.) in sapropels (Szajdak unpublished data)

No	Location	District, country	IAA
Organic sapropels			
1	Sominskoje	Brest, Belarus	189.9 ± 7.2
2	Kwietino	Minsk, Belarus	192.2 ± 7.0
3	Maloje Swino	Vitebsk, Belarus	373.9 ± 13.1
4	Lukowo	Brest, Belarus	292.5 ± 9.9
5	Homel	Homel, Belarus	168.93 ± 8.1
6	Homel	Homel, Belarus	161.01 ± 7.5
Mineral sapropels			
1	P1 95–99 cm	Khanty-Mansiysk, Russia	65.33 ± 2.9
2	P 16 100 cm	Khanty-Mansiysk, Russia	14.52 ± 0.7
3	P 16 150 cm	Khanty-Mansiysk, Russia	19.8 ± 0.9
4	P 17 150–200 cm	Khanty-Mansiysk, Russia	14.78 ± 0.8

μg kg^{-1} of IAA ±95 % confidence interval

of self-purification. This raw material plays a vital role by stimulating seed germination and root growth of cultivated plants. Sapropel as a fertilizer increases:

- Grain crops' yields, vegetables, fruits, and root crops
- Protein, carotene, sugar, and starch quantity in cultivated plants

Sapropel has been found to be the most efficient on light sandy and stony soils which are oversaturated with mineral fertilizers.

When mineral fertilizers are applied simultaneously with the organic sapropel, the amount of heavy metals decreases. Tubers also remain clean and do not accumulate any heavy metals.

Sapropel modifies and improves soil structure, physical properties, soil aeration, viscosity, and capillary rise. It has a positive impact on the hydrophilic–hydrophobic properties of fertilized soils by activating the water movement and air mode in soils.

Sapropel exhibits a strong ability to consume and retain water, increases the content of humus in soils on the second and third year of use, and activates soil processes.

Sapropel increases humus content and participates in the cycle of nitrogen, phosphorus, sulfur, and microelements in soils. Sapropel is also used in the preparation of composts.

The rich content of inorganic and organic chemicals contributes to the fact that sapropel has a variety of usages. It can be used:

(i) As an organic fertilizer
(ii) As a supplement for animals and poultry

(iii) In mud therapy
(iv) As an adsorbent for organic and mineral compounds
(v) In veterinary
(vi) In organic pelletized stock production
(vii) For yeast assimilation

Expert Group for Technical Advice on Organic Production (EGTOP) (2011) of European Commission Directorate-General for Agriculture and Rural Development in the final report on Fertilizers and Soil Conditioners (EGTOP/2/2011) during the third plenary meeting on June 29–30, 2011, concluded that, according to the dossier, sapropel could be added to the soil with the aim of increasing soil organic matter content. In addition, sapropel as well as other similar organic sediments from fresh water bodies is in line with the objectives, criteria, and principles of organic farming and should be included in Annex I of the final report on Fertilizers and Soil Conditioners (EGTOP/2/2011), however, with the following restrictions:

- Only organic sediments, which are by-products of water body management and which are extracted in ways that cause minimal negative impact on the aquatic ecosystem, should be used.
- Same limits for heavy metals as given in Annex I for household waste should be applied.
- Sediments rich in contaminants such as petrol-like substances should not be used.

Sapropel is also used in the preparation of composts. Sapropels which are rich in the salts of calcium, iron, and phosphorus contain no sand and hardly any clay. They are added to the rations given to agricultural animals as a mineral supplement; daily sapropel supplement ration reaches 2 kg for hogs, 3 kg for cows, and 10–15 g for hens.

The term "organic fertilizer" refers to any soil enhancer derived from natural sources that guarantees to contain at least a minimum percentage of sulfur, nitrogen, phosphorus, and potassium, as well as a low content of heavy metals and other contaminations, low biomass weeds, etc.

Sapropel contributes to an increase in crop yields. Examples of organic fertilizers include plant and animal by-products, rock powders, seaweed, inoculants, and conditioners.

As peat and sapropel have a high content of organic matter and biologically active substances, they are a rich source of organic fertilizers for agriculture, horticulture, and pomology. The use of organic fertilizers, particularly nitrogen, is the largest single factor responsible for the increase in the potential productivity of agriculture in the world.

The introduction of fertilizers in agriculture has not only enhanced the productivity of many good soils but it has also transformed many regions of inherently low productivity into agriculturally effective regions. The application of fertilizers influences land transformation in two ways. First, the judicious use of fertilizer improves the quality of good arable soils over time by increasing the organic return from crop residues, thus stimulating microbiological activity. Secondly, increased productivity achieved with fertilizers on good arable lands reduces the need to crop lower-quality areas better left in permanent grass or forest cover.

Understanding the processes and mechanisms of the degraded materials added to soils is a prerequisite for understanding the availability and cycling of such nutrients as the derivatives of nitrogen, carbon, sulfur, and phosphorus. Research on the structure and properties of organic substances such as peat, sapropel, and brown coal reveals essential differences between them, mostly predetermined by their genesis. It allows us to estimate their potential agroecological efficiency and prospective directions of technological processing and use. Soil organic matter represents an equilibrium system which plays a major role in supplying nutrients to plants grown thereon. The transformation of fresh organic matter to stable humic compounds affects the cation and anion exchange capacity. It is known that under the influence of enzymes, the macromolecules of complex organic compounds, which are secreted by microorganisms, are being destructed. The microorganisms degrade the heterocyclic compounds and produce low- and high-molecular organic substances such as carbohydrates, lignin, as well as peptides (Schnitzer and Khan 1978; Bambalov et al. 2000; Nieder et al. 2003).

Changes in the quality and quantity of soil organic matter caused by the application of a different organic fertilizer occur very slowly. Numerous long-term field experiments have shown that, with realistic and feasible improvement, soil organic matter in fertilized plots increased only by mere 30 % over decades (Sauerbeck 1993).

The challenges in determining nutrient availability in cropping systems that are managed to accumulate soil organic matter include assessing the interaction of added nutrients (via organic fertilizers or organic residues) with soil organic matter nutrient pools. As an organic fertilizer is added and soil organic matter formation proceeds, soil biomass of fauna and soil microbial pools increases (Hassink et al. 1994; Paul and Clark 1996). These components of the active soil organic matter are crucial for promoting nutrient mineralization in agroecosystems. Microbial and faunal biomass mediates the N mineralization (Hassink 1995; Wilson et al. 2001).

The use of organic fertilizers in cultivated fields leads to quantitative and qualitative variations of different organic substances.

The tested types of "Balanced Organic-Mineral Fertilizers" (BOMF) are:

- $BOMF_P$—peat, cow manure, and NPK
- $BOMF_S$—sapropel, cow manure, and NPK
- $BOMF_{BC}$—brown coal, cow manure, and NPK

They were applied in the cultivation of potato in crop rotation system.

The following materials were used: fen *sedge* peat deposit with the 30 % degree of organic matter decomposition, siliceous sapropel, brown coal, and semifluid cattle manure (Szajdak and Sokolov 1997; Bambalov and Sokolov 1998; Sokolov et al. 2008).

The organic fertilizers were applied in an equal dose of 60 ton $\cdot$ ha^{-1} to the soil plowed in April, and the potatoes were planted in May. The addition of the fertilizer

N invariably leads to the increases in the amount of soil N taken up by the plants. Reasons for the increased consumption include:

(i) The increased uptake is a special feature of the mineralization-immobilization process.
(ii) The fertilizer N causes enhanced mineralization of native humus N through a "priming" action.
(iii) Plants growing on treated soil develop a more extensive root system, thereby permitting better utilization of untagged soil N by the plant (Stevenson 1986).

Soil organic matter affects biochemical, chemical, biological, and physical soil properties that control soil microbial activity (Flaig 1971; Stevenson 1986). The degradation of organic matter and the autolysis of microorganisms in soils cause the release of amino acids, which typically amount to 40–60 % of the total organic N present in soil (Świętochowski and Miklaszewski 1965; Sowden 1966a, b; Miklaszewski 1968; Umarov and Aseeva 1971; Parsens and Tinsley 1975; Ktsoyev 1977; Goh and Edmeades 1979; Stevenson 1985; Kunnas and Eronen 1994; Szajdak et al. 1998). Amino acids in soils can undergo mineralization, migration down the soil profile, plant uptake, soil adsorption, and humification (Gupta and Reuszer 1967; Sørensen 1967; Ivarson and Sowden 1969; Sørensen 1972; Holtzclaw et al. 1980; Szajdak 1996; Kuzyakov 1997; Szajdak et al. 2003, 2004b). The total amount of bound amino acids in reference soil equaled 519 mg kg^{-1} (Table 10.6). All used fertilizers increased the total amount of bound amino acids in soils.

$BOMF_{BC}$ fertilizer almost doubled the amount of bound amino acids, causing an increase by 97.7 %. $BOMF_S$ had the smallest impact on the amount of amino acids and increased their quantity only by 57.5 %. $BOMF_P$ lad to 69.5 % increase in the total amount of amino acids in soil. In all samples amino acids with the neutral net charge were the most abundant (Życzyńska-Bałoniak and Szajdak 1993). The samples, on the other hand, had the lowest concentration of acid amino acids. The organic fertilizers exerted the highest influence on the amount of basic amino acids, which increased from 105.1 % to 182.4 % compared with the reference soil (Tables 10.6 and 10.7).

Basic amino acids react with reducing sugars and quinones at considerably higher rates than neutral and acidic amino acids. Thus, if these reactions occurred in soils, the more basic compounds, such as lysine, would be affected to a greater extent than other amino acids. Another factor to consider is that the accessory amino group of basic amino acids, when present in peptides, is capable of combining with carbonyl-containing substances, whereas the amino group of neutral and acidic amino acids is inaccessible because of participation in peptide linkages (Stevenson 1982).

Glutamic acids predominated among acidic amino acids, while glycine and leucine, in the neutral amino acids.

In addition, the study found that the sample soils had a higher quantity of β-alanine after treating them with different organic fertilizers than in the reference soil. $BOMF_P$ and $BOMF_S$ increased the concentration of β-alanine in the sample soils by 67.1 % and 20.2 %, respectively. $BOMF_P$ and $BOMF_S$ increased the concentration

Table 10.6 Bound amino acid quantities and nitrogen in amino acids (italic) in soils treated with different kinds of fertilizers in $mg \cdot kg^{-1}$ of soils

Amino acids	Reference soil	Soil treated BOMF$_{BC}$	BOMF$_P$	BOMF$_S$
Acidic				
Taurine	1.13±0.03	1.25±0.04	1.45±0.05	4.33±0.15
	0.13±0.01	*0.14±0.01*	*0.16±0.01*	*0.48±0.02*
Cysteic acids	9.28±0.32	11.16±0.50	14.20±0.49	12.37±0.43
	0.76±0.03	*0.92±0.03*	*1.18±0.05*	*1.02±0.04*
Phosphoethanolamine	5.31±0.19	0.54±0.02	6.72±6.72	2.51±0.09
	0.53±0.02	*0.05±0.01*	*0.67±0.03*	*0.25±0.01*
Glutamic acids	37.14±1.29	41.27±1.44	67.31±2.34	57.74±2.01
	3.53±0.14	*3.93±0.16*	*6.41±0.26*	5.49±*0.22*
α-Aminoadipic acid	26.16±0.92	44.00±1.53	28.31±0.98	37.42±1.31
	2.27±0.09	*3.82±0.15*	*2.46±0.09*	*3.56±0.16*
Neutral				
Proline	23.69±0.83	2.37±0.08	5.96±0.20	6.13±0.20
	2.88±0.12	*0.29±0.01*	*0.73±0.01*	*0.75±0.02*
Glycine	71.14±2.48	97.13±3.36	103.4±13.61	99.18±3.46
	13.26±*0.53*	18.12±*0.62*	19.21±*0.67*	18.50±*0.55*
Alanine	29.46±2.48	45.42±1.59	50.28±1.75	39.46±1.38
	4.63±0.18	*7.13±0.21*	*7.90±0.23*	*6.20±0.15*
α-Aminobutyric acid	–	0.27±0.01	8.95±0.31	5.92±0.21
	–	*0.01±0.001*	*1.22±0.05*	*0.80±0.03*
Valine	33.02±1.15	138.12±4.82	47.54±1.67	53.78±1.87
	3.95±0.19	*16.52±0.32*	*5.68±0.24*	*0.68±0.03*
Methionine	26.14±0.91	32.32±1.13	40.74±1.42	28.49±0.96
	2.45±0.15	*3.07±0.18*	*3.82±0.22*	*2.67±0.11*
Cystathionine	2.18±0.08	6.51±0.22	6.04±0.21	4.37±0.15
	0.28±0.01	*0.82±0.03*	*0.76±0.03*	*0.55±0.02*
Leucine	71.45±2.49	104.74±3.65	119.6±14.17	96.76±3.38
	7.63±0.31	*11.13±0.45*	*12.77±0.51*	*10.28±0.42*
Tyrosine	7.17±0.25	21.34±0.75	13.02±0.45	11.24±0.39
	0.55±0.02	*1.65±0.08*	*1.01±0.04*	*0.87±0.03*
β-Alanine	32.71±1.14	32.11±1.12	54.83±1.91	39.33±1.36
	5.14±0.21	*5.05±0.24*	*8.61±0.31*	*6.18±0.27*
β-Aminoisobutyric acid	0.92±0.03	0.29±0.01	1.02±0.03	1.38±0.05
	0.13±0.01	*0.04±0.01*	*0.14±0.01*	*0.19±0.01*
γ-Aminobutyric acid	10.34±0.36	19.40±0.67	6.18±0.21	7.61±0.27
	1.40±0.06	*2.63±0.11*	*0.84±0.04*	*1.03±0.04*
Basic				
Ornithine	5.44±0.19	6.98±0.24	8.12±0.28	6.60±0.23
	1.15±0.05	*1.48±0.05*	*1.72±0.06*	*1.40±0.05*

(continued)

Table 10.6 (continued)

Amino acids	Reference soil	Soil treated		
		$BOMF_{BC}$	$BOMF_P$	$BOMF_S$
Lysine	37.34 ± 1.30	52.60 ± 1.83	52.24 ± 1.89	44.86 ± 1.57
	7.15 ± 0.29	*10.08 ± 0.41*	*10.00 ± 0.43*	*8.59 ± 0.34*
Histidine	13.82 ± 0.48	20.65 ± 0.72	22.08 ± 0.77	14.46 ± 0.51
	3.74 ± 0.15	*5.59 ± 0.22*	*5.98 ± 0.25*	*3.92 ± 0.16*
1-Methylhistidine	11.53 ± 0.40	188.55 ± 6.51	111.47 ± 3.89	129.02 ± 4.50
	2.86 ± 0.12	*46.81 ± 1.88*	*27.67 ± 1.11*	*32.03 ± 1.28*
3-Methylhistidine	70.64 ± 2.46	138.20 ± 4.81	88.92 ± 3.10	96.94 ± 3.38
	17.54 ± 0.71	*34.31 ± 1.36*	*22.07 ± 0.88*	*24.07 ± 0.96*
Arginine	13.38 ± 0.46	22.68 ± 0.74	22.24 ± 0.77	18.33 ± 0.64
	4.30 ± 0.17	*7.29 ± 0.32*	*7.15 ± 0.34*	*5.89 ± 0.25*
Total amount of amino acids	519.39 ± 22.8	1027.90 ± 45.02	880.51 ± 38.30	818.23 ± 36.2
Total amount of nitrogen in amino acids	86.26 ± 3.43	180.87 ± 7.22	148.16 ± 5.92	137.41 ± 5.48

Szajdak and Sokolov (1997) and Sokolov et al. (2008)
$BOMF_P$, peat, cow manure, NPK; $BOMF_S$, sapropel, cow manure, NPK; $BOMF_{BC}$, brown coal, cow manure, NPK. The initial materials were taken from fen sedge peat deposit with the degree of organic matter decomposition of 30 %, siliceous sapropel from Chervonoe Lake, brown coal from Khandinskoe deposit, semifluid cattle manure (x ± 95 % confidence interval)

Table 10.7 Total amount of bounded amino acids and total nitrogen amount in amino acids (italic) in soils treated with different kind of fertilizers in $mg \cdot kg^{-1}$ of soils

Amino acids	Reference soil	Soil treated		
		$BOMF_{BC}$	$BOMF_P$	$BOMF_S$
Acidic	79.02 ± 5.5	98.22 ± 6.4	117.99 ± 7.7	114.37 ± 7.4
	7.22 ± 0.28	*8.85 ± 3.53*	*10.88 ± 0.43*	*10.80 ± 0.49*
Neutral	308.22 ± 22.0	500.02 ± 32.5	457.45 ± 29.8	393.65 ± 25.8
	42.30 ± 1.65	*66.46 ± 2.65*	*62.69 ± 2.63*	*50.71 ± 2.32*
Basic	132.15 ± 8.7	429.66 ± 28.2	305.07 ± 19.8	310.21 ± 20.5
	36.74 ± 1.43	*105.56 ± 4.49*	*74.59 ± 3.14*	*75.90 ± 2.86*

Szajdak and Sokolov (1997) and Sokolov et al. (2008)
(x ± 95 % confidence interval)

of β-alanine 67.1 % and 20.2 %, respectively. β-alanine is a typical constituent of bacterial cell walls (Stevenson 1972; Durska and Kaszubiak 1980a, b, c). Thus, soils treated with organic fertilizers can have a higher β-alanine microbial biomass.

A similar phenomenon of the accumulation of β-alanine was observed in the studies of soils under continuous cropping of rye and crop rotation as well as under conventional and no-tillage management (Życzyńska and Szajdak 1993; Szajdak 1996; Szajdak et al. 2003; 2004b).

The study also revealed a higher content of lysine in the fertilized soils. The soils treated with the fertilizers had on average 34 % higher concentration of lysine than the reference soils. A higher concentration of lysine may be correlated with a higher microbial biomass in the fertilized soils. Lysine is transformed in soils from α,ε-diaminopimelic acid (Durska and Kaszubiak 1980a, b, c). The occurrence of the α,ε-diaminopimelic acid in soils is of great interest to researchers as this compound is confined to bacteria's cell wall. The highest quantity of lysine was found in the soils treated with $BOMF_{BC}$. The quantity equaled to 52.60 mg kg^{-1} of soil, and it was 40.9 % higher than the quantity of lysine in the reference soils (Nieder et al. 2003).

In addition, the results showed that a significant amount of organic nitrogen, included in amino acids structures, is supplied to soils. Amino acids represent a form of organic nitrogen which can be easily hydrolyzed by chemicals and enzymes and are available to plants and soil microorganisms (Maciak et al. 1977). $BOMF_{BC}$ supplied the sample soils with the biggest amount of nitrogen. The amount of nitrogen was 93.7 % higher than the amount found in the reference soil. Two other fertilizers—$BOMF_P$ and $BOMF_S$—supplied, respectively, 64.1 % and 56.3 % higher amount of nitrogen than the amount which was found in the reference soil. The highest supply was observed in case of $BOMF_{BC}$. This fertilizer supplied the soils with 93.7 % higher amount of nitrogen than the amount found in the reference soil. Two other fertilizes—$BOMF_P$ and $BOMF_S$—caused a 64.1 % and 56.3 %, respectively, increase in the amount of nitrogen.

The study showed also that the concentration of proline was lower in the fertilized soils than in the reference soil. On average in the fertilized soils, the quantity of proline was 79.6 % lower than in the reference soil. The smallest amount of proline was determined in soils treated with $BOMF_{BC}$ and it equaled 2.37 mg kg^{-1}. This amount was 89.9 % lower than the amount of proline in the reference soil. A decrease in the amount of proline is beneficial for soils. Proline is a secondary amine, which may form N-nitrosamines with nitrate ions (under acidic conditions). Therefore, it is a potent toxin which can have a carcinogenic, mutagenic, and teratogenic effect on biota in soils (Kofoed et al. 1981; Larsson et al. 1990).

The results showed a significant addition of nitrogen included in amino acids to soils (Table 10.7).

The distribution of amino acids in soils may be affected by a variety of factors, which include the synthesis and the degradation of nitrogen organic substances by the indigenous biota and chemical agents, the adsorption of amino acids by clay minerals, and reactions of amino acids with quinones and reducing sugars. The efficiency of BOMF in soils under crop rotation (corn–barley–oats–buckwheat) was rather high and stable up to 3 years after they had been used.

On average these fertilizers ensured yield increase of 18.2–21.6 % in comparison with their mineral equivalent.

The studies revealed a significant impact of organic fertilizers prepared on the basis of peat, brown coal, and sapropel on the amino acid content in soils. The highest amount of bound amino acids was determined in the soil treated with the organic fertilizer produced on the basis of brown coal. The lowest content of bound amino acids was found in the soil treated with the fertilizer containing sapropel. Neutral amino acids were abundant in the soils treated with all organic fertilizers, whereas acidic amino acids were uncommon. Overall, all organic fertilizers supplied the soils with the organic form of nitrogen included in amino acids. These substances influence nitrogen-containing compounds, including amino acids of root exudates, proteins, nitrates, and other nitrogen forms.

References

Aeshad M, Frankenburger WT Jr (1991) Microbial production of plant hormones. Plant Soil 133:1–8

Aldesuquy HS, Mansour FA, Abo-Hamed SA (1998) Effect of the culture filtrates of Streptomyces on growth and productivity of wheat plants. Folia Microbiol 43:465–470

Bambalov N, Sokolov G (1998) New soil improving agents for accelerated cultivation of soils with low fertility or damaged. J Int Agrophysc 12:357–360

Bambalov N, Smychnik T, Maryganova V, Strigutsk V, Dite M (2000) Peculiarities of chemical composition and the molecular structure of peat humic substances. Acta Agroph 26:149–177

Bandurski RS, Schulze A (1977) Concentration of indole-3-acetic acid and its derivatives in plants. Plant Physiol 60:211–213

Belkevitch PI (1962) Chimia i genesis torfa i sapropel. Publishing of Belarus Academy of Sciences, Minsk, pp 3–319 (in Russian)

Bender DA (1989) The kynurenine pathway of tryptophan metabolism. In: Stone TW (ed) Quinolinic acid and the kynurenines. CRC Press, Boca Raton, pp 3–38

Beyeler M, Keel C, Micheaux P, Haas D (1999) Enhanced production of indole-3-acetic acid by a genetically modified strain of *Pseudomonas fluorescens* CHAO affects root growth in cucumber, but does not improve protection of the plant against *Pythium root rot*. FEMS Microbiol Ecol 28:225–233

Bhavdish N, Johri A, Sharma J, Virdi S (2003) Rhizobacterial diversity in India and its influence on soil and plant health. Adv Biochem Eng Biotechnol 84:49–89

Brown ME (1972) Plant growth substances produced by microorganisms of soil and rhizosphere. J Appl Bacteriol 35:443–451

Chalvignac MA, Mayaudon J (1971) Extraction and study of soils enzymes metabolizing tryptophan. Plant Soil 34:25–31

Chen Y, De Nobili M, Aviad T (2004) Stimulatory effects of humic substances on plant growth. In: Magdoff RR, Weil R (eds) Soil organic matter in sustainable agriculture. CRC Press, Boca Raton, pp 103–129

Cohen JD, Bandurski RS (1982) Chemistry and physiology of the bound auxin. Annu Rev Plant Physiol 33:403–430

Dahm H, Sitek JM, Strzelczyk E (1997) Synthesis of auxins by bacteria isolated from roots pine seedlings inoculated with rusty forest soil. Pol J Soil Sci 10:131–137

Durska G, Kaszubiak H (1980a) Occurrence of α, ε-diaminopimelic acid in soil. II. Usefulness of α, ε-diaminopimelic acid determination for calculations of the microbial biomass. Pol Ecol Stud 6(2):195–199

Durska G, Kaszubiak H (1980b) Occurrence of α, ε-diaminopimelic acid in soil. I. The content of α, ε-diaminopimelic acid in different soils. Pol Ecol Stud 6(2):189–193

Durska G, Kaszubiak H (1980c) Occurrence of α, ε-diaminopimelic acid in soil. III. α, ε-diaminopimelic acid as the nutritional component of soil microorganisms. Pol Ecol Stud 6(2):201–206

El-Tarabily KA (2008) Promotion of tomato (Lycopersicon esculentum Mill.) plant growth by rhizosphere competent 1-aminocyclopropane-1-carboxylic acid deaminase-producing streptomycete actinomycetes. Plant Soil 308:161–174

Flaig W (1971) Organic compounds in soil. Soil Sci 111:19–33

Frankenberger WT, Arshad M (1991a) Yield response of watermelon and muskmelon to L-tryptophan applied in soil. HortSci 26:35–37

Frankenberger WT, Arshad M (1991b) Yield response of *Capsicum annuum* to the auxin precursor, L-tryptophan applied to soil. Plant Growth Reg Soc Am Q 19:231–240

Gelinas DA (1973) Proposed model for the peroxidase-catalyzed oxidation of indole-3-acetic acid in the presence of the inhibitor ferulic acid. Plant Physiol 51(5):967–972

Goh KM, Edmeades DC (1979) Distribution and partial characterization of acid hydrolysable organic nitrogen in six New Zealand soil. Soil Biol Biochem 11:127–132

Goldsmith MHM (1977) The polar transport of auxin. Annu Rev Plant Physiol 28:439–478

Goodfellow M, Simpson KE (1987) Ecology of streptomycetes. Front Appl Microbiol 2:97–125

Grapelli A, Rossi W (1979) The effect herbicides, respiration and IAA biosynthesis in *Arthrobacter sp.* Chemosphere 26:341–377

Gupta UC, Reuszer HW (1967) Effect of plant species on the amino acid content and nitrification of soil organic matter. Soil Sci 104:395–400

Harley FR, Smith SE (1983) *Mycorrhizal symbiosis*. Academic Press. A subsidiary of Harcourt Brace Jovanovich, Publishers, London/New York. pp 304–316

Hassink J (1995) Decomposition rate constant of size and density fractions of soil organic matter. Soil Sci Soc Am J 59:1631–1635

Hassink J, Neutel M, De Ruiter PC (1994) C and N mineralization in sandy and loamy grassland soils: the role of microbes and microfauna. Soil Biol Biochem 26:1565–1571

Holtzclaw KM, Schauberd G, LeVesque-MadoreS, Sposito G, Heic JA, Johnston CT (1980) Analytical properties of the soluble, metal-complexing fractions in sludge-soil mixtures: V. Amino acids, hexosamines, another carbohydrates in fulvic acid. Soil Sci Soc Am J 44:736–740

Ilic B, Konstantinovic SS, Todorovic ZB, Lazic ML, Veljkovic VB, Jokovic N, Radovanovic BC (2007) Characterization and antimicrobial activity of the bioactive metabolites in streptomycete isolates. Microbiology 76:421–428

Ilnicki P (ed) (2002) Torfowiska i torf (Peatlands and peat). Handbook. Publishing Agriculture University, Poznań, pp 1–606 (in Polish)

Ivarson CK, Sowden JF (1969) Free amino acid composition of the plant root environment under field conditions. Can J Soil Sci 49:121–127

Kampert M, Strzelczyk E (1975) Synthesis by auxin by fungi isolated from the roots of pine seedlings (*Pinus silvestris L.*) and from soil. Acta Microbiol Pol 7:223–230

Kaunat H (1969) Bidung von indolderivaten durch rhizospha reenspezifisch Bakterien und Aktinomyzeten. Zentr Bakt Abt II 123:501–515

Kennedy AC, Stubbs TL, Schillinger WF (2004) Soil and crop management effect on soil microbiology. In: Magdoff RR, Weil R (eds) Soil organic matter in sustainable agriculture. CRC Press, Boca Raton, pp 295–326

Kireicheva LV, Khokhlov OB (2000) Elemental composition of different fraction from the sapropel organic matter. Eur Soil Sci 33(9):947–949. Translated from Pochvovedenie 9:1083–1085

Kofoed D, Nemming O, Brunfeldt K, Nebelin E, Thomsen J (1981) Investigations on the occurrence of nitrosamines in some agricultural products. Acta Agric Scand 31:40–48

Kravchenko V, Borovkov AV, Pshikvil Z (1991) The possibility of auxin biosynthesis in wheat rhizosphere by associated nitrogen-fixing bacteria. Microbiology 60:927–931

Krylov SN, Krylova SM, Chebotarev IG, Chebotareva AB (1994) Inhibition enzymatic indole-3-acetic acid oxidation by phenols. Phytochemistry 36(2):253–267

Ktsoyev BK (1977) Free amino acids in soils of northern Caucasus. Pochvovedenie 5:50–53

Kunnas AV, Eronen M (1994) Identification of free amino acids in peat by gas chromatography and mass spectrometry. Eur J Soil Sci 45:387–392

Kuraishi S, Sakurai N (1988) Auxin induced coleoptilar growth of dwarf barley. Plant Physiol 86:69

Kurzo B (2010) Features of accumulation and sapropel structure in lake-mire complexes (on an example of the western areas of Belarus). In: Szajdak LW, Karabanov AK (eds) Physical, chemical and biological processes in soils. Prodruk, Poznań, pp 323–332

Kutacek M, Kefeli V (1970) Biogenesis of indole compounds from D- and L- tryptophan in segments of etiolated seedlings od cabbage, maize and pea. Biol Plant 12:145–158

Kuzyakov YV (1997) The role of amino acids and nucleic bases in turnover of nitrogen and carbon in soils humid fractions. Eur J Soil Sci 48:121–130

Larsson BK, Österdahl BG, Regner S (1990) Polycyclic aromatic hydrocarbons and volatile N-nitrosamines in some dried agricultural products. Swed J Agric Res 20:49–56

Law DM (1987) Gibberellin-enhanced indole-3-acetic acid biosynthesis: D-tryptophan as the precursor of indole-3-acetic acid. Physiol Plant 70:626–632

Lebuhn MT, Heulin T, Hartmann A (1995) Production of auxin and other indolic and phenolic compounds by Paenibacillus polymyxa strains isolated from different proximity to plant roots. FEMS Microbiol Ecol 22:325–334

Lishtvan II (2004) Peat and its role in solving the environment protection issues. In: Päivänen J (ed) Wise use of peatlands, vol 2. Proceedings of the 12th international peat congress. Tampere, Finland, pp 1129–1133

Lüttig G (1986) Plants to peat: the process of humification. In: Fuchsman CH (ed) Peat and water. Aspect of water retention and dewatering in peat. Elsevier Applied Science Publishers, London, pp 9–19

Lynch JM (1990) Introduction: some consequences of microbial rhizosphere competence for plant and soil. In: Lynch JM (ed) The rhizosphere. Wiley, Chichester, pp 1–10

Maciak F, Söchtig H, Flaig W (1977) Composition of amino acids in peat-forming plants and in peats. In: Soil organic matter studies, vol II. Proceedings of a symposium Braunschweig, 6–10 September 1976, International Atomic Energy Agency, Vienna, pp 343–357

Mahadevan B, Crawford DL (1997) Properties of the chitinase of the antifungal biocontrol agent Streptomyces lydicus WYEC108. Enzym Microb Technol 20:489–493

Martens DA, Frankenberger WT Jr (1993) Metabolism of tryptophane in soil. Soil Biol Biochem 25:1679–1687

Martens A, Frankenberger WT Jr (1994) Assimilation of exogenous 2′-14C-indole acetic acid and 3′-14C-tryptophan exposed to the roots of three wheat varieties. Plant Soil 166:281–290

Martin JK, Kemp JR (1980) Carbon loss from roots of wheat cultivars. Soil Biol Biochem 12:551–554

Maryganova V, Szajdak L (2007) Content of indol-3-acetic acid in peat. Vesti Natl Acad Sci Belarus Chem Ser 4:111–114

McQueen-Mason SJ, Hamilton RH (1989) The biosynthesis of indole-3-acetic acid from D-tryptophan in Alaska pea plastids. Plant Cell Physiol 30:999–1005

Merckx R, Dijkra A, Hartog AD, Veen JAV (1987) Production of root-derived material and associated microbial growth in soil at different nutrient levels. Biol Fertil Soils 5:126–132

Miklaszewski S (1968) Dynamic of free amino acids under different cultivated sandy soils. Zesz Probl Post Nauk Rol 77a:151–158 (in Polish)

Müller M, Deigele C, Ziegler H (1989) Hormonal interactions in the rhizosphere of maize (Zea mays L.) and their effects on plant development. Zeit Pfl Bod 152:247–254

Muscolo L, Cutrupi S, Nardi S (1998) IAA detection in humic substances. Soil Biol Biochem 30:1199–1201

Musculo A, Nardi S (1997) Auxin or auxinlike activity of humic matter. In: Drozd J, Gonet SS, Senesi N, Weber J (eds). The role of humic substances in the ecosystems and in environmental protection. Proceedings of the 8th meeting of the International Humic Substances Society, Wrocław, Poland, 9–14 September 1997, PTSH – Polish Society of Humic Substances, Polish Chapter of the International Humic Substances Society, pp 997–992

Nieder R, Benbi DK, Isermann K (2003) Soil organic matter. In: Bendi DK, Nieder R (eds) Handbook of processes and modeling in the soil-plant system. Food Products Press, The Haworth Reference Press Imprints of the Haworth Press, Inc., New York, pp 345–408

Parsens JW, Tinsley J (1975) Nitrogenous substances. In: Gieseking JE (ed) Soil components. Organic components, vol 1. Springer, Berlin, pp 263–304

Paul EA, Clark FE (1996) Soil microbiology and biochemistry. Academic, San Diego, p 275

Rossi W, Grapelli A, Pietrosantu W (1984) Phytohormones in soil after atrazine application. Folia Microbiol 29:325–329

Rovira AD (1965) Plant root exudates and their influence upon soil microorganisms. In: Baker KF, Synder WC (eds) Ecology of soil-borne plant pathogens. University of California Press, Berkeley, pp 170–186

Sanderson KJ, Jameson PE, Zabkiewicz JA (1987) Auxin in a seaweed extract: Identification and quantitation of indole-3-acetic acid by gas chromatography-mass spectrometry. J Plant Physiol 129:363–367

Sarwar M, Arshad M, Frankenberger WT Jr (1992) Tryptophan-dependent biosynthesis of auxins in soil. Plant Soil 47:207–215

Sauerbeck D (1993) CO_2 – emissions from agriculture: sources and mitigation potentials. Water Air Soil Pollut 70:381–388

Schnitzer M, Khan SU (1978) Soil organic matter. Elsevier Scientific Publishing Company, Amsterdam, pp 261–281

Shahab S, Ahmed N, Khan NS (2009) Indole acetic acid production and enhanced plant growth promotion by indigenous PSBs. Afr J Agric Res 4(11):1312–1316

Smith WH (1976) Character and significance of forest tree root exudates. Ecology 57:324–331

Smolander A, Kitunen V (2002) Soil microbial activities and characteristics of dissolved organic C and N in the relation to tree species. Soil Biol Biochem 34:651–660

Final Report on Fertilizers and Soil Conditioners (2011) In: Expert Group for Technical Advice on Organic Production – EGTOP. Directorate-General for Agriculture and Rural Development. European Commission. pp 12–13

Sokolov GA, Bambalov NN (2000) Peat-sapropel material as soil improvers. In: Rochefort L, Daigle JY (eds) Sustaining our peatlands. Proceedings of the 11th international peat symposium, Quebec, Canada, International Peat Society, pp 461–465

Sokolov G, Szajdak L, Simakina I (2008) Changes in the structure of nitrogen-containing compounds of peat, sapropel, brown coal and organic fertilizers on their basis. Agron Res 6(1):141–149

Sørensen LH (1967) Duration of amino acid metabolites formed in soil during decomposition of carbohydrates. Soil Sci 104:234–241

Sørensen LH (1972) Stabilization of newly formed amino acid metabolites in soil by clay minerals. Soil Sci 114:109–114

Sowden FJ (1966a) Nature of amino acid compounds of soil: I. Isolation and fractionation. Soil Sci 102(3):202–207

Sowden FJ (1966b) Nature of amino acid compounds of Soil: II. Amino acids and peptides produced by partial hydrolysis. Soil Sci 102(4):264–271

Stevenson G (1972) Biology of fungus, microbes and viruses. PWRiL, Warszawa, p 280 (in Polish)

Stevenson FJ (1982) Organic forms of soil nitrogen. In: Humus chemistry. A Wiley-Interscience Publication/Wiley, New York, pp 55–92

Stevenson FJ (1985) Amino acids. In: Humus chemistry, genesis, composition, reactions. Wiley, New York, pp 65–78

Stevenson FJ (1986) The micronutrient cycle. In: Cycles of soil, carbon, nitrogen, phosphorus, sulfur, micronutrients. A Wiley-Interscience Publication/Wiley, New York, pp 321–367

Stonier T, Yoneda Y (1967) Auxin destruction and growth in Japanese morning glory stems. Ann N Y Acad Sci 144:129–135

Stonier T, Singer R, Yang H-M (1970) Studies on auxin protectors. X. Inactivation of certain protectors by polyphenol oxidase. Plant Physiol 46:454–457

Strzelczyk E, Pokojska-Burdziej A (1984) Production of auxin and gibberellin-like substances by mycorrhizal fungi, bacteria and actinomycetes isolated from soil and mycorrhizosphere of pine (*Pinus silvestris L.*). Plant Soil 81:185–194

Strzelczyk E, Dahm H, Pachlewski R, Różycki R (1997) Production of indole compounds by the ectomycorrhizal fungus *Cantharellus cibarius* Fr. Pedobiologia 41:402–411

Suzuki S, Yamamoto K, Okuda T, Nishio M, Nakanishi N, Komatsubara S (2000) Selective isolation and distribution of *Actinomadura rugatobispora*. strains in soil. Actinomycetologica 14:27–33

Świętochowski B, Miklaszewski S (1965) Free amino acids as a test of biological processes in the soil. Zesz Nauk Wyż Szk Rol Wroc Rolnictwo XIX 55:97–107 (in Polish)

Szajdak L (1996) Impact of crop rotation and phenological periods on rhodanese activity and sulphur amino-acids concentrations in brown soils under continuous rye cropping and crop rotation. Environ Int 22(5):563–569

Szajdak L (2004) Substancje aktywne biologicznie w kompostach z odpadów komunalnych na tle innych podłoży organicznych. In: Drozd J (ed) Komposty z odpadów komunalnych, produkcja, wykorzystanie i wpływ na środowisko. (Composts from waste materials, production, utilization and the impact on the environment). Polskie Towarzystwo Substancji Humusowych, Wrocław. pp 186–196

Szajdak L, Maryganova M (2007) Occurrence of auxin in some organic soils. Agron Res 5(2):175–187

Szajdak L, Maryganova V (2008) Quantities of indole-3-acetic acid in secondary transformed peat-moorsh soils. Tomsk Pedagogical State University Herald. Biol Ser Sci 4(78):34–39

Szajdak L, Sokolov GA (1997) Impact of different fertilizers on the bound amino acids content in soils. Int Peat J 7:29–32

Szajdak L, Matuszewska T, Gawlik J (1998) Effect of secondary transformation state of peat-muck soils on their amino acids content. Int Peat J 8:76–80

Szajdak L, Jezierski A, Cabrera ML (2003) Impact of conventional and no-tillage management on soils amino acids, stable and transient radicals and properties of humic and fulvic acids. Org Geochem 34:693–700

Szajdak L, Maryganova V, Bambalov N (2004a) Concentration of indole-3-acetic acid in different kinds of peat and sapropel. In: Päivänen J (ed) Wise use of peatland, vol 2. International Peat Society, Tampere, pp 1152–1155

Szajdak L, Życzyńska-Bałoniak I, Meysner T, Blecharczyk A (2004b) Bound amino acids in humic acids from arable cropping systems. J Plant Nutr Soil Sci 167:562–567

Szajdak L, Brandyk T, Szatyłowicz J (2007) Chemical properties of different peat-moorsh soils from the Biebrza River Valley. Agron Res 5(2):165–174

Szajdak LW, Szatylowicz J, Kõlli R (2011) Peat and peatlands, physical properties. In: Gliński J, Horabik J, Lipiec J (eds) Encyclopedia of agrophysics, Encyclopedia of earth sciences series. Springer, Dordrecht, pp 551–555

Takizawa M, Colwell RR, Hill RT (1993) Isolation and diversity of actinomycetes in the Chesapeake Bay. Appl Environ Microbiol 59:997–1002

Tena M, Pinilla JA, Magallanes M (1986) L-phenylalanine deaminating activity in soil. Soil Biol Biochem 18(3):321–325

Tokala RK, Strap JL, Jung CM, Crawford DL, Salove MH, Deobald LA, Bailey JF, Morra MJ (2002) Nove. l plant-microbe rhizosphere interaction involving Streptomyces lydicusW-YEC108 and the pea plant (*Pisum sativum*). Appl Environ Microbiol 68:2161–2171

Tresner HD, Davies MC, Backus EJ (1961) Electron microscopy of Streptomyces spores morphology and its role in species differentiation. J Bacteriol 81:70–80

Tsavkelova EA, Klimova SY, Cherdyntseva TA, Netrusov AI (2006) Microbial producers of plant growth stimulators and their practical use: a review. Appl Biochem Microbiol 42:117–126

Tsurusuki K, Watanabe S, Sakuri N, Kuraishi S (1990) Conversion of D-tryptophan to indole-3-acetic acid in coleoptiles of normal and semidwarf barley (*Hordeum vulgare*) strain. Physiol Plant 79:221–225

Umarov MM, Aseeva IV (1971) Free amino acids in some soils of the USSR. Pochvovedenie 10:108–111

Uosukainen H, Pihlaja K (2006) Peat in balneology and therapy. Terraviva Oy, Kempele, pp 1–96

Van Dijk H (1971) Colloid chemical properties of humic matter. In: McLaren AD, Skujins J (eds) Soil biochemistry, vol II. Marcel Dekker, New York, p 21

Wheeler CT, Crozier A, Sandberg G (1984) The biosynthesis of indole-3-acetic acid by Frankia. Plant Soil 78:99–104

Wilson T, Paul EA, Harwood RR (2001) Biologically active soil organic matter fractions in sustainable cropping systems. Appl Soil Ecol 16:63–76

Życzyńska-Bałoniak I, Szajdak L (1993) The content of the bound amino acids in soil under rye monoculture and Norfolk crop rotation in different periods development. Pol J Soil Sci 20:111–117

Chapter 11
Cranberry: A Plant Growing on Organic Soils with a Broad Spectrum of Pharmaceutical and Medical Use

Lech Wojciech Szajdak and Lydia I. Inisheva

Abstract Cranberry is a significant example of a plant which is growing on peat soils and has a high content and a rich spectrum of biologically active substances. The organic compounds in the plant are responsible for its wide medical and pharmaceutical use. The beneficial impact of cranberries on human health is caused by the presence of the following substances in the berries: anthocyanins, proanthocyanidins (condensed tannins), flavonol glycosides, low-molecular-weight phenolic acids, organic acids, and sugars. Cranberry juice and fruits are reported to display a number of health benefits including: potent antioxidant activity, cholesterol reduction, vasorelaxant effects, the prevention of urinary tract infections, the reduction of biofilm formation, and in vivo anticancer effects.

Keywords Cranberry • Biologically active substances • Health benefits

11.1 Introduction

Cranberry was introduced to the Western civilization by the pilgrims of Plymouth Colony. For numerous centuries it was used as a staple ingredient of pemmican—a mixture of dried venison, fat, and cranberries—by the indigenous peoples of America inhabiting the Northern Atlantic Coastal Plain. Growers have breeded a group of high-quality cultivars of the original *Vaccinium macrocarpon*. Cranberry is a unique fruit requiring a high level of horticulture. It demands an acid peat soil (pH 4–6) and grows in natural forest pockets. It can be easily damaged by trampling or

L.W. Szajdak (✉)
Institute for Agricultural and Forest Environment, Polish Academy of Sciences, Str. Bukowska 19, 60-809 Poznań, Poland
e-mail: lech.szajdak@isrl.poznan.pl

L.I. Inisheva
Tomsk State Pedagogical University, 634041 Tomsk, Russia
e-mail: inisheva@mail.ru

L.W. Szajdak (ed.), *Bioactive Compounds in Agricultural Soils*,
DOI 10.1007/978-3-319-43107-9_11

improper drainage. These factors make cranberry growing a difficult and a high-risk endeavor.

Unlike most other crop systems, cranberry growing is a form of swamp agriculture. Successful water management is the *sine qua non* of cranberry breeding. Bog or marsh sites are chosen with consideration of air drainage, water drainage, and water availability. The establishment of an entire complex of reservoirs, dikes, ditches, trunks, and bogs has only one aim: a precise control of water distribution (Brick 1980).

Cranberry—*Vaccinium macrocarpon Ait*—is a creeping evergreen shrub which is both cultivated and wild harvested. Cranberries are characteristic of boreal landscape and organic soils—peat soils—where climate is severe. The plants are native to acidic bogs and peat wetlands as they favor acidic conditions.

Cranberry can grow and survive only when certain conditions are met. These include acidic peat soil, an adequate fresh water supply, and a growing season extending from April to November. Cranberries grow on low-lying vines in beds layered with sand, peat, gravel, and clay. These beds were originally created by glacial deposits and are commonly known as bogs or marshes. Commercial bogs use a system of wetlands, uplands, ditches, flumes, ponds, and other water bodies that provide a natural habitat for a variety of plant and animal life (Brick 1980).

In 2004, according to FAO report, the world production of cranberry amounted to 757 million pounds. Some analysts predict that the world production will rise by ca, 40 % over the period of 2013–2018, surpassing the one million tons mark by 2018.

The production of cranberry is mostly limited to the Northern USA and Canada, which are at the top of world production of this plant—82 % and 14 %, respectively. The production of cranberries in other countries is significantly lower than in the USA or Canada: Latvia 2 %, Belarus <1 %, Azerbaijan <1 %, Ukraine <1 %, and Estonia <1 %.

Cranberries are one of the very few crops that are native to the USA. The USDA National Agricultural Statistics Service released its projected yields for the 2014 season. The total USA cranberry production was estimated to be 8.72 million barrels (a barrel is 100 pounds of cranberries). Wisconsin, which produces 57 % of the country's yields, was predicted to yield 4.5 million barrels, up 2 % from 2011. Massachusetts was forecasted to produce 2.10 million barrels, down 9 % from 2011. New Jersey was going to yield 542,500 barrels; Oregon, 400,000 barrels; and Washington, 142,000 barrels. The average yield in 2014 was 211.6 barrels per acre nationwide, an increase of 0.2 barrels per acre in comparison with 211.4 barrels in 2013. Approximately ten million barrels are produced globally, including the production in the USA, Canada, and Chile.

11.2 Methods

pH Instrument, pH meter; reagent, 0.1 N KCl

pH of peat was assayed potentiometrically in 1 N KCl (1:2.5, v/v) suspensions using combination electrode.

The Determination of Total Nitrogen The total nitrogen was estimated by Kjeldahl method.

The Determination of Total Organic Carbon The total organic carbon was estimated by TOC 5050A with Solid Sample Module, SSM-5000A, Shimadzu, Japan.

The Determination of Dissolved Organic Carbon For the estimation of dissolved organic carbon (DOC), soil samples in redistilled water were heated at 100 °C during 2 h under reflux condenser. Extracts were separated with the mean filter paper and analyzed on TOC 5050A a facilities (Shimadzu, Japan).

The method of the determination of IAA is described in Chap. 10.

11.3 Conclusions

Peat soils contain a higher concentration of mineral and organic compounds of a well-known and unknown structure than mineral soils. The amount of amino acids, hemicellulose, cellulose, lignin, pectin, bitumens, lipids, waxes, resins, carbohydrates, phenolic compounds, amines, amides, terpenoids, amino sugars, nucleic acids, vitamins, nitrogenous material, non-saturated and saturated fatty acids, organic sulfur compounds, starch compounds, ethereal oils, balsam, antibiotics (streptomycin and penicillin), hormones (derivatives of estrogen), and enzymes is considerable higher in peat soils than in mineral soils. Therefore, peat as a raw material with a rich spectrum of biologically active substances is widely used in agriculture, horticulture, veterinary, pharmacy, and medicine.

Biologically active substances of well-known and unknown structure of peat and peat extracts influence on organisms (Banaszkiewicz and Drobnik 1994; Beer et al. 2003). Peat is involved in the stimulation of digestion as it causes:

(i) The reduction of pH in the intestines
(ii) The contraction of smooth muscles in the gastrointestinal tracts
(iii) The improvement of nutrient uptake and conversion efficiency
(iv) The formation of chelate complexes with heavy metals, toxins produced by pathogenic bacteria, pesticides, etc.
(v) Lipid and protein metabolism

Antibiotics formed by strains of eubacteria and actinomycetes in peat influence the composition of intestinal microflora (Huck et al. 1991). Peat extracts protect

Table 11.1 pH, N-total, dissolve organic carbon (DOC), total organic carbon (TOC), and indole-3-acetic acid (IAA) in soil under cultivation of cranberry (Szajdak, unpublished data)

Time of sampling	pH	N-total	DOC g kg^{-1}	TOC g kg^{-1}	IAA μg kg^{-1}
March 6	6.16	2.10	21.79	475.7	273.20
April 10	5.95	2.29	24.92	434.6	249.45
May 7	5.22	1.80	30.93	433.1	225.69
June 19	5.09	2.40	32.52	450.0	231.63
July 14	5.49	2.01	24.99	473.4	201.93
August 8	5.38	2.30	25.25	472.8	195.99
September 10	5.29	2.22	22.23	451.3	225.69
October 8	5.34	2.45	23.10	450.1	237.57

DOC, dissolved organic carbon; TOC, the total organic carbon; IAA, indole-3-acetic acid

from viral diseases (Klocking and Sprossig 1972; Thiel et al. 1977; Schiller et al. 1979; Sydow et al. 1986; Schols et al. 1991; Jankowski et al. 1993; Laub 2000), activate the immune system (Riede et al. 1991), reduce or minimalize infection in patients (Kuhnert et al. 1982; Pukhova et al. 1987), decrease the total lipid content in liver, and increase leg muscle strength (Stepchenko et al. 1991).

pH of peat soils directly affects the life and growth of plants. pH-dependent processes are important in terms of hydrogen ion equilibrium, stability, ligand interactions, assembly, and dynamics. Therefore, pH affects the availability of all plant nutrients. The nutrients have to be soluble and have to successfully move through the soil solution toward the roots. Nitrogen, for example, has its greatest solubility between soil pH 4–8. Soil microorganisms are involved in the conversion and transformation of chemical and biochemical compounds in soils.

During the entire vegetation seasonal of cranberry, pH values of peat ranged from 5.29 to 6.16 (Table 11.1) (Szajdak, unpublished data).

The concentration of total nitrogen, DOC, and TOC ranged from 1.80 to 2.40 % (21.79–32.53 g kg^{-1} and 433.1–475.7 g kg^{-1}, respectively). The highest concentration of IAA was noted in March when it equaled to 273.2 ng kg^{-1}. The content of IAA decreased later during the period of intense plant development. The lowest amount of IAA was determined in August. In autumn the concentration of IAA increased again (Table 11.2). The amount of IAA in Table 11.2 concerns the free form of this compound. Free forms are very unstable and migrate very quickly in soils, IAA physiological function, which expressed by the interaction of IAA with IAA receptor in plants. Therefore, the research on the free form of IAA in fact concerns the biochemical and biological processes in soils.

Cranberry grows in very comfortable conditions of peatlands (Figs. 11.1, 11.2, and 11.3).

The fruit of cranberry has gained significant attention for its putative human health benefits. However, research has mostly been focused on the phenolic acids, their derivatives, and flavonoids due to their high chemical, biological, physiological, and pharmaceutical activity. In vitro chemical analyses showed that cranberries, out of 21 fruits, had the highest antioxidant values (Vinson et al. 2001; Sun et al. 2002).

Table 11.2 Structures of phenolic acids

Benzoic acid	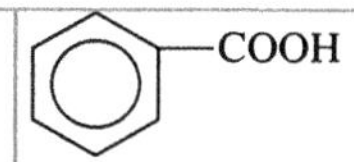
o-Hydroxybenzoic acid	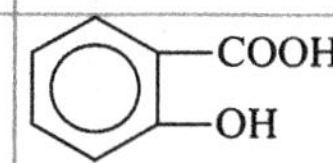
m-Hydroxybenzoic acid	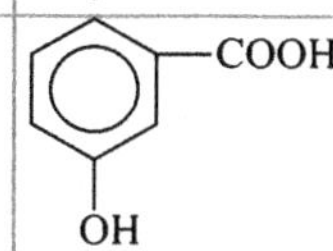
p-Hydroxybenzoic acid	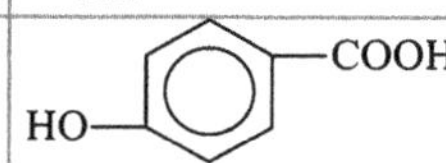
Phthalic acid	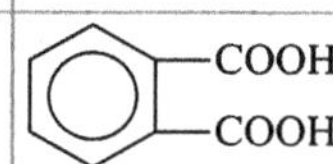
Cinnamic acid	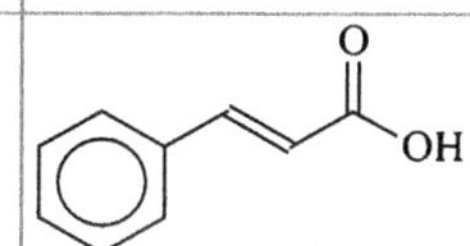
o-Hydroxycinnamic acid	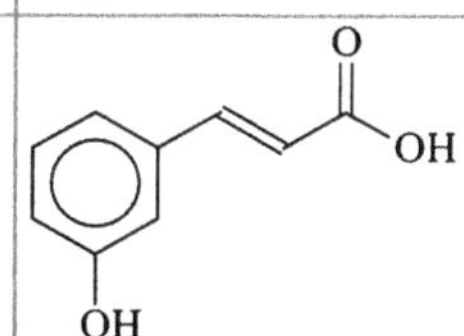
2,3-Dihydroxybenzoic acid	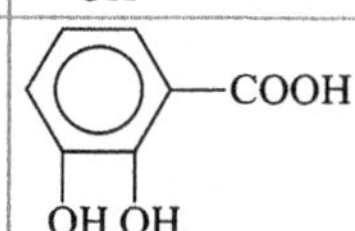
4-Hydroxyphenylacetic acid	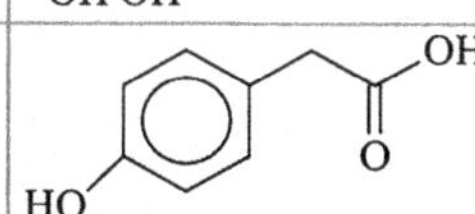
Vanillic acid	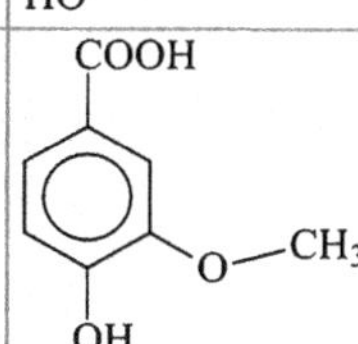

Table 11.2 (continued)

2,4-Dihydroxybenzoic acid	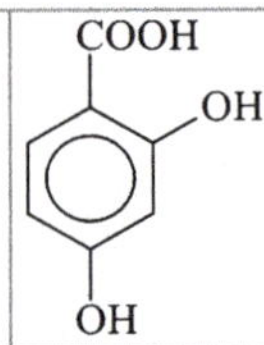
p-Coumaric acid	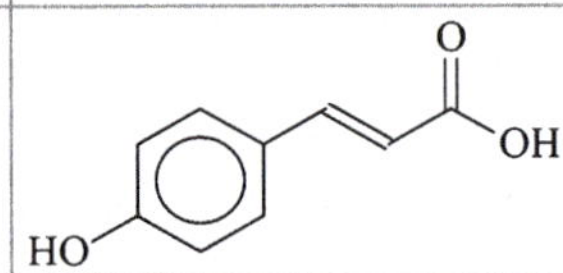
Ferulic acid	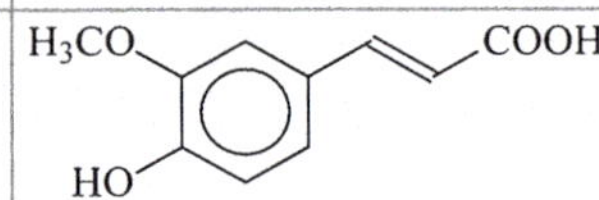
Caffeic acid	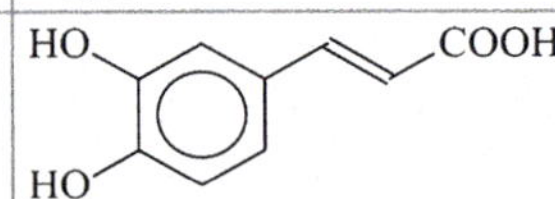
Sinapic acid	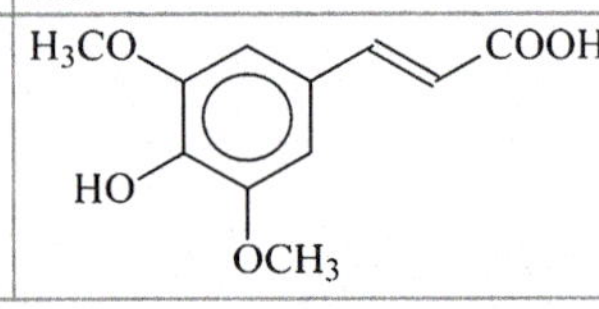

Fig. 11.1 Cranberry cultivation

The overall phenolic content appeared to correlate with the level of antioxidant activity (Vinson et al. 2001; Sun et al. 2002). The phenolic classes identified in cranberry include some phenolic acids (Marwan and Nagel 1982; Heimhuber et al. 1990; Zheng and Shetty 2000; Zuo et al. 2002): anthocyanins (Hong and Wrolstad 1986, 1990), flavonols, procyanidins, proanthoycyanidins, and the derivatives of flavon-3-ols. The derivatives of flavon-3-ols are represented by both monomers and the polymer classes of proanthoycyanidins (Foo and Porter 1981; Foo et al. 2000a, b).

Cranberries are epigynous or "false" berries. The fruit are bright red with waxy bloom at maturity, giving dark red to black appearance. The color changes from green to white and then red during its development. The fruit matures in 60–120

Fig. 11.2 Creeping shrub of cranberry

Fig. 11.3 Mature fruits

days after fertilization, depending on the cultivar and weather. Red color is the primary determinant of harvest maturity and fruit quality. The color increases over time; therefore, harvest is delayed as long as possible to allow color development.

About 90–95 % of the cranberry crop is processed into juices and sauces. Recently, juice blends have become more popular. Cranberry juice and fruits are reported to display a number of health benefits including:

(i) Potent antioxidant activity
(ii) Cholesterol reduction
(iii) Vasorelaxant effects
(iv) The prevention of urinary tract infections
(v) The reduction of biofilm formation
(vi) In vivo anticancer effects

The beneficial impact of cranberries on human health is caused by the presence of the following substances in the berries:

(a) Anthocyanins (Hong and Wrolstad 1990; Mazza and Miniati 1993)
(b) Proanthocyanidins (condensed tannins) (Foo and Porter 1981; Foo et al. 2000a, b) and flavonol glycosides (Puski and Francis 1967)

(c) Low-molecular-weight phenolic acids (Zuo et al. 2002)
(d) Organic acids (Heimhuber et al. 1990)
(e) Sugars (Hong and Wrolstad 1986)

Therefore, the knowledge of bioavailability, the mechanism of uptake, and the consequences of biotransformation of these biologically active substances after cranberry consumption is urgently needed. Incorporating cranberries in a balanced diet rich in fruits, vegetables, and whole grains is recommended for the prevention of cardiovascular disease (CVD). CVD is a major cause of death in most industrialized countries (Chu et al. 2002; Chu and Liu 2005). In the USA, the annual CVD death rate exceeds one million with an annual economic cost evaluated at over 350$ billion, surpassing 3 % of the US GDP (gross domestic product) for a single disease (American Heart Association 2002). Wise diet is said to be an effective method for reducing the formation of artherosclerotic lesions. An increase in fruits and vegetables intake of one daily serving decreases the risk of CVD from 7 to 4 % (Joshipura et al. 2001). Oxidized low-density lipoprotein (LDL) plays a significant role in the initiation and the acceleration of artherosclerotic process. Cardiovascular diseases can be prevented through lowering LDL oxidation.

Cranberries exhibit a very high phytochemical and antioxidant activity (Lusis 2000; Chu et al. 2002; Sun et al. 2002; Liu 2003). Just a single intake of cranberry juice leads to a significant increase in plasma antioxidant level for up to 7 h after the intake. Moreover, it also causes changes in high-density lipoprotein (HDL) in hypercholesterolemic human subjects when consumed for an extended period of time (Miller et al. 1998; Viason et al. 2003). In vitro fresh cranberry extracts inhibited low-density lipoprotein (LDL) oxidation (Wilson et al. 1998), which is a critical point in artherosclerotic conversion (Lusis 2000). The antioxidative activity of 100 g cranberries against LDL oxidation is equivalent to that of 1000 mg vitamin C (ascorbic acid) or 3700 mg vitamin E (tocopherol). Cranberry extracts also significantly induce the expression of hepatic LDL receptors and increase the intracellular uptake of cholesterol in HepG2 cell in vitro in dose-dependent manner. This may suggests that cranberries can clear excessive plasma cholesterol from the vascular system. The pharmacokinetic mechanism is based on the properties of biologically active substances in cranberries. These compounds are responsible for the inhibition of LDL oxidation. They induce the expression of LDL receptors and increase the uptake of cholesterol in hepatocytes (Chu and Liu 2005).

The derivatives of phenolic compounds found in cranberries represent a rich source of natural antioxidants and reveal an inhibitory effect on mutagenesis and carcinogenesis (Rice-Evans et al. 1996; Zuo et al. 2002). Flavonoids and phenolic acids exist in berries predominantly in combined forms, such as glycosides and esters. A total of 400 mg of flavonoids and phenolic compounds (44 %, phenolic acids; 56 %, flavonoids) was found in a freshly squeezed cranberry juice sample. Fifteen benzoic and phenolic acids (benzoic, *o*-hydroxybenzoic, cinnamic, *m*-hydroxybenzoic, *p*-hydroxybenzoic, *p*-hydroxyphenyl acetic, phthalic, 2,3-dihydroxybenzoic, vanillic, *o*-hydroxycinnamic, 2,4-dihydrobenzoic, *p*-coumaric, ferulic, caffeic, and sinapic) were identified in cranberry fruit in their free and bound forms. Cranberry fruit contained a high amount of benzoic and phenolic acids (5.7 g/kg fresh weight) with benzoic acid being the most abundant (4.7 g/kg fresh weight) (Table 11.2).

Benzoic and phenolic acids occur mainly in bound forms, while only about 10 % of them occur as free acids. What is more, a cranberry fruit included *p*-coumaric (0.25 g/kg fresh weight) and sinapic acids (0.21 g/kg fresh weight). Quercetin and myricetin were the most common flavonoids found in the freshly squeezed cranberry juice. In addition, the derivatives of flavonol glycoside conjugates were represented by quercetin-3-α-arabinopyranoside, quercetin-3-*O*-(6″-*p*-coumaroyl)-*β*-galactoside, 3′-methoxyquercetin-3-α- xylopyranoside, myricetin-3-*β*-xylopyranoside, and quercetin-3-*β*-glucoside (Vvedenskaya et al. 2004) (Table 11.3).

It has been well established that complex biologically active substances in cranberry fruit can provide protective health benefits mainly through a combination of additive and/or synergistic effects. These substances can have a complementary and overlapping influence on oxidative stress, the immune system, gene expression in cell proliferation and apoptosis, and hormone metabolism. They can also have a direct antibacterial and antiviral effect. In addition, cranberries inhibit the development of tumor cells in oral, prostate, and colon cancer patients (Yan et al. 2002; Vaisberg and Neto 2003; Seeram et al. 2004).

Moreover, recent studies indicate that cranberry extracts exhibit antimicrobial activity against several food-borne and human pathogens (Cavanagh et al. 2003; Vattem et al. 2005). They also inhibit the adhesion of *Helicobacter pylori* to the gastric mucus. *Helicobacter pylori* is a "Gram-negative" microaerophilic bacterium that lives in the stomach and duodenum. Infections caused by *Helicobacter pylori* are generally recognized as one of the etiological agents of gastritis, peptic ulcer, gastric cancer, mucosa-associated lymphoid tissue lymphoma (Uemura et al. 2001; Fox and Wang 2001), and cardiovascular diseases (Pellicano et al. 2003). Helicobacter pylori is indigenous to the stomach of more than 50 % of the entire population, reaching 80 % in some countries (Dunn et al. 1997; Vattem et al. 2005). Most chronic infections caused by *Helicobacter pylori* are asymptomatic, and only if the colonization of the bacteria persists, symptoms appear in 15–20 % of the infected population (Parsonnet et al. 1991). Dietary management of *Helicobacter pylori* infection by consuming fruits of cranberries and their products could be an effective strategy due to likelihood of high compliance and absence of side effects.

Clinical studies also confirm that cranberries have a beneficial impact on the prevention of urinary tract infections (Kontiokari et al. 2001; Stothers 2002; Howell et al. 1998, 2005). The adhesion of microorganisms to the uroepithelium is the initial step in the development of mammalian urinary tract infections. Cranberry may inhibit the adhesion of P-fimbriated uropathogenic strains of *Escherichia coli* to uroepithelial cells (Ofek et al. 1991). *Escherichia coli* strains that express P-fimbriae are linked to both cystitis and pyelonephritis (Roberst et al. 1989). The majority of P-fimbriated *Escherichia coli* that cause a urinary tract infection bind glycosphingolipid receptor sites on the uroepithelium that are similar in structure to the P blood group antigens on the surface of A_1, Rh+human red blood cells (Kallenius et al. 1980).

Table 11.3 Structures of quercetin and myricetin and flavonol glycosides isolated from cranberry (Vvedenskaya et al. 2004)

Quercetin	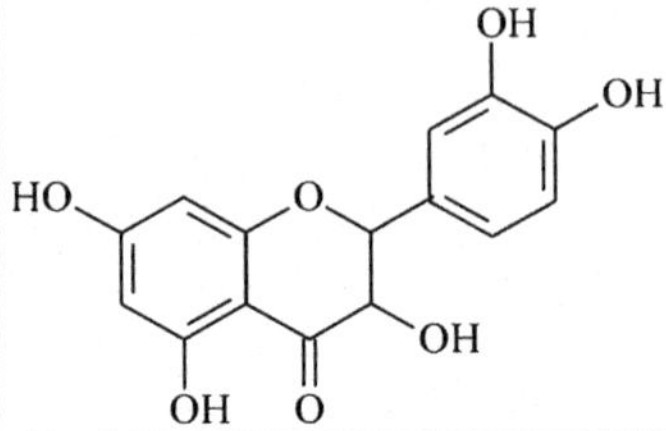
Myricetin	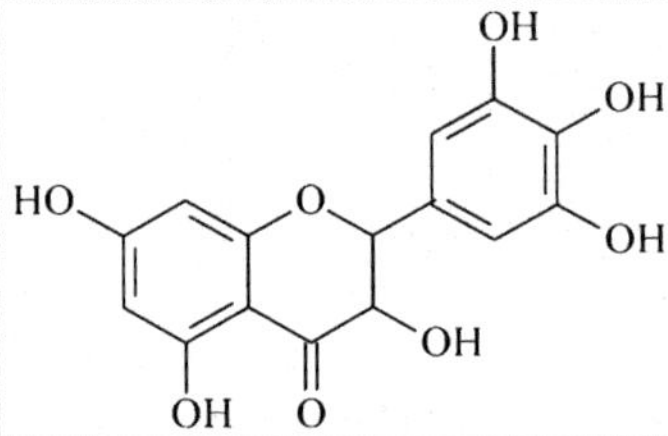
Myricetin-3-β-xylopyranoside	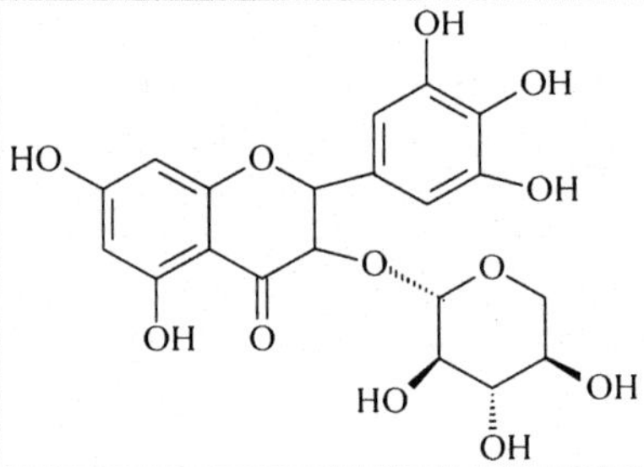
Quercetin-3-β-galactoside	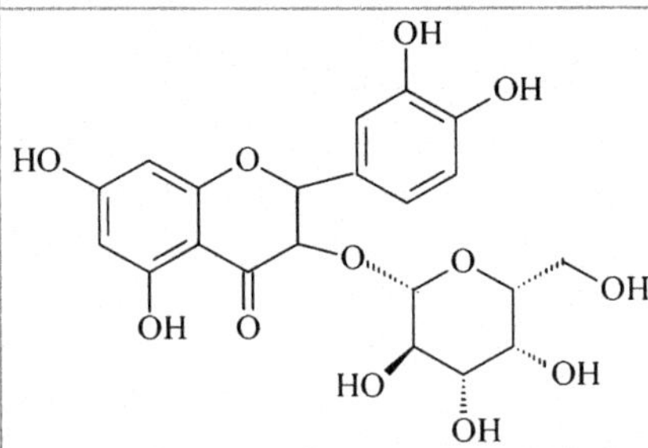
Quercetin-3-β-glucoside	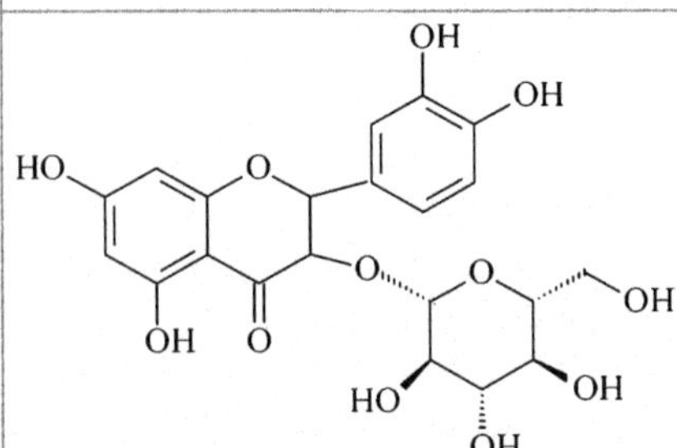

(continued)

Table 11.3 (continued)

Quercetin-3-α-arabinopyranoside	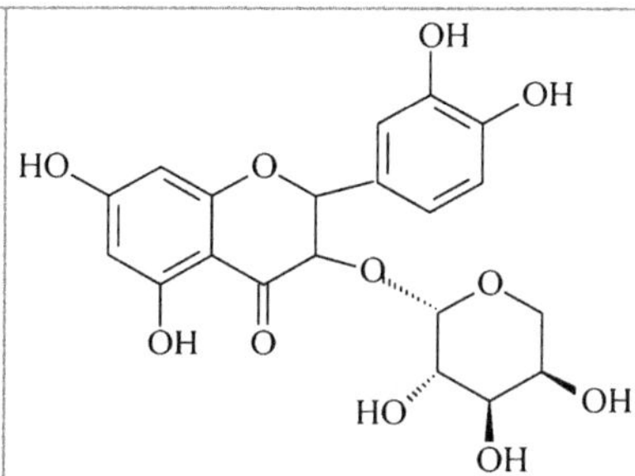
Quercetin-3-α-arabinofuranoside	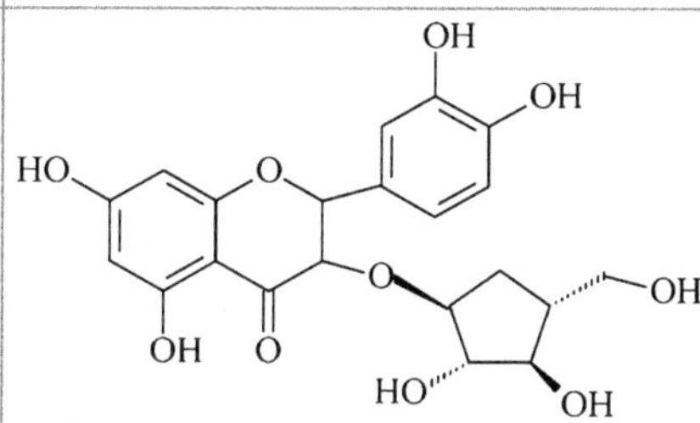
3′-Methoxyquercetin-3-α-xylopyranoside	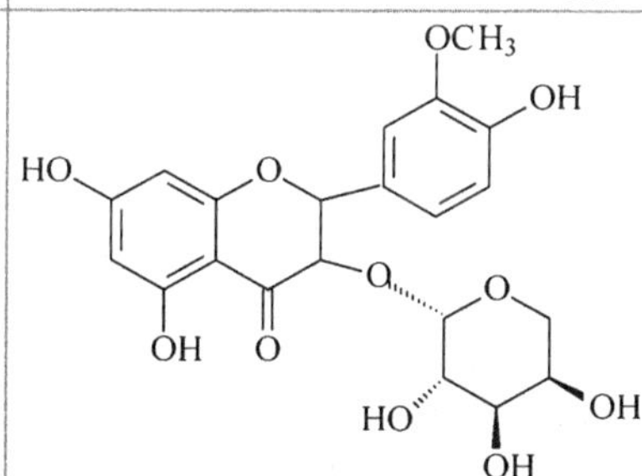
Quercetin-3-O-(6″-p-coumaroyl)-β-galactoside	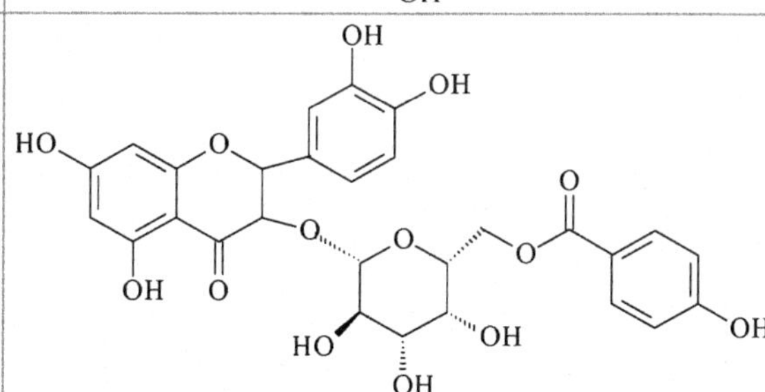
Quercetin-3-O-(6″-benzoyl)-β-galactoside	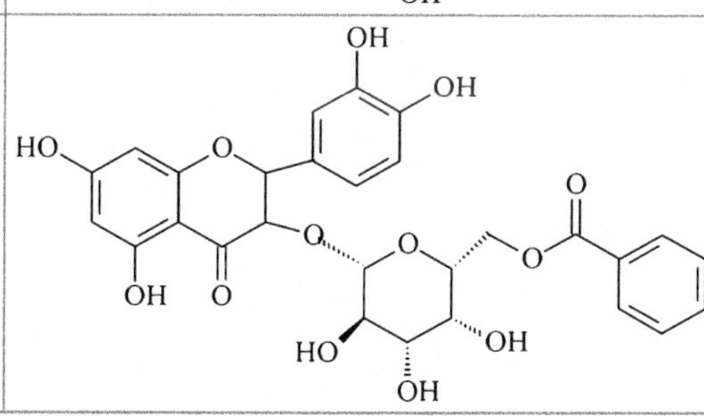

Proanthocyanidins isolated from cranberry inhibit P-fimbrial adhesion in vitro and, thus, may be able to prevent urinary tract infections (Howell et al. 1998, 2005).

Cranberries are also an excellent dietary source of a wide array of phytochemicals of physiological significance, which include:

(i) Flavonol glycosides
(ii) Anthocyanins
(iii) Proanthocyanidins
(iv) Organic and phenolic acids

Cranberry juice and fruit exhibit various health benefits. Numerous studies have shown that cranberries display potent antioxidant activity. Moreover, they reduce cholesterol and biofilm formation and prevent urinary tract infections. Furthermore, cranberries are reported to have vasorelaxant and in vivo anticancer effect.

Cranberry is a significant example of a plant which is growing on peat soils and has a high content and rich spectrum of biologically active substances. The organic compounds in the plant are responsible for its wide medical and pharmaceutical use.

References

American Heart Association (2002) Heart disease and stroke statistics-2003 update. American Heart Association, Dallas. http://www.americanheart.org/downloadable/

Banaszkiewicz W, Drobnik M (1994) The influence of natural peat and isolated humic acid solution on certain indices of metabolism and of acid-base equilibrium in experimental animals (in Polish). Rocz Państ Zak Hig 45:353–360

Beer AM, Juginger HE, Lukanov J, Sagorchev P (2003) Evaluation of the permeation of peat substances through human skin in vitro. Int J Pharm 253:169–175

Brick AR Jr (1980) Energy inputs in cranberry production. In: Pimental D (ed) Handbook of energy utilization in agriculture. CRC Press, Boca Raton, pp 315–325

Cavanagh HM, Hipwell M, Wilkinson JM (2003) Antibacterial activity of berry fruits used for culinary purposes. J Med Food 61(1):57–61

Chu YF, Liu RH (2005) Cranberries inhibit LDL oxidation and induce LDL receptor expression in hepatocytes. Life Sci 77:1892–1901

Chu YF, Sun J, Wu XZ, Liu RH (2002) Antioxidant and antiproliferative activities of common vegetables. J Agric Food Chem 50:6910–6916

Dunn BE, Cohen H, Blaser MJ (1997) Helicobacter pylori. Clin Microbiol Rev 10(4):720–741

Foo LY, Porter LJ (1981) The structure of tannins of some edible fruits. J Agric Food Chem 32:711–716

Foo LY, Lu Y, Howell AB, Vorsa N (2000a) The structure of cranberry proanthocyanidins which inhibit adherence of uropathogenic P-fimbriated *Escherichia coli* in vitro. Phytochemistry 5:173–181

Foo LY, Lu Y, Howell AB, Vorsa N (2000b) A-type proanthocyanidins trimers from cranberry that inhibit adherence of uropathogenic P-fimbriated *Escherichia coli*. J Nat Prod 63:1225–1228

Fox JG, Wang TC (2001) *Helicobacter pylori*-not a good bug after all. N Engl J Med 345(11):829–832

Heimhuber B, Wray V, Galensa R, Hermann K (1990) Benzoylglucoses from two *Vaccinium* species. Phytochemistry 29:2726–2729

Hong V, Wrolstad RE (1986) Cranberry juice composition. J Assoc Off Anal Chem 69:199–207

Hong V, Wrolstad RE (1990) Use of HPLC separation photodiode array detection for characterization of anthocyanins. J Agric Food Chem 38:708–715

Howell AB, Vorsa N, DerMarderosian A, Foo LY (1998) Inhibition of adherence of P-fimbriated *Escherichia coli* to uroepithelial-cell surfaces by proanthocyanidin extracts from cranberries. N Engl J Med 339:1085–1086

Howell AB, Reed JD, Krueger CG, Winternbottom R, Cunningham DG, Leahy M (2005) A-type cranberry proanthocyanidins and uropathogenic bacterial anti-adhesion activity. Phytochemistry 66:2281–2291

Huck TA, Porter N, Bushell ME (1991) Positive selection of antibiotic-producing soil isolates. J Gen Microbiol 137(10):2321–2329

Jankowski A, Nienartowicz B, Polanska B, Lewandowicz-Uszyńska A (1993) A randomised, double-blind study on the efficacy of Tołpa Torf Preparation (TTP) in the treatment of recurrent respiratory tract infections. Arch Immunol Ther Exp (Warsz) 41:95–97

Joshipura KJ, Hu FB, Manson JE, Stampfer MJ, Rimm EB, Spezer FE, Colditz G, Ascherio A, Rosner B, Spiedelman D, Willen WC (2001) The effect of fruit and vegetable intake on risk for coronary heart disease. Ann Intern Med 134:1106–1114

Kallenius G, Molby R, Svenson SB, Winberg J, Lunblad A, Svenson S, Cedergree B (1980) The P-antigen as receptor for the haemagglutinin of pyelonephritis *Escherichia coli*. Microbiol Lett 7:297–302

Klocking R, Sprossig M (1972) Antiviral properties of humic acids. Experientia 28:607–608

Kontiokari T, Sundquist K, Nuutinen M, Pokka T, Koskela M, Uhari M (2001) Randomized trial of cranberry-lingonberry juice and Lactobacillus GG drink for the prevention of urinary tract infections in women. Br Med J 322:1571–1573

Kuhnert M, Fuchs V, Golbs S (1982) Chemical characterization and pharmacologico-toxicological peculiarities of humic acid (in German). Archiv Exp Vet 36:169–177

Laub R (2000) Laub developing humate with anti-HIV, HSV, HPV and other antiviral activity. Biotechnology Information Institute. Antiviral Drug and Vaccine Development Information, 12, ISBN 0897–9871. www.laubbiochem.com/pdf/AABullVol13reprint.pdf

Liu RH (2003) Health benefits of fruit and vegetables are from additive and synergistic combination of phytochemicals. Am J Clin Nutr 78:5175–5208

Lusis AJ (2000) Atherosclerosis. Nature 407:233–241

Marwan AG, Nagel CW (1982) Identification of the hydroxycinnamic acid derivatives in cranberries. J Food Sci 47:774–782

Mazza G, Miniati E (1993) Anthocyanins in fruits, vegetables and grains. CRC Press, Boca Raton, pp 144–199

Miller ER, Appel LJ, Risby TH (1998) Effect of dietary patterns on measures of lipid peroxidation results from a randomized clinical trial. Circulation 98:2390–2395

Ofek I, Goldhar J, Zafriri D, Lis H, Adar R, Sharon N (1991) Anti-*Escherichia coli* adhesion activity of cranberry and blueberry juices. N Engl J Med 324:1599

Parsonnet J, Friedman GD, Vandersteen DP, Chang Y, Vogelman JH, Orentreich N et al (1991) *Helicobacter pylori* infection and the risk of gastric carcinoma. N Engl J Med 325:1127–1131

Pellicano R, Fagoonee S, Rizetto M, Ponzetto A (2003) *Helicobacter pylori* and coronary heart disease which directions for future studies? Crit Rev Microbiol 29(4):351–359

Pukhova GG, Druzhina NA, Stepchenko LM, Chebotarev EE (1987) Effect of sodium humate on animals irradiated with lethal doses (in Russian). Radiobiologii 27:650–653

Puski D, Francis F (1967) Flavonol glycosides in cranberries. J Food Sci 32:527–530

Rice-Evans CA, Miller NJ, Paganga G (1996) Structure-antioxidant activity relationships of flavonoids and phenolic acids. Free Radic Biol Med 20(7):933–956

Riede UN, Zeck-Kapp G, Freudenberg N, Keller HU, Seubert B (1991) Humate induced activation of human granulocytes. Virchows Arch B Cell Pathol Incl Mol Pathol 60:27–34

Roberst JA, Kaack MB, Fussell EN (1989) Bacterial adherence in urinary tract infections: Preliminary studies in a primate model. Infection 17:401–404

Schiller F, Klocking R, Wutzler P, Farber I (1979) Results of an oriented clinical trial of ammonium humate for the local treatment of herpesvirus hominis (HVH) infections (in German). Dermatol Monatsschr 165:505–509

Schols D, Wutzler P, Klocking R, Helbig B, De Clercq E (1991) Selective inhibitory activity of polyhydroxy carboxylates derived from phenolic compounds against human immunodeficiency virus replication. J Acquir Immune Defic Syndr 4:677–685

Seeram NP, Adams LS, Hardy ML, Heber D (2004) Total cranberry extract versus its phytochemical constituents: antiproliferative and synergistic effects against human tumor cell lines. J Agric Food Chem 52:2512–2517

Stepchenko LM, Zhorina LV, Kravtsova LV (1991) The effect of sodium humate on metabolism and resistance in highly productive poultry (in Russian). Nauchnye Doki Vyss Shkoly Biol Nauki 10:90–95

Stothers L (2002) A randomized trial to evaluate effectiveness and cost effectiveness of naturopathic cranberry products as prophylaxis against urinary tract infection in women. Can J Urol 9:1158–1162

Sun J, Chu YF, Wa XZ, Liu RH (2002) Antioxidant and antiproliferative activities of common fruits. J Agric Food Chem 50:7449–7454

Sydow G, Wunderlich V, Klocking R, Helbig B (1986) The effects of phenolic polymers on retroviruses. Pharmazie 41(12):865–868 (in German)

Thiel KD, Klocking R, Schweizer H, Sprossig M (1977) In vitro studies of the antiviral activity of ammonium humate against Herpes simplex virus type 1 and type 2 (in German). Zentralbl Bakteriol Orig A 239:304–321

Uemura N, Okamoto S, Yamamoto S, Matsumura N, Yamaguchi S, Yamakido M et al (2001) *Helicobacter pylori* infection and the development of gastric cancer. N Engl J Med 345(11):784–789

Vaisberg AJ, Neto CC (2003) Identification of triterpene hydroxycinnamates with in vitro antitumor whole cranberry fruit (*Vaccinium macrocarpon*). J Agric Food Chem 51:3541–3545

Vattem DA, Lin YT, Ghaedian R, Shelly K (2005) Cranberry synergies for dietary management of *Helicobacter pylori* infections. Process Biochem 40:1583–1592

Viason JA, Al Khrmat H, Proch J, Sammonn N (2003) Single dose and supplementation studies with cranberry juice relevant to its role as an antioxidant and heart disease. FASEB J 17:A1097–A1098

Vinson JA, Su X, Zubik L, Bose P (2001) Phenol antioxidant quantity in foods: Fruits. J Agric Food Chem 49:5315–5321

Vvedenskaya IO, Rosen RT, Guido JE, Russel DJ, Mills KA, Vorsa N (2004) Characterization of flavonols in cranberry (*Vaccinium macrocarpon*) powder. J Agric Food Chem 52:188–195

Wilson T, Porcari JP, Harbin D (1998) Cranberry extract inhibits low density lipoprotein oxidation. Life Sci 62(24):PL381–PL386

Yan X, Murphy BT, Hammond GB, Vinson JA, Neto CC (2002) Antioxidant activities and antitumor screening of extracts from cranberry fruit (*Vaccinium macrocarpon*). J Agric Food Chem 50:5844–5849

Zheng Z, Shetty K (2000) Solid-state bioconversion of phenolics from cranberry pomace and role of *Lentinus edodes* β-glucosidase. J Agric Food Chem 48:895–900

Zuo Y, Wang C, Zhan J (2002) Separation, characterization, and quantization of benzoic and phenolic antioxidants in American cranberry fruit by GC-MC. J Agric Food Chem 50:3789–3794

Chapter 12
The Importance of Horticultural Growing Media and Biochemical Processes

Lech Wojciech Szajdak, Katarzyna Styła, Wioletta Gaca, Teresa Meysner, Marek Szczepański, and Jacek Stanisław Nowak

Abstract The effect of the IAA content in the different media on the rooting of the cuttings of ornamental plants: chrysanthemum *Dendranthema grandiflora* "Zembla White," poinsettia *Euphorbia pulcherrima* "Prestige Early Red," and hydrangea *Hydrangea* L. was investigated The rooting of the cuttings was carried out in cell trays with the use of six rooting substrates (three commercial substrates and three self-prepared substrates based on Polish neutralized white peat mixed with perlite). All substrates contained four concentrations of IAA (natural concentration, 200, 300, and 400 μg kg^{-1}). The natural content of IAA in the studied substrates was as follows: (i) commercial growing media for rooting of cuttings "Klasmann Steck Medium," 142.52 μg kg^{-1} d.m.; (ii) commercial growing media for rooting of cuttings "Substrate for rooting cuttings of ornamental plants AURA," 114.82 μg kg^{-1} d.m.; (iii) commercial growing media for rooting of cuttings CERES, 158.36 μg kg^{-1} d.m.; (iv) white peat (H3–H4) from Northwestern Poland, 133.63 μg kg^{-1} d.m.; (v) white peat (H3–H4) from Northern Poland, 109.88 μg kg^{-1} d.m.; and (vi) white peat (H3–H4) from the Northeastern Poland, 123.54 μg kg^{-1} d.m.

In addition, the activity of the enzymes which participate in the nitrogen cycle and redox processes was measured in growing media during cultivation period. The experiments were in the line with EPPO norms [European and Mediterranean Plant Protection Organization—Guideline for the efficiency evaluation of plant growth regulators, Rooting of cuttings, PP 1/186(2)].

Keywords Growing media • Rooting cuttings • Auxin • Enzymatic processes

---

L.W. Szajdak (✉) • K. Styła • W. Gaca • T. Meysner • M. Szczepański
Institute for Agricultural and Forest Environment, Polish Academy of Sciences, Str. Bukowska 19, 60-809 Poznań, Poland
e-mail: lech.szajdak@isrl.poznan.pl

J.S. Nowak
Research Institute of Horticulture, Str. Pomologiczna 18, 96-100 Skierniewice, Poland
e-mail: jacek.nowak@inhort.pl

L.W. Szajdak (ed.), *Bioactive Compounds in Agricultural Soils*,
DOI 10.1007/978-3-319-43107-9_12

12.1 Introduction

A growing medium should maintain a plant and provide it with water, nutrients, and the right conditions for gas exchange in the root zone. The root system needs a sufficient amount of oxygen and free space between the solid particles of a substrate. The choice of a substrate is one of the decisive factors influencing the ornamental plant cultivation. The main component of a horticultural substrate is peat. Depending on its origin and the depth of a bed, the peat exhibits different physicochemical and biological properties.

Growing media used for the cultivation of plants include, beside peat coir, compost, bark, perlite, vermiculite, mineral wool, sand, etc. The choice of a medium component used depends on the availability of materials, cultivation methods, and the size and type of a container used for rooting. The process of root formation sometimes takes a few months; therefore, a growing medium should be sterile, airy, highly permeable, and free of disease pathogens or weeds. It ought to have a good moisture content and it should provide air at the base of shoots (Bhekithemba and Wahome 2010). Water and air properties have a significant influence on the formation of a root system, the growth of young roots, and the content of biologically active compounds responsible for biochemical, chemical, and physical processes.

White peat mixed in different proportions with permeable components such as perlite, vermiculite, and sand meets these requirements (Matysiak and Nowak 2008; Shadparvar et al. 2011). Coir (coco fiber, coconut) has similar properties and exhibits many characteristics that make it an equally good component of growing substrates.

Peat is a rich source of humic substances and it contains a low concentration of ash. Therefore, peats yield large amounts of humic substances per unit volume with desirable low ash contents. According to a number authors (Cresswell 1992; Meerow 1994, 1995; Martinez et al. 1997), the physical properties of a coir substrate put it on a par with *Sphagnum* peat or even make it superior to the peat. Verdonck et al. (1983) reported that coir had a very high carbon content, had degraded slowly, and was similar to peat moss in look and structure. In addition, coir is suggested as a substitute for peat as it had excellent drainage, physical resilience, and wettability, and it was free of weeds and pathogens. Furthermore, it was readily renewable (Smith 1995) and, in comparison with peat, it also had a lower water-holding capacity, a higher total porosity, and a very high air content (Martinez et al. 1997; Noguera et al. 1997; Stamps and Evans 1999). Therefore, coir has been successfully been used in cuttings and seed propagation (Farnsworth and Guam 1995; Stoven and Kooima 1999; Rose and Haase 2000). The use of the mixture of coco fibers, peat, sand, and perlite (60 % peat, 15 % coconut, 10 % perlite, 15 % sand) or the mixture of coco fibers and just peat (peat 65 %, 35 % coconut) led to a higher percentage of rooting and the formation of better root systems than the use of a common growing medium comprised of 70 % peat, 20 % sand, and 10 % perlite (Matysiak and Nowak 2008). The peat industry, quality assurance organizations, and growers are responsible for the quality of growing media. Therefore the most important parameters: pH, electrochemical conductivity (EC), bulk density, organic matter, total pore space, water and air capacity, and shrinkage have to be analyzed according to the European Standards in order to provide consumers with a greater transparency Fisher et al. (2013). Growing media constituents are the basic components of the mixtures, which are generally formulated on a percentage volume basis. Peats maintain its leading position as the best ingredient of growing media. They have been the most widely used constituent of growing media for commercial horticulture and floriculture.

The properties of peats are mainly determined by the composition of plant communities and the degree of humification of its remains. Bogs with a low degree of decomposition are very acidic and mostly lack the available forms of both macronutrients and micronutrients. Mineral materials, i.e., perlite, vermiculite, pumice, expanded clay, and lava, are being used in relatively small amounts. Their usage is not expected to increase dramatically for reason of price competitiveness and less favorable overall properties (Table 12.1).

In the past century, horticulture has seen a tremendous development in the industrialized countries and it is now an economic force. To some extent these changes will also affect the peat industry (Huttunen and Reinikainen 2000; Schmilewski and Falkenberg 2000; Pudelski 2002; Schmilewski 2008).

The development of alternatives for peat substrates is necessary for the following reasons: (i) the resources of peat are limited, (ii) the pressure for using waste coming from human or industrial activities increases rapidly, and (iii) the economic

Table 12.1 Estimated amounts of materials used for the production of growing media in Germany and the Netherlands for the professional and hobby market in 1999

	Germany		The Netherlands	
Material	(m^3)	(%)	(m^3)	(%)
Black peat	6,000,000	63	1,500,000	39
White peat	3,000,000	32	1,600,000	42
Wood fibers	190,000	2	5000	<1
Clay (fresh and dried)	170,000	2	/	/
Composed biogenic waste	90,000	1	/	/
Composed bark/bark	30,000	<1	40,000	1
Mineral wood	20,000	<1	500,000	13
Perlite	9000	<1	50,000	1
Sand	8000	<1	/	/
Coir (fibers and dust)	5000	<1	100,000	3
Rice hulls	3000	<1	10,000	<1
Pumice	400	<1	20,000	<1
Expanded clay	400	<1	20,000	<1
Others (lava, vermiculite, synthetic–organic material, etc.)	7000	<1	5000	<1
Total	*9,522,800*	*100*	*3,850,000*	*100*

Schmilewski and Falkenberg (2000)

Table 12.2 Peat classification based on the degree of decomposition

Class	Degree of decomposition
Light peat	H1–H3
Dark peat	H4–H6
Black peat	H7–H10

Raviv et al. (2002)

necessity to use locally produced waste products is pressing. Different types of peat vary in terms of the degree of decomposition. Plant species, climate, and quality of water affect characteristics of peat. The classification of peat by von Post, which is based on the degree of decomposition, is shown in Table 12.2 (Raviv et al. 2002).

About 90 % of *Sphagnum* leaf is made up of long, empty, thin-walled hyaline cells. As a consequence, the leaf behaves like a sponge and readily soaks up water. The function of 5-keto-D-mannuronic acid is hypothesized to be similar to α-ketoglutaric acid in the roots of higher plants in that they are both responsible for the extraction of ammonium from the environment (Painter 1998). This should occur via the reversible nucleophilic addition of ammonium to a carbonyl group (C=O) in the α-keto carboxylic acid group to yield a Schiff base (C=N). In higher plants, this Schiff base (formed with α-ketoglutaric acid) is first reductively aminated (C–N) to form glutamic acid and then aminated to form glutamine. Sphagnan is covalently cross-linked to other cell wall glycans and is an integral part of the structure of the holocellulose. This product is completely insoluble in water and available to the roots of cultivated plants (Painter 1998). Sphagnan binds and deactivates enzymes and limits microbial growth. It is reactive and unstable in both its liberated soluble form and its insoluble leaf-bound form (Painter 1998; Børsheim et al. 2001).

12.2 Conclusions

12.2.1 Phytohormones in Growing Media

The physiological activity of plants is regulated by many biochemical compounds called hormones. Since the discovery of the first phytohormone—auxin—in 1926, scientists have identified several types of phytohormones including auxins, cytokinins, abscisic acid, gibberellins, and ethylene. There are two types of growth regulators: (i) endogenous regulators (natural), which are produced by plants, and (ii) exogenous gulators, which are artificially introduced into plants in the form of synthetic preparations. The phytohormones can be used in cultivation in order to obtain a higher yield in agriculture, horticulture, pomiculture, and moriculture (Jankiewicz 1997; Ma et al. 2008; Zhen et al. 2008).

The use of plant growth regulators is very important for the rooting of cutting in ornamental plant production. Currently, ready-made substances containing auxins can be used for the rooting of plant cuttings. They come in the form of a powder in which the base of a cutting is immersed before insertion (Szajdak and Nowak 2013).

The first known phytohormone was auxin—indole-3-acetic acid (IAA). Auxins chemically represent derivatives of indole ring (IAA; IBA, indole-3-butyric acid; NAA, naphthalene-1-acetic acid) (formula 12.1).

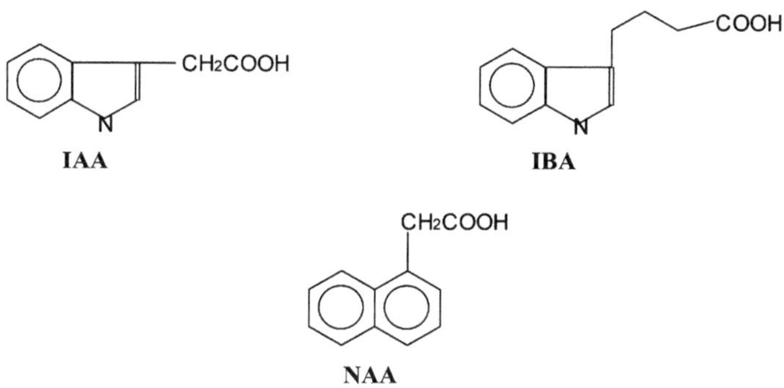

(12.1)

The chemical structure of auxins expresses the relationship between the quantity and the activity. Auxins are naturally produced in buds and leaves of plants from where they are later transported to the base of shoots. These substances in low quantities move within the plant from the site of production to the site of action; however, the exact mechanism remains unknown. Plant growth regulators (phytohormones) are organic substances which in low concentrations promote, inhibit, or modify the growth and development of plants (Halda-Alija 2003; Martinez-Morales et al. 2003; Szajdak and Maryganowa 2007; Szajdak et al. 2011a, c; Szajdak and Nowak 2013). These compounds occur in plants in free and bound forms. The free form of IAA is responsible for the biological activity (Szajdak 2004).

IAA is formed in mineral and organic soils (peat, moorsh, and sapropels) from tryptophan (TRP) by enzymatic conversion (Arshad and Frankenberger 1991; Beyeler et al. 1999; Kamnev et al. 2001; Halda-Alija 2003; Szajdak 2004; Szajdak and Maryganowa 2007; Szajdak et al. 2011a, c).

The application of chemical compounds exhibiting a hormonal effect or a hormonal-like effect has become an important agricultural production practice, particularly in horticulture, agriculture, pomology, and floriculture. IAA regulates several critical processes including the differentiation of cells, apical dominance, root initiation, parthenocarpy, abscission callus formation, and respiration (Halda-Alija 2003; Martinez-Morales et al. 2003; Szajdak and Maryganowa 2007; Szajdak et al. 2011a, c; Szajdak and Nowak 2013). It is vital that the level of endogenous IAA and its homeostasis is strictly regulated from biosynthesis and metabolism (Kamnev et al. 2001; Kawano et al. 2001).

Plant growth-promoting rhizobacteria (PGPR) are free-living soilborne bacteria, which stimulate the growth of plants either directly or indirectly (Kloepper et al.

1980; Glick 1995). The direct mechanisms involve nitrogen fixation; phosphorus solubilization; HCN production of phytohormones such as auxins, cytokinins, and gibberellins; and the lowering of ethylene concentration (Kloepper et al. 1989; Glick 1995; Glick et al. 1999). *Azotobacter*, *Arthrobacter*, *Nocardia*, *Chromobacterium*, *Pseudomonas fluorescens*, *Rhizobium*, and actinomycetes *Frankia* are responsible for the synthesis of IAA (Dahm et al. 1977; Garcia-Rodriguez et al. 1986; Beyeler et al. 1999). According to Ali et al. (2009), bacterial strains of *Bacillus*, *Pseudomonas*, *Escherichia*, *Micrococcus*, and *Staphylococcus* genera isolated from rhizosphere, histoplane, and phyllosphere of different plant species were found to increase endogenous IAA content and stimulate the growth of *Triticum aestivum* var. Inqalab-91. These authors found out that IAA in the presence of L-TRP increased the number bacterial strains. Hence, microbial strains associated with different plant species could be effectively used to increase the growth and yield of important crops (Ali et al. 2009). Also, Beyeler et al. (1999) reported that part of the plant growth-promoting effect of *Azospirillum brasilense* may have been caused by IAA formation. Moreover, the application of TRP on the rhizosphere could stimulate vegetative growth of corn, and this effect was ascribed to the conversion of TRP to IAA by rhizosphere bacteria (Beyeler et al. 1999). Bacteria belonging to the following genera: *Azospirillum*, *Pseudomonas*, *Xanthomonas*, *Rhizobium*, *Alcaligenes faecalis*, *Enterobacter cloacae*, *Acetobacter diazotrophicus*, and *Bradyrhizobium japonicum* produce auxins which stimulate plant growth (Patten and Glick 1996). The addition of IAA to *Azotobacter paspali* significantly increased the dry weight of leaves and roots of several plant species (Barea and Brown 1974). Tien et al. (1979) suggested that the addition of TRP (1–100 $\mu g\ ml^{-1}$) to *Azospirillum brasilense* Sp13t and SR2 increased the production of IAA and ILA (indole-3-lactic acid) (Tien et al. 1979). The inoculation of wheat seedlings with *Azospirillum brasilense* increased the number and length of lateral roots (Barbieri et al. 1986). The inoculation of canola seeds with *Pseudomonas putida* GR12-2, which produces low levels of IAA, resulted in a two- to a threefold increase in the length of seedling roots (Glick et al. 1986; Caron et al. 1995). The plant growth regulators produced by *Pseudomonas* species could also have an impact on plant growth (Karnwal 2009).

The synthetic auxins such as naphthalene-1-acetic acid (NAA), indole-3-butyric acid (IBA), and 2,4-dichlorophenoxyacetic acid (2,4-D) stimulate the formation of adventitious roots in horticulture and agriculture (Ludwig-Müller et al. 1995; Jankiewicz 1997; Chao et al. 2001; Szydło 2003). The most prominent synthetic derivative of auxin is IBA. IBA is rapidly metabolized in the plant tissue. IBA promotes and accelerates the root formation of plant clippings and reduces transplant shock of nonfood ornamental nursery stock. IBA is classified as a biochemical pesticide because it has a similar structure and function to the naturally occurring plant growth hormone IAA (United States Environmental Protection Agency 1992).

Many soil compounds exhibit strong auxin-like activity and differ in their IAA synthesizing capacity depending on the fertility status and organic matter content (Sarwar et al. 1992). Szajdak and Maryganova (2007, 2009) and Szajdak et al. (2013) evaluated the content of IAA in a commercial growing medium, raw peat

Table 12.3 The contents of chemical compounds in commercial growing medium before and after rooting of cuttings of *Euphorbia pulcherrima*

Commercial growing medium	pH (KCl)	IAA (g kg^{-1})	TOC (g kg^{-1})	C_{HWE} (g kg^{-1})	NH_4^+ (mg kg^{-1})	NO_3^- (mg kg^{-1})	N_{total} (g kg^{-1})	C/N
Before rooting of cuttings of *Euphorbia pulcherrima*								
Control (natural concentration of IAA)	4.58	134.47	391.30	10.47	25.59	54.20	8.96	44
Substrate + IAA (200 μg kg^{-1})	4.65	142.54	393.70	10.96	43.56	43.37	9.52	41
Substrate + IAA (300 μg kg^{-1})	4.72	128.29	388.10	10.94	40.84	42.35	9.56	41
Substrate + IAA (400 μg kg^{-1})	4.72	134.47	395.20	7.33	27.01	37.77	9.62	42
After rooting of cuttings of *Euphorbia pulcherrima*								
Control (natural concentration of IAA)	5.01	158.37	425.20	14.52	7.89	3.94	10.08	42
Substrate + IAA (200 μg kg^{-1})	5.16	190.06	421.70	12.81	4.40	4.93	10.36	41
Substrate + IAA (300 μg kg^{-1})	5.27	182.12	426.35	14.47	2.43	2.19	10.92	39
Substrate + IAA (400 μg kg^{-1})	5.41	170.25	433.70	13.81	3.24	1.11	11.20	39

Szajdak et al. (2013)

(high-moor peat and low-moor peat), sapropels, old and young shelterbelts, and adjoining cultivated fields. The content of IAA was the lowest before the rooting of cuttings (from 128.29 to 142.54 μg kg^{-1}) and increased after the plants were cultivated (from 158.37 to 190.06 μg kg^{-1}) reaching a maximum amount of 200 μg kg^{-1} of IAA additions (Table 12.3) Szajdak et al. (2013).

Plant roots release many compounds including inorganic ions and organic substances: amino acids, amides, sugars, aliphatic acids, aromatic acids, vitamins, peptides, proteins, enzymes, ketones, urea, phytoalexins, and plant hormones like IAA (Dundek et al. 2011). Szajdak and Maryganova (2009) suggested that soil flora also produced significant concentrations of auxin.

Certain types of peats and sapropels are commonly used in substrates and growing media for horticulture and pomology as well as in organic fertilizers for agriculture (Szajdak and Nowak 2013). Szajdak et al. (2004) and Szajdak and Maryganova (2007) measured the contents of IAA in raw peat (high-moor peat and low- moor peat) and sapropels. The authors observed that the amount of IAA in sapropels was higher (from 189.9 to 373.9 μg kg^{-1}) than in peats (from 57.9 to 210.2 μg kg^{-1}). In high-moor peat, the contents of IAA ranged from 124.4 to 210.0 μg kg^{-1} and in low-moor peat the quantity of IAA ranged from 69.3 to 186.1 μg kg^{-1}.

In addition, Szajdak and Maryganova (2009) showed that the concentrations of IAA in the soil under a 200-year-old shelterbelt (contains mainly *Robinia pseudo-acacia* and a small admixture of *Quercus robur and Larix decidua*) ranged from 64.4 to 241.5 μg kg^{-1} and were much higher than in the soil under a 15-year-old shelterbelt and adjoining cultivated field (from 5.8 to 21.8 μg kg^{-1}). Furthermore the contents of IAA in moorsh ranged from 69.3 to 186.1 μg kg^{-1}.

Falkowski and Szydło (2005) reported that the time of transplanting and the way of applying an auxin solution (450 ppm NAA, 25 ppm IBA) had an influence on the growth of 14 species of ornamental plants. The use of a solution containing 450 ppm NAA and 25 ppm IBA revealed a positive impact on the growth of Thuja plants which were transplanted late. In addition, soaking the roots of deciduous plants in a solution containing 450 ppm, 25 ppm NAA, and IBA before replanting beneficially influenced the length of shoots of *Hypericum kalmianum* "Gemo" and *Physocarpus opulifolius* "Dart's Gold." Moreover, watering the plants with this solution stimulated the growth of *Berberis thunbergii* "Coronita," *Hypericum kalmianum* "Gemo," *Physocarpus opulifolius* "Dart's Gold," and *Rosa rugosa* (Table 12.4).

Świstowska and Hetman (2004) proved that auxins: IAA, IBA, and NAA in vitro in concentrations 5, 10, 20, and 40 μM influenced the growth and development of *Columnea hirta* in a greenhouse. Auxins used at the stage of the rooting of the cuttings of plants had an impact on the diversity, adaptability, and development of plants in greenhouse conditions. *Columnea hirta* plants, cultivated in spring, adapted, grew faster, and consequently produced overground part and roots of the highest quality. Furthermore, NAA used in vitro during the rooting of the cuttings of plants had an adverse effect on plant growth in the same conditions. Auxins have an influence on plant growth for a long time after their application. Their impact is dependent on the concentration of auxins and the metabolic activity of plants. The most favorable results are obtained with the concentration of IBA at 20 μM (Table 12.5).

Table 12.4 Effect of auxins and the way of their application on shoot length of deciduous plants (cm)

	Plants		
Species and cultivar	Control	Watered with auxin solution	Soaked in auxin solution
Berberis thunbergii "Coronita"	137	209	–
Ginko biloba	21.9	21.0	23.0
Hypericum kalmianum "Gemo"	215	247	273
Pachysandra terminalis "Green carpet"	30.0	29.3	–
Physocarpus opulifolius "Dart's Gold"	152	184	210
Rosa rugosa	100.7	132.7	–

Falkowski and Szydło (2005)

Table 12.5 Average values of length of *Columnea hirta* (cm) roots in relation to kind and concentration of auxin and term and kind of the medium

Term	Auxin	*Sphagnum* peat (T)					Coconut fiber (K)					Mean		
		Concentration of auxin (μM)												
		0	5	10	20	40	0	5	10	20	40	T	K	Mean
Spring	IAA	8.59	8.89	8.47	9.17	9.03	11.37	11.74	11.55	11.70	11.10	8.83	11.50	10.16
	IBA	8.59	9.11	11.95	10.92	10.92	11.37	12.16	12.24	11.61	10.30	10.29	11.53	10.91
	NAA	8.59	6.94	5.37	8.19	2.97	11.37	8.99	6.34	6.41	–	6.41	6.62	6.51
Mean		8.59	8.31	8.59	9.42	7.64	11.37	10.96	10.04	9.90	7.13	8.51	9.88	9.19
NIR–LSD (p=0.05)														
Auxin					1.15					0.93				
Concentration					n.i.					n.i.				
Interaction					3.55					3.08				
Summer	IAA	3.81	4.32	4.72	5.16	4.76	6.04	6.04	5.36	6.54	6.54	4.55	6.10	5.32
	IBA	3.81	5.12	4.52	4.34	4.36	6.04	5.05	5.49	6.21	5.78	4.43	5.71	5.07
	NAA	3. 81	3.81	2.87	2.87	1.75	6.04	3.34	3.16	2.52	–	3.02	3.01	3.01
Mean		3.81	4.41	4.03	4.12	3.62	6.04	4.81	4.67	5.09	4.10	3.99	4.94	4.46
NIR–LSD (p=0.05)														
Auxin					0.54					0.61				
Concentration					n.i.					n.i.				
Interaction					3.03					5.30				

Świstowska and Hetman (2004)

Szajdak and Nowak (2013) evaluated the effect of IAA content in different media on the rooting of the cuttings of ornamental plants: chrysanthemum *Dendranthema grandiflora* "Zembla White," poinsettia *Euphorbia pulcherrima* "Prestige Early Red" and hydrangea *Hydrangea* L. From April 2011 to August 2012, the rooting of the cuttings was carried out in cell trays with the use of six rooting substrates (three commercial substrates and three self-prepared substrates based on Polish neutralized white peat mixed with perlite). All contained four concentrations of IAA (natural concentration, 200, 300, and 400 μg kg^{-1}). The natural content of IAA in the studied substrates was as follows:

1. Commercial growing media for rooting of cuttings "Klasmann Steck Medium"—142.52 μg kg^{-1} d.m.
2. Commercial growing media for rooting of cuttings "Substrate for rooting cuttings of ornamental plants AURA"—114.82 μg kg^{-1} d.m.
3. Commercial growing media for rooting of cuttings CERES—158.36 μg kg^{-1} d.m.
4. White peat (H3–H4) from Northwestern Poland—133.63 μg kg^{-1} d.m.
5. White peat (H3–H4) from Northern Poland—109.88 μg kg^{-1} d.m.
6. White peat (H3–H4) from Northeastern Poland—123.54 μg kg^{-1} d.m.

The experiments were in the line with EPPO norms [European and Mediterranean Plant Protection Organization—Guideline for the efficiency evaluation of plant growth regulators, Rooting of cuttings, PP 1/186(2)].

The evaluation of cuttings was established after their rooting. The following features were evaluated: the number of rooted cuttings (%), the length of roots (cm), the number of new shoots growing from the cutting, and the assessment of root system in the scale of 1–5, according to EPPO norms (1, very good rooting, produced bulk; 2, good rooting, roots did not produce bulk; 3, weak rooting, tips of roots or individual roots are visible; 4, absence of roots or callus visible at the base of cutting's shoot; 5, absence of roots, necrosis of the base of cutting's shoot).

Commercial substrates and the mixtures of peat and perlite had a significant impact on the cuttings. The weakest cuttings were obtained with the peat-based substrate H3–H4 from the Northwestern Poland. The concentration of IAA had hardly any effect on the characteristics of rooted cuttings except for H3–H4 peat from the Northern Poland. A higher concentration of IAA in this substrate led to the increase in the length and fresh weight of chrysanthemum's roots. The addition of IAA had a positive effect on the rooted cuttings. In most of the substrates, the largest rooted cuttings were obtained when IAA was added to the substrate. The natural content of IAA in peat and commercial substrates was also found to have a significant effect on the quality and the rooting of cuttings of ornamental plants. The best rooted cuttings were obtained in the rooting of the cuttings of "Klasmann Steck Medium" in the commercial growing media of "Substrate for rooting cuttings of ornamental plants AURA" and in the peat originating from the Northern and Northeastern Poland (Szajdak and Nowak 2013).

These results indicate that the source of peat has a significant influence on the investigated parameters of rooted cuttings of ornamental plants. The source of peats

Table 12.6 Physical properties in investigated growing media

Growing media	C_{OM} (v/v)	D_{BD} (kg m^{-3})	$S_{\%}$ (v/v)	P_s (v/v)	W_V—10 cm H_2O (v/v)	A_V—10 cm H_2O (v/v)
Klasmann Steck Medium	87.6	119.6	38.1	92.7	72.0	20.6
Substrate for rooting cuttings of ornamental plants AURA	79.0	142.7	30.8	91.6	70.0	21.6
CERES—coir dust with perlite	79.5	85.3	32.9	95.0	71.8	23.1
White peat (H3–H4) from the Northwestern Polish regions mixed with perlite (3:1)	51.3	150.8	21.5	92.2	60.3	31.9
White peat (H3–H4) from the Northern Polish regions mixed with perlite (3:1)	52.5	134.5	19.8	93.0	62.6	30.5
White peat (H3–H4) from the Northeastern Polish regions mixed with perlite (3:1)	64.8	188.5	19.2	89.6	62.8	26.8
Optimal level[a]	>40	60–250	<35	85–95	35–80	15–65

Szajdak and Nowak (2013)
[a]Kipp et al. (2000)
C_{OM}, organic matter content; D_{BD}, dry bulk density; $S_{\%}$, shrinkage; Ps, total pore space; Wv, water volume; Av, air volume

has an impact on the content of IAA and various physical and chemical parameters such as organic matter content, porosity, and air content, which all ensure good conditions for the development of root systems (Table 12.6).

Szajdak et al. (2013) showed that different doses of IAA added to the commercial growing medium for the rooting of the cuttings of "Klasmann Steck Medium" had a significant influence on the tested characteristics of rooted cuttings of poinsettia "Prestige Early Red" (Table 12.7).

All of the cuttings (100 %) were rooted when IAA in the amount of 200 and 300 μg kg^{-1} was added to the substrate. The longest root and the highest fresh and dry weight of the roots were found when IAA was added to the substrate in the amount of 400 μg kg^{-1}. All poinsettia cuttings had a good quality and were suitable for further cultivation irrespective of the content of IAA in the medium (Photos 12.1 and 12.2).

12.2.2 Enzymatic Processes in Growing Media

Soil microorganisms, the living component of soil, usually occupy less than 1 % of the soil volume; however their number and efficiency are very high. The number and the activity of soil microorganisms depends on crop type, soil type, soil treatment, soil cultivation, as well as on the macro- and microclimate at each location (Dalal and Mayer 1986). They are important for catalyzing innumerable reactions necessary for life processes of microorganisms in soils, the decomposition of

Table 12.7 The effect of commercial growing medium for rooting of cuttings of *Euphorbia pulcherrima*

Commercial growing medium	Assessment of root system (1–5 scale, according to EPPO)	Length of roots (cm)	Fresh matter of roots (g plant^{-1})	Dry matter of roots (g plant^{-1})	% of rooted cuttings
Control (natural concentration of IAA)	2.48 cd	4.70 abc	0.6179 abcd	0.0603 abc	97.5
Substrate + IAA (200 μg kg^{-1})	2.30 bc	5.30 bc	0.7530 abcde	0.0639 abcd	100.0
Substrate + IAA (300 μg kg^{-1})	2.05 ab	5.70 c	1.1904 ef	0.1198 bcde	100.0
Substrate + IAA (400 μg kg^{-1})	2.45 bc	7.90 d	1.3926 f	0.1269 cdef	92.5

Szajdak et al. (2013)
In columns, values followed by the same letter(s) are not significantly different at $\alpha=0.05\%$

Photos 12.1 and 12.2 The effect of commercial substrate for rooting of cuttings of *Euphorbia pulcherrima* "Prestige Early Red" (photo J.S. Nowak). Explanations: *1/1*, natural IAA content; *1/2*, 200 μg kg^{-1} IAA; *1/3*, 300 μg kg^{-1} IAA; *1/4*, 400 μg kg^{-1} IAA with peat substrate

organic residues, the cycling of nutrients, and the formation of organic matter and soil structure (Deng and Tabatabai 1997).

Soil enzymes are a quantitatively minute but a very important fraction of soil organic matter because all biochemical actions are dependent upon or related to them. The frequently poor correlation between overall metabolic activity and activity of a particular enzyme is probably the result of stabilization of extracellular enzymes by association with soil organic matter and clay surfaces (Stevenson and Cole 1999). It is well known that the role of enzymes in coupling reactions leading to polymerization, polyaddition, and polycondensation is limited to the oxidation of substrates. Both microorganisms, as well as their metabolites—enzymes—participate actively in the organic matter decomposition, the detoxification of mineral xenobiotics (heavy metals), as well as in soil-forming processes.

Microbial metabolism within the unusual physical conditions of the peat environment leads to the release of soluble humic substances and organic compounds of a small molecular weight: amino acids, alkaloids, vitamins, fates, antibiotics, phenol compounds, enzymes, etc. These substances activate several biochemical pathways. Phenolics of a well-known and unknown structure inhibit the activity of the enzymes. In addition, the evolution of carbon dioxide, methane, nitrous oxide, and dinitrogen is observed (Freeman et al. 2004; Matocha et al. 2004; Makoi and Ndakidemi 2008).

Soil enzymes have a key biochemical function in the process of organic matter decomposition in the soil system (Nannipieri et al. 2002). They catalyze biochemical reactions necessary for the life processes of microorganisms in soils, the stabilization of soil structure, the decomposition of organic waste, and the organic matter formation and nutrient cycling. Extracellular enzyme activity is used as a sensitive indicator of changes that occur in carbon cycling associated with management practices. Therefore, the enzyme activity is treated as an important determinant of soil quality. Soil enzyme activities are indicators of soil degradation since they integrate information about microbial status from soil physicochemical conditions. They are used as sensors in the studies on the influence of soil treatments on soil fertility (Linch 1976; Matocha et al. 2004; Ralte et al. 2005; Sardans and Peñuelas 2005; Makoi and Ndakidemi 2008).

Enzymes are generally substrate specific and each enzyme normally catalyzes only a single type of reaction. They are required to catalyze the processing of high-molecular-weight organic matter into small subunits, thus enabling heterotrophic bacteria to obtain suitable substrates. The acquisition of nutrients released by extracellular hydrolysis depends to a large extent on the quality of the decomposing detritus. This degradation requires the involvement of many hydrolytic enzymes. The soil enzymes include a wide spectrum of oxido-reductases, transferases, hydrolases, and lyases. Xanthine oxidase, phenol oxidase, peroxidase, dehydrogenases, catalase, phosphatase, amylase, cellulase, pectinase, saccharase, protease, urease, arginine deaminase, nitrate reductase, etc. are mostly found in soils. These enzymes are generally of a bacterial or fungal origin and only a small fraction of enzymes is excreted by animals or plants (Rejmánkova and Sirová 2007; Singh and Kumar 2008). Enzymes are constantly being synthesized, accumulated, inactivated, and/or decomposed in the soil; hence they are playing an important role in agriculture, particularly in nutrient cycling. Enzymes in soils undergo complex biochemical processes including immobilization and enzyme stability. In this regard, all soils contain a group of enzymes that determine soil metabolic processes which, in turn, depend on the physical, chemical, microbiological, and biochemical processes and mechanisms in soils (Makoi and Ndakidemi 2008). Specifically, the assessment of the activities of hydrolases can provide information on the status of key reactions that participate in the rate-limiting steps of the decomposition organic matter and transformation of nutrients in soils (Trasar-Cepeda et al. 1998).

12.2.3 Xanthine Oxidase Activity

Xanthine oxidase [E.C. 1.2.3.2] plays an important role in redox processes and in the cycle of nitrogen in soils. This enzyme is a metal-flavoprotein containing FAD, molybdenum, and iron in the ratio of 2:2:8. Xanthine oxidase catalyzes the oxidation of hypoxanthine to xanthine and next to uric acid. It can be illustrated with the following figure:

$$\text{Xanthine} + O_2 + H_2O \xrightarrow{\text{XO}} \text{Uric acid} + H_2O_2 \qquad (12.2)$$

XO – xanthine oxidase

Furthermore, this enzyme is the last one in the pathway of the degradation of purine derivatives from nucleic acids, and it is assumed to be a rate-limiting step in purine metabolism. It is also believed to take part in alcohol metabolism and to play a role in the incorporation of iron into ferritin. This enzyme is widespread in nature, having been found in vertebrates, invertebrates, higher plants, green algae, fungi, and bacteria (Fried and Fried 1983; Hille and Massey 1985; Montalbini 1992; Fujimoto et al. 2000; Masuoka and Kubo 2004).

Szajdak et al. (2013) measured the activity of this enzyme in growing media before and after the rooting of the cuttings of poinsettia with a different dose of IAA. The results of this research indicated a high activity of xanthine oxidase before the rooting of the cuttings of poinsettia. In all analyzed substrate samples xanthine oxidase activity ranged from 4.03 to 8.88 μmol h^{-1} g^{-1} (Table 12.8).

The study revealed a decrease in the activity of xanthine oxidase from 15 to 24% when IAA was added to the substrate before the rooting of the cuttings of plants (from 200 μg kg^{-1} to 400 μg kg^{-1} IAA). In addition, in other studies Szajdak et al. (2011b) determined a higher activity of xanthine oxidase in the peat sample of Tagan peatland (from 22.93 to 54.73 μmol h^{-1} g^{-1}) in comparison with its activity in a growing medium.

12.2.4 Phenol Oxidase Activity

Phenol oxidase [E.C. 1.10.3.1] is an enzyme which catalyzes the oxidation of phenolic compounds to quinines and participates in the formation of humic acids. Moreover, it indicates the capacity of the microflora degrade recalcitrant organic substances like phenolic materials. It can be illustrated with the figure below:

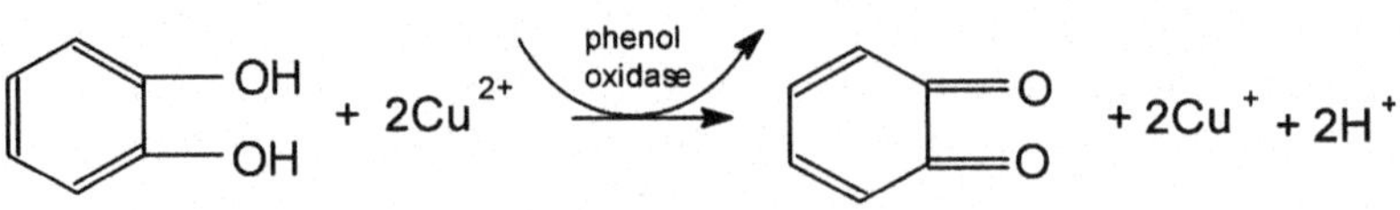

(12.3)

Table 12.8 Enzyme activity in commercial growing medium before and after rooting of cuttings of *Euphorbia pulcherrima*

Commercial growing medium	XOA (μmol h^{-1} g^{-1})	POA (nmol h^{-1} g^{-1})	PA (μmol h^{-1} g^{-1})	NRA (μgN $24h^{-1}$ g^{-1})	UA (μmol h^{-1} g^{-1})
Before rooting of cuttings of *Euphorbia pulcherrima*					
Control (natural concentration of IAA)	6.13	8.06	1.34	0.29	9.36
Substrate + IAA (200 μg kg^{-1})	8.88	9.62	1.22	0.31	9.76
Substrate + IAA (300 μg kg^{-1})	7.23	9.20	2.59	0.40	9.12
Substrate + IAA (400 μg kg^{-1})	6.77	9.96	1.26	0.44	8.32
After rooting of cuttings of *Euphorbia pulcherrima*					
Control (natural concentration of IAA)	4.13	10.81	0.95	4.21	15.13
Substrate + IAA (200 μg kg^{-1})	5.85	12.99	1.01	5.04	16.65
Substrate + IAA (300 μg kg^{-1})	4.03	9.62	0.54	4.32	19.29
Substrate + IAA (400 μg kg^{-1})	5.41	13.11	0.59	5.51	16.81

Szajdak et al. (2013)
XOA, xanthine oxidase activity; POA, phenol oxidase activity; PA, peroxidase activity; NRA, nitrate reductase activity; UA, urease activity

The group of enzymes includes o-diphenol oxidase (tyrosinase), p-diphenol oxidase (laccases), and polyphenol oxidases (Freeman et al. 2004; Benitez et al. 2006). Phenol oxidase enzymes catalyze polyphenol oxidation in the presence of oxygen (O_2) by removing phenolic hydrogen or hydrogens from radicals or quinines. According to Benitez et al. (2004) and Matocha et al. (2004), phenoloxidase enzymes are directly involved in humus formation, and therefore higher levels of their activity could contribute to an increase in humification processes. The activity of phenol oxidase can be limited by low oxygen pressure, low temperature, and acidic pH. A number of microbial species such as fungi, bacteria, and actinomycetes are able to produce phenol oxidase (Freeman et al. 1996; Freeman et al. 2001; Sinsabaugh et al. 2008; Sun et al. 2010).

In agricultural soil, phenol oxidase appears to act as an "enzymatic latch" which regulates soil organic carbon storage. The activity of phenol oxidase was found to be negatively related to soluble polyphenolics and to dissolve soil organic carbon. The reduction in phenol oxidase decreased the decomposition of soluble polyphenols, which in turn inhibited the decomposition of dissolved organic carbon (Shi 2011).

Many studies have focused on the effect which soil temperature and peat nitrogen status have on soil organic matter dynamics. During the decomposition of peat organic matter, secondary compounds are released by leaching and microbial breakdown or synthesized by microbial activity. Plant root–soil–microbe interactions are an important aspect of plant nutrient uptake strategies. Karsisto et al. (2004) showed a negative relationship between total nitrogen, temperature sum, and phenolic acid components. They suggested that peat nitrogen affected p-coumaric, 3,4-dihydroxybenzoic, vanillic, ferulic, and 2,5-dihydroxybenzoic acids.

Polyphenols are a very large and important group of compounds naturally occurring in nature. The following groups can be distinguished according to the chemical structure of the basic carbon skeleton: phenolic acids, flavonoids, stilbenes, and lignans. Phenolic acids are simple phenolic compounds of the non-flavonoid family. They are synthesized through the shikimic acid pathway and may occur in the bound or free form. Phenolic compounds are an integral component of plants. They participate in a variety of protective mechanisms, such as the formation of natural barriers on the way of infection. Phenolic compounds stimulate the growth inhibitors. Plant phenolics also act as modulators of IAA catabolism and increase the rigidity of plant cell walls, acting as molecular bridges between cell wall components (Savio et al. 2011). The content and composition of phenolic compounds depend on the plants that predominate in phytocoenotic microcomplex as well as the characteristics of the microhabitat (Djurdjevic' et al. 2003; Tarnawski et al. 2006; Tomson et al. 2010). The content of total phenolics in the peat profile in the pristine peatland of the Tuchola Forest National Park (Pomerania, Northern Poland) was measured by Szajdak and Styła (2012). The authors noted that the content of total phenolics increased with the depth of a profile. The highest concentrations of phenolic acids were observed at the depth of 50–70 cm and 70–100 cm (from 4.76 to 5.27 mg g^{-1}) and the lowest—at the depth 0–20 cm and 20–50 cm (from 3.95 to 2.86 mg g^{-1}, respectively). The typically low oxygen conditions at the depth of 50–100 cm inhibit phenol oxidase activity which allows the phenolic compounds to accumulate (Fenner et al. 2005).

Szajdak et al. (2013) measured phenol oxidase activity in growing media before and after the rooting of the cuttings of poinsettia when different doses of IAA were added. Overall higher activity was observed in peat substrates after the rooting of the cuttings of poinsettia. Phenol oxidase activity ranged from 8.06 to 9.96 μmol h^{-1} g^{-1} before the cultivation of poinsettia and ranged from 9.62 to 13.11 μmol h^{-1} g^{-1} after planting (Table 12.8).

Yang et al. (2011) postulated that the microorganisms influence the exudation of organic materials into the soil. Phenolic acids are released by plant roots and microorganisms as metabolites or biotransformation products in the rhizosphere and the nutrient solution (Linch, 1976). The study of Szajdak et al. (2011b) revealed a higher phenol oxidase activity in Tagan peatlands in Russia (from 6.18 to 46.01 μmol h^{-1} g^{-1}) than in a growing medium (from 8.06 to 13.11 μmol h^{-1} g^{-1}). Szajdak and Styła (2012) also showed that a comparable level of activity of phenol oxidase (from 10.92 to 54.70 μmol h^{-1} g^{-1}) was found in peatland at lobelia Lake Nierybno located in the National Tuchola Park in Poland and in the Tagan peatland in Russia (Szajdak et al. 2011b).

12.2.5 Peroxidase Activity

Peroxidase [E.C. 1.11.1.7] catalyzes the oxidation of a large number of aromatic structures (such as phenols and aromatic amines, hydroquinones and hydroquinoid amines, especially benzidine derivatives) at the expanse of H_2O_2. It can be illustrated with the following figure:

$$SH_2 + H_2O_2 \xrightarrow{\text{peroxidase}} S + 2H_2O \tag{12.4}$$

S – the substrate in the oxidized stateSH2 – the substrate in the reduced state

Peroxidases can be divided into two superfamilies, namely, the mammalian peroxidase superfamily, which includes enzymes such as lactoperoxidase and myeloperoxidase, and the plant superfamily. These enzymes occur in plants, animals, and microorganisms (Nakayama and Amachi 1999; Conesa et al. 2002). Peroxidase is involved, among many other processes, in the redox processes taking place in soils. Its specificity and biological functions vary depending on the source of the enzyme. As is well known, the role of peroxidases in coupling reactions leading to polymerization is limited to the oxidation of the substrates. Peroxidases mediate key ecosystem functions of lignin degradation, humification, carbon mineralization, dissolved organic carbon export, and the biological availability of soil nitrogen compounds (Johnsen and Jacobsen 2008).

Peroxidases also play a role in the synthesis of humic acids (HA) and fulvic acids (FA). Peroxidases are present in natural soil and they may originate from microorganisms, plants, or other organisms. Metalloenzyme contains iron heme prosthetic groups. This enzyme is involved in the oxidation of high-redox potential phenols and aromatic amines in the presence of hydrogen peroxide as an electron acceptor in the reactions. The release of carboxyl and methoxyl groups from phenolic substrates is mainly ascribed to microbial activity and it may lead to CO_2 production in soil (Criquet et al. 2000; Dec et al. 2003).

Szajdak et al. (2013) showed that the activity of peroxidase was higher before the rooting of the cuttings of poinsettia than after the rooting. Peroxidase activity ranged from 1.22 to 2.59 nmol h^{-1} g^{-1} before the rooting of the cuttings of poinsettia, while after planting it ranged from 0.54 to 1.01 nmol h^{-1} g^{-1}. The highest peroxidase activity was found at the dose of 300 μg kg^{-1} of IAA before the cultivation of plants (Table 12.8). Moreover, Szajdak et al. (2012a, b) documented a higher peroxidase activity in raw peat (peatlands Kusowo bog, from 5.41 to 8.22 nmol h^{-1} g^{-1} and fen of Stążka Mire, from 4.22 to 22.34 nmol h^{-1} g^{-1}) than in a growing medium (from 0.54 to 2.59 nmol h^{-1} g^{-1}).

12.2.6 Nitrate Reductase Activity

Denitrification plays an important role in the nitrogen cycle. It occurs in the anaerobic conditions with the participation of denitrification bacteria (*Paracoccus denitrificans*, *Paracoccus halodenitrificans*, *Thiobacillus denitrificans*, *Bacillus licheniformis*, *Pseudomonas aeruginosa*, and *Pseudomonas denitrificans*). This process accelerates the pathways of easily decayed organic matter and alkalinity or neutrality of soil. The potential for denitrification in soils is a complex interaction among aeration, nitrate availability, carbon substrate availability, and other intrinsic soil factors (Firestone 1982).

It is well known that the absence of O_2 or reduced O_2 availability is required for both the synthesis and the activity of denitrification enzymes. Quantification of O_2 availability and related rates of denitrification in soil is complicated due to dynamic relationships between aeration potential (O_2 flux) and microbial oxygen use (Klemedtsson et al. 1988). Water content, matric potential, and water-holding capacity all serve as relative predictors of microbial activity in soil (Sommers et al. 1981).

One of the enzymes participating in the denitrification process is the nitrate reductase [E.C. 1.6.6.3]. The prosthetic group of the enzyme creates a flavoprotein (FAD) containing molybdenum. The nitrogen oxides act as terminal electron acceptors in the absence of oxygen. In the anaerobic conditions, nitrate ions are reduced to nitrite ions and this enzyme catalyzes this conversion (Szajdak et al. 2005; Szajdak and Gaca 2010). In the anaerobic conditions, nitrate ions are reduced to nitrite ions and nitrate reductase is the catalyst of this reaction. Subsequently, the formed NO_2^- anions are reduced with the participation of nitrite reductase to N_2O which easily reacts with oxygen. The reduction reaction of N_2O to molecular nitrogen is catalyzed with the nitrous oxide reductase (Fu and Tabatabai 1989; Ma 2000). It can be illustrated with the following figure:

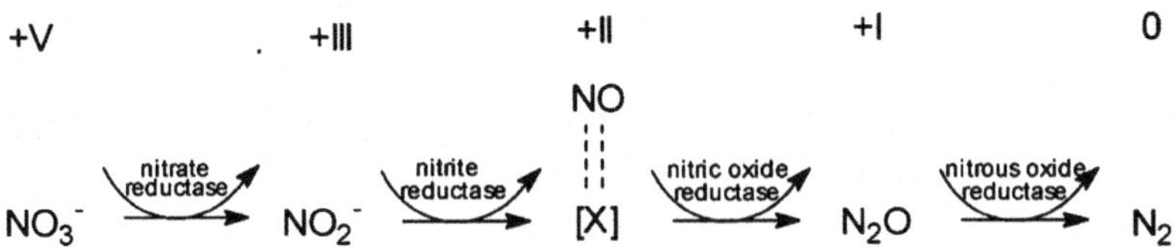

(12.5)

N_2O is a greenhouse gas contributing to the destruction of the Earth's stratospheric ozone layer (Groffman et al. 1992; Robertson and Klemedtsson 1996; Carnol and Ineson 1999; Murray and Knowles 1999). In addition, the nitrogen can be also removed from the soil in the form of NO_3^-. The nitrates are absorbed with the sorption complex, and when it is raining, they may migrate to the deeper soil layers and next to the groundwaters.

Szajdak et al. (2013) conducted the study of nitrate reductase activity in growing media and raw peat (raised bog and fen). The obtained results indicated that the activity of nitrate reductase in commercial substrates for rooting (from 0.60 to 0.76 μgN $24h^{-1}g^{-1}$) was at a similar level as the activity in peat from Kusowo bog (0.53 μgN $24h^{-1}$ g^{-1}) (Table 12.8). The highest nitrate reductase activity was found in peat from Stążka Mire (2.32 μgN $24h^{-1}$ g^{-1}) and the lowest in raised bog from a commercial company (from 0.02 to 0.03 μgN $24h^{-1}$ g^{-1}) (Szajdak et al. 2013). Amha and Bohne (2011) hypothesized that the rate of denitrification from the horticultural peats was influenced by the respective peat-forming environment. They suggested that the rate of denitrification from a given horticultural peat was also affected by the availability of an easily decomposable carbon source to heterotrophic microorganisms.

12.2.7 Urease

Urease [E.C. 3.5.1.5] is an important enzyme involved in the organic matter mineralization in soil and in N cycle. This enzyme releases $N\text{-}NH_4^+$ through urea hydrolysis. It is essential in the chain of hydrolysis of amino compounds, which are supplied to the soil by plants and to a smaller extent by animals and microorganisms. It can be illustrated with the following figure:

$$NH_2CONH_2 + H_2O \xrightarrow{\text{urease}} CO_2 + 2\,NH_3 \tag{12.6}$$

Alkaline and aerobic conditions together with the ammonium accumulation during the urease activity assays tend to enhance the nitrification process in soils (Chaperon and Sauve 2007). Soil ureases are microbial products that can accumulate in cell-free forms in the soil because they are highly resistant to environmental degradation. Soil urease originates mainly from plants and microorganisms found as both intra- and extracellular enzyme. Urease activity in soils is influenced by cropping history, organic matter content, soil depth, soil amendments, heavy metals, and temperature (Makoi and Ndakidemi 2008; Sardans et al. 2008).

The stability of urease enzyme in the system is affected by a number of factors. Extracellular urease associated with soil organo-mineral complexes is more stable than urease in the soil solution. Those humus–urease complexes extracted from soil are highly resistant to denaturing agents such as extreme temperature and proteolytic attack. On the other hand, urease extracted from plants or microorganisms is rapidly degraded in soil by proteolytic enzymes. This suggests that a significant fraction of ureolytic activity in soil is carried out by extracellular urease, which is stabilized by the immobilization of organic and mineral soil colloids (Marzadori et al. 2000; Makoi and Ndakidemi 2008).

Szajdak et al. (2013) found out that the urease activity in fen and raised bog was lower (13.78 and 12.18 μmol h^{-1} g^{-1}) than in the mixture of commercial substrates and IAA (from 25.29 to 45.38 μmol h^{-1} g^{-1}). Similarly, the concentrations of ammonium ions were higher in the raised bog from a commercial company with different

IAA additions than in the commercial substrates for rooting and peat soils. It was related to a higher urease activity in the raised bog from the commercial company (71 %) than in peat soils and commercial substrates for rooting (41%). The results suggest that the urease activity can vary depending on a different growing media (Szajdak et al. 2013). Furthermore, Styła and Sawicka (2009) evaluated the activity of this enzyme in a mineral soil under apple-tree cultivation in a replanted orchard. Urease activity ranged from 1.78 to 3.35 μmol h^{-1} g^{-1}.

The study of Yang et al. (2011) indicated that enzymes in the rhizosphere of soil played an essential role in soil processes such as nutrient cycling and energy transformation by catalyzing numerous chemical, physical, and biological reactions. They are mainly exuded by roots and microorganisms and their activity can also have significant influence on nutrients.

References

Ali B, Sabri AN, Hasnain S (2009) Indole-3-acetic acid production by plant associated bacteria: potential to alter endogenous IAA content and growth of *Triticum aestivum* L. New Biotechnol 25S. doi:10.1016/j.nbt.2009.06.883

Amha Y, Bohne H (2011) Denitrification from the horticultural peat: effects of pH, nitrogen, carbon, and moisture contents. Biol Fertil Soils 47:293–302

Arshad M, Frankenberger WT Jr (1991) Microbial production of plant hormones. Plant Soil 133:1–8

Barbieri P, Zanelli T, Galli E, Zanetti G (1986) Wheat inoculation with *Azospirillum brazilance* Sp6 and some mutants altered in nitrogen fixation and indole-3-acetic acid production. FEMS Microbiol Lett 36:87–90

Barea JM, Brown ME (1974) Effects on plant growth by *Azotobacter paspali* related to synthesis of plant growth regulating substances. J Appl Bacteriol 37:583–593

Benitez E, Melgar R, Nogales R (2004) Estimating soil resilience to a toxic organic waste by measuring enzyme activities. Soil Biol Biochem 36:1615–1623

Benitez E, Nogales R, Campos M, Ruano F (2006) Biochemical variability of olive-orchard soils under different management systems. Appl Soil Ecol 32:221–231

Beyeler M, Keel C, Michau P, Haas D (1999) Enhanced production of indole-3-acetic acid by a genetically modified strain of *Pseudomonas fluorescens* CHA0 affects root growth of cucumber, but does not improve protection of the plant against *Pythium* root rot. FEMS Microbiol Ecol 28:225–233

Bhekithemba M, Wahome PK (2010) Propagation of geranium (*Pelargonium hortorum*) using different medium components. Am Eurasian J Agric Environ Sci 7(5):497–500

Børsheim KY, Christensen BE, Painter TJ (2001) Preservation of fish by embedment in *Sphagnum* moss, peat or holocellulose: experimental proof of the oxopolysaccharidic nature of the preservative substance and of its antimicrobial and tanning action. Innov Food Sci Emerg Technol 2:63–74

Carnol M, Ineson P (1999) Environmental factors controlling NO_3^- leaching, N_2O emissions and numbers of NH_4^+ oxidisers in a coniferous forest soil. Soil Biol Biochem 31:979–990

Caron M, Patten CL, Ghosh S (1995) Effects of plant growth promoting rhizobacteria *Pseudomonas putida* GR-122 on the physiology of canola roots. Proc Plant Growth Reg Soc Am 7:18–20

Chao IL, Cho CL, Chen L-M, Liu Z-H (2001) Effect of indole-3-butyric acid on the endogenous indole-3-acetic acid and lignin contents in soybean hypocotyl during adventitious root formation. J Plant Physiol 158:1257–1262

Chaperon S, Sauve S (2007) Toxicity interaction of metals (Ag, Cu, Hg, Zn) to urease and dehydrogenase activities in soils. Soil Biol Biochem 39:2329–2338

Conesa A, Punt PJ, Van Den Hondel CAMJJ (2002) Fungal peroxidases: molecular aspects and applications. J Biotechnol 93:143–158

Cresswell GC (1992) Coir dust – a viable alternative to peat. In: Proceedings of the Australian potting mix manufacturers conference. 2–3 April 1992, Sydney, Australia, pp 1–5

Criquet S, Farnet AM, Tagger S, Le Petit J (2000) Annual variations of phenoloxidase activities in an evergreen oak litter: influence of certain biotic and abiotic factors. Soil Biol Biochem 32:1505–1513

Dahm H, Sitek JM, Strzelczyk E (1977) Synthesis of auxins by bacteria isolated from the roots of pine seedlings inoculated with rusty forest soil. Pol J Soil Sci 10:131–137

Dalal RC, Mayer RJ (1986) Long-term trends in fertility of soils under continuous cultivation and cereal cropping in southern Queensland. V. Rate of loss of total nitrogen from the soil profile and changes in carbon-nitrogen ratios. Aust J Soil Res 24:493–504

Dec J, Haider K, Bollag JM (2003) Release of substituents from phenolic compounds during oxidative coupling reactions. Chemosphere 52:549–556

Deng SP, Tabatabai MA (1997) Effect of tillage and residue management on enzyme activities in soils: III. phosphatases and arylsulfatase. Biol Fertil Soils 24:141–146

Djurdjevic' L, Dinic' A, Mitrovic' M, Pavlovic' P, Teševic V (2003) Phenolic acids distribution in a peat of the relict community with Serbian spruce in the Tara Mt. forest reserve (Serbia). Eur J Soil Biol 39:97–103

Dundek P, Holík L, Rohlík T, Hromádko L, Vranová V, Rejšek K, Formánek P (2011) Methods of plant root exudates analysis. Rev Acta Univ Agric et Silv Mend 59(3):241–246

Falkowski G, Szydło W (2005) Wpływ terminu przesadzania i sposobu zastosowania auksyn na wzrost wybranych gatunków drzew i krzewów ozdobnych. (The effect of transplanting date and the way of applying auxins on the growth of selected ornamental trees and shrubs). Zesz Nauk Inst Sad Kwiac 13:111–117 (in Polish)

Farnsworth K, Guam RH (1995) Root initiation in *Ocotea bullata* (Burch.) Baill. Cuttings. South Afr For J 173:31–33

Fenner N, Frejman C, Reynolds B (2005) Observations of a seasonally shifting thermal optimum in peatland carbon-cycling processes; implications for the global carbon cycle and soil enzyme methodologies. Soil Biol Biochem 37:1814–1821

Firestone MK (1982) Biological denitrification. In: Stevenson FJ (ed) Nitrogen in agricultural soils. American Society of Agronomy, Madison, pp 289–326

Fisher PR, Dickson RW, Mohammad-Pour G, Huang J (2013) Effect of the solution electrical conductivity (EC) and pre-plant nutrient form on the pH of a peat-perlite substrate. In: Book of abstracts. The international symposium on growing media and soilless cultivation. 17–21 June 2013, Leiden, Netherlands, p 55

Freeman C, Liska G, Ostle NJ, Lock MA, Reynolds B, Hudson J (1996) Microbial activity and enzymic decomposition processes following peatland water table drawdown. Plant Soil 180:121–127

Freeman C, Ostle N, Kang H (2001) An enzymic 'latch' on a global carbon storage. Nature 409:149

Freeman C, Ostle NJ, Fener N, Kang H (2004) A regulatory role for phenol oxidase during decomposition in peatlands. Soil Biol Biochem 36:1663–1667

Fried R, Fried LW (1983) Xanthine oxidase (Xanthine dexydrogenase). In: Bergmeyer HU (ed) Methods of enzymatic analysis, vol 2. Wiley, New York, pp 644–649

Fu MH, Tabatabai MA (1989) Nitrate reductase activity in soils: effects of trace elements. Soil Biol Biochem 21:943–946

Fujimoto Y, Sakuma S, Tagami T, Ichikawa R, Fujita T (2000) N-ethylmaleimide inhibits xanthine oxidase activity with no detectable change in xanthine dehydrogenase activity in rabbit liver. Life Sci 68:517–524

Garcia-Rodriguez T, Alvarez C, Peréz-Silva J (1986) Indole-3-acetic acid production by cell-free extracts of *Rhizobium trifolii*. Pol J Soil Sci 17:59–65

Glick BR (1995) The enhancement of plant growth by free-living bacteria. Can J Microbiol 41:109–117

Glick BR, Brooks HE, Pasternak JJ (1986) Physiological effects of plasmid DNA transformation of *Azotobacter vinelandii*. Can J Microbiol 32:145–148

Glick BR, Patten CL, Holguin G, Penrose DM (1999) Biochemical and genetic mechanisms used by plant growth promoting bacteria. Imperial College Press, London, p 270

Groffman PM, Tiedje JM, Mokma DL, Simkins S (1992) Regional scale analysis of denitrification in north temperate forest soils. Landsc Ecol 7:45–53

Halda-Alija L (2003) Identification of indole-3-acetic acid producing freshwater wetland rhizosphere bacteria associated with *Juncus effusus* L. Can J Microbiol 49:781–787

Hille R, Massey V (1985) Molybdenum – containing hydroxylases: xanthine oxidaes, aldehyde oxidase, and sulfite oxidase. In: Thomas S (ed) Molybdenum enzymes. Wiley, New York, pp 443–518

Huttunen J, Reinikainen O (2000) Peat growing media in modern vegetable production in greenhouses. In: Rochefort L, Daigle JY (eds) Sustaining our peatlands. Proceedings of the 11th international peat congress, vol II. Quebec, Canada, p 522

Jankiewicz LS (1997) Przegląd regulatorów roślinnych. (Overview of plant regulators). In: Jankiewicz LS (ed) Regulatory wzrostu i rozwoju roślin. Właściwości i działanie. (Regulators of plant growth and development. Properties and action). Part 1. Wydawnictwo Naukowe PAN, Warszawa. pp 17–37 (in Polish)

Johnsen AR, Jacobsen OS (2008) A quick and sensitive method for the quantification of peroxidase activity of organic surface soil from forests. Soil Biol Biochem 40:814–821

Kamnev AA, Shchelochkov AG, Perfiliev Y D, Tarantilis PA, Polissiou MG (2001) Spectroscopic investigation of indole-3-acetic acid interaction with iron (III). J Mol Struct 563–564:565–572

Karnwal A (2009) Production of indole acetic acid by fluorescent *Pseudomonas* in the presence of L-Tryptophan and rice root exudates. J Plant Pathol 91(1):61–63

Karsisto M, Hartman M, Potila H, Sarjala T (2004) Phenolic compounds and metabolic profiles of microbial communities in peat; effects of temperature and peat nitrogen status. In: Päinvänen J (ed) Wise use of peatlands. Proceedings of the 12th international peat congress vol I. Tampere, Finland, p 306

Kawano T, Kawano N, Hosoya H, Lapeyrie F (2001) Fungal auxin antagonist hypaphorine competitively inhibits indole-3-acetic acid dependent superoxide generation by horseradish peroxidase. Biochem Biophys Res Commun 288:546–551

Kipp JA, Wever G, De Kreij C (2000) International substrate manual. Elsevier International Business Information, Doetinchem, pp 1–94

Klemedtsson L, Svensson BH, Rosswall T (1988) A method of selective inhibition to distinguish between nitrification and denitrification as sources of nitrous oxide in soil. Biol Fertil Soils 6:112–119

Kloepper JW, Leong J, Teintze M, Schroth MN (1980) Enhanced plant growth by siderophores produced by plant growth promoting rhizobacteria. Nature 286:885–886

Kloepper JW, Lifshitz R, Zablotowicz RM (1989) Free-living bacterial inocula for enhancing crop productivity. Trends Biotechnol 7:39–44

Linch JM (1976) Products of soil micro-organisms in relation to plant growth. CRC Crit Rev Microbiol 5:67–107

Ludwig-Müller J, Hilgenberg W, Epstein E (1995) The in vitro biosynthesis of indole-3-butyric acid in maize. Phytochemistry 40(1):61–68

Ma RX (2000) Effects of allelochemicals on activity of nitrate reductase. J Environ Sci 12:125–128

Ma Z, Ge L, Lee ASY, Yong JWH, Tan SN, Ong ES (2008) Simultaneous analysis of different classes of phytohormones in coconut (*Cocos nucifera* L.) water using high-performance liquid chromatography and liquid chromatography – tandem mass spectrometry after solid-phase extraction. Anal Chim Acta 610:274–281

Makoi JHJR, Ndakidemi PA (2008) Selected soil enzymes: examples of their potential roles in the ecosystem. Afr J Biotechnol 7(3):181–191

Martinez FX, Sepò N, Valero J (1997) Physical and physicochemical properties of peat-coir mixes and the effect of clay-material addition. Acta Hortic 450:39–46

Martinez-Morales LJ, Soto-Urzúa LS, Baca BE, Sánchez-Ahèdo JA (2003) Indole-3-butyric acid (IBA) production in culture medium by wild strain *Azospirillum brasilense*. FEMS Microbiol Lett 228:167–173

Marzadori C, Francioso O, Ciavatta C, Gessa C (2000) Influence of the content of heavy metals and molecular weight of humic acids fractions on the activity and stability of urease. Soil Biol Biochem 32:1893–1898

Masuoka N, Kubo I (2004) Characterization of xanthine oxidase inhibition by anacardic acids. Biochim Biophys Acta 1688:245–249

Matocha CJ, Haszler GR, Grove JH (2004) Nitrogen fertilization suppresses soil phenol oxidase enzyme activity in no-tillage systems. Soil Sci 169(10):708–714

Matysiak B, Nowak JS (2008) Coir substrates for rooting of ornamental ericaceous plants. Prop Orn Plants 8(2):76–80

Meerow AW (1994) Growth of two subtropical ornamentals using coir (coconut mesocarp pith) as a peat substitute. HortSci 29(12):1484–1486

Meerow AW (1995) Growth of two tropical foliage plants using coir dust as a container medium amendment. HortTechnology 5:237–239

Montalbini P (1992) Changes in xanthine oxidase activity in bean leaves induced by *Uromyces phaseoli* infection. J Phytopathol 134:63–74

Murray RE, Knowles R (1999) Chloramphenicol inhibition of denitrifying enzyme activity in two agricultural soils. Appl Environ Microbiol 65:3487–3492

Nakayama T, Amachi T (1999) Fungal peroxidase: its structure, function, and application. J Mol Catal B Enzym 6:185–198

Nannipieri P, Kandeler E, Ruggiero P (2002) Enzyme activities and microbiological and biochemical processes in soil. In: Burns RG, Dick RP (eds) Enzymes in the environment. Activity, ecology, and applications. Marcel Dekker Inc., New York, pp 1–33

Noguera P, Abad M, Puchades R, Noguera V, Macquieira A, Martinez J (1997) Physical and chemical properties of coir waste and their relation to plant growth. Acta Hortic 450:365–369

Painter TJ (1998) Carbohydrate polymers in food preservation: an integrated view of the Maillard reaction with special reference to discoveries of preserved foods in *Sphagnum*-dominated peat bogs. Carbohydr Polym 36:335–347

Patten C, Glick BR (1996) Bacterial biosynthesis of indole-3-acetic acid. Can J Microbiol 42:207–220

Pudelski T (2002) Torf i wyroby z torfu w ogrodnictwie. In: Ilnicki P (ed) Torfowiska i torf. (Peatlands and peat). Wydawnictwo. AR im. Augusta Cieszkowskiego, Poznań, pp 458–467 (in Polish)

Ralte V, Pandey HN, Barik SK, Tripathi RS, Prabhu SD (2005) Changes in microbial biomass and activity in relation to shifting cultivation and horticultural practices in subtropical evergreen forest ecosystem of north-east India. Acta Oecol 28:163–172

Raviv M, Wallach R, Silber A, Bar-Tal A (2002) Substrates and their analysis. In: Savvas D, Passam H (eds) Hydroponic production of vegetables and ornamental. Embrio Publications, Athens, pp 25–101

Rejmánkova E, Sirová D (2007) Wetland macrophyte decomposition under different nutrient conditions: Relationships between decomposition rate, enzyme activities and microbial biomass. Soil Biol Biochem 39:526–538

Robertson K, Klemedtsson L (1996) Assessment of denitrification in organogenic forest soil by regulating factors. Plant Soil 178:49–57

Rose R, Haase D (2000) The use of coir as a containerized growing medium for Douglas-fir seedlings. Nat Plants J 2:107–111

Sardans J, Peñuelas J (2005) Drought decreases soil enzyme activity in a Mediterranean *Quercus ilex* L. forest. Soil Biol Biochem 37:455–461
Sardans J, Penñuelas J, Estiarte M (2008) Changes in soil enzymes related to C and N cycle and in soil C and N content under prolonged warming and drought in a Mediterranean shrubland. Appl Soil Ecol 39:223–235
Sarwar M, Arshad M, Martens DA, Frankenberger WT Jr (1992) Tryptophan-dependent of auxins in soil. Plant Soil 147:207–215
Savio LEB, Astarita LV, Santarém ER (2011) Secondary metabolism in micropropagated *Hypericum perforatum* L. grown in non-aerated liquid medium. Plant Cell Tiss Org Cult. doi: 10.1007/s11240-011-0058-9
Schmilewski G (2008) The role of peat in assuring the quality of growing media. Mires and peat 3/02. http://www.mires-and-peat.net/, ISSN 1819-754X. pp 1–8
Schmilewski G, Falkenberg H (2000) Production and processing of peat-based growing media – a precondition for sustainable horticulture in Europe. In: Rochefort L, Daigle JY (eds) Sustaining our peatlands. Proceedings of the 11th international peat congress, vol II. Quebec, Canada, pp 533–541
Shadparvar V, Torkashvand MA, Alamshiri AH (2011) Effect of IBA and soil mixture on rooting of *Hibiscus rosa* – sinensis. Eur J Exp Biol 1(4):142–146
Shi W (2011) Agricultural and ecological significance of soil enzymes: soil carbon sequestration and nutrient cycling. In: Shukla G, Varma A (eds) Soil enzymology. Chapter 3. Springer-Verlag, Berlin/Heidelberg, pp 43–60
Singh DK, Kumar S (2008) Nitrate reductase, arginine deaminase, urease and dehydrogenase activities in natural soil (ridges with forest) and in cotton soil after acetamiprid treatments. Chemosphere 71:412–418
Sinsabaugh RL, Lauber CL, Weintraub MN, Ahmed B, Allison SD, Crenshaw C, Contosta AR, Causack D, Frey S, Gallo ME, Gartner MN, Ahmed TB, Hobbie SE, Holland K, Keeler BL, Powers JS, Stursova M, Takacs-Vesbach C, Wallenstein MD, Zak DR, Zeglin LH (2008) Stoichiometry of a soil enzyme activity at global scale. Ecol Lett 11:1252–1264
Smith C (1995) Coir: a viable alternative to peat for potting. Horticulturist 4:24–28
Sommers LE, Gilmour CM, Wildrung RE, Beck SM (1981) The effect of water potential on decomposition processes in soils. In: Parr JF (ed) Water potential relationships in soil microbiology, Special Publication 9. Soil Science Society of America, Madison, pp 97–117
Stamps RH, Evans MR (1999) Growth of *Dracaena marginata* and *Spathiphyllum* 'Petite' in *Sphagnum* peat and coconut coir dust-based growing media. J Environ Hortic 17:49–52
Stevenson FJ, Cole MA (1999) Cycles of soil. Wiley, New York, p 448
Stoven J, Kooima H (1999) Coconut coir-based media versus peat-based media for propagation of woody ornamentals. Comb Proc Int Plant Prop Soc 49:373–374
Styła K, Sawicka A (2009) Biochemical activity of soil in apple tree (*Malus domestica*) orchard after replantation. Agron Res 7(2):855–864
Sun X, Xiang W, He L, Zhao Y (2010) Impacts of hydrological conditions on enzyme activities and phenolic concentrations in peatland soil: an experimental simulation. Front Earth Sci Chin 4(4):463–470
Świstowska A, Hetman J (2004) Wpływ auksyn na ukorzenianie mikrosadzonek i adaptację roślin *Columnea mirta* Klotzsch et Haust. Cz. II. Następczy wpływ w uprawie szklarniowej. (The influence of auxins on the rooting of microcuttings and acclimatization of plants of *Columnea hirta* Klotzsch et Hanst. Part II. The consequent influence in the greenhouse cultivation). Acta Scient Polon Hort Cultus 3(2):239–248 (in Polish)
Szajdak L (2004) Substancje aktywne biologicznie w kompostach z odpadów komunalnych na tle innych podłoży organicznych. (Biological active of substances in compost from municipal waste against a background organic growing media). In: Drozd J (ed) Komposty z odpadów komunalnych. Produkcja, wykorzystanie i wpływ na środowisko. (Municipal solid waste composts, production, utilization and influence on the environment). Polish Humic Substances Society, Wrocław, pp 186–196 (in Polish)

Szajdak LW, Gaca W (2010) Nitrate reductase activity in soil under shelterbelt and adjoining cultivated field. Chem Ecol 26(4):123–134

Szajdak L, Maryganowa V (2007) Occurrence of IAA auxin in some organic soil. Agron Res 5(2):175–187

Szajdak L, Maryganowa V (2009) Impact of age and composition of shelterbelts plant on IAA content as allelochemical in soils. Allelopath J 23(2):461–468

Szajdak LW, Nowak JS (2013) Impact of peat substrates with different concentrations of indole-3-acetic acid on ornamental plant cultivation. Peatlands Int 1:25–27

Szajdak LW, Styła K (2012) Phenol oxidase activity and the concentrations of total phenolic in peat profile of peatland by Nierybno Lake in Tuchola Forest National Park. In: Szajdak LW, Gaca W, Meysner T, Styła K, Szczepański M (eds) Necessity of peatlands protection. Wydawnictwo Prodruk, Poznań, pp 77–86

Szajdak L, Maryganowa V, Bambalov N (2004) Concentration of indole-3-acetic acid in different kinds of peat and sapropel. In: Päinvänen J (ed) Wise use of peatlands. Proceedings of the 12th international peat congress, vol II. Tampere, Finland, pp 1152–1155

Szajdak L, Gaca W, Karg M (2005) Impact of the age of shelterbelts and the composition of plants on the dissimilatory nitrate reductase activity in soils. Pol J Soil Sci 38(2):135–144

Szajdak LW, Gaca W, Meysner T, Styła K, Maryganova V (2011a) Enzymes activity and IAA contents in soils. In: Narwal SS, Pavlovic P, John J (eds) Forestry and agroforestry, vol 2, Research methods in plant sciences. Studium Press LLC, Houston, pp 207–230

Szajdak LW, Inisheva LI, Meysner T, Gaca W, Styła K (2011b) Activities of enzymes participating in redox potential in the two depths of Tagan peatland. Tomsk Pedagogical State University Herald 8(110):68–75

Szajdak LW, Meysner T, Styła K (2011c) Biochemical and chemical characterization of soils under shelterbelts and adjoining cultivated fields. In: Szajdak LW (ed) Shelterbelts: efficient element of the landscape. LAP Lambert Academic Publishing, Saarbrücken, pp 33–53

Szajdak LW, Gaca W, Styła K, Meysner T (2012a) Changes of enzyme activities in peat profile of Kusowo bog. In: Szajdak LW, Gaca W, Meysner T, Styła K, Szczepański M (eds) Necessity of peatlands protection. Wydawnictwo Prodruk, Poznań, pp 47–60

Szajdak LW, Styła K, Meysnr T, Gaca W (2012b) Choice enzymes participating in oxydoreduction properties in peat profile of Stążka Mire. In: Szajdak LW, Gaca W, Meysner T, Styła K, Szczepański M (eds) Necessity of peatlands protection. Wydawnictwo Prodruk, Poznań, pp 61–75

Szajdak LW, Nowak JS, Gaca W, Meysner T, Styła K, Szczepański M (2013) Mixture of growing medium with IAA, biochemical and chemical properties in *Euphorbia pulcherrima* cultivation. ProEnvironment 6(14):254–261

Szydło W (2003) Auksyny w rozmnażaniu drzew i krzewów ozdobnych przez sadzonki. Charakterystyka auksyn. Cz. I. (Auxin in the propagation of ornamental trees and shrubs by cuttings. Characteristics of auxin. Part 1). Szkółkarstwo 4:7–9 (in Polish)

Tarnawski M, Depta K, Grejciun D, Szelepin B (2006) HPLC determination of phenolic acids and antioxidant activity in concentrated peat extract-a natural immunomodulator. J Pharm Biomed Anal 41:182–188

Tien TM, Gaskins MH, Hubbell DH (1979) Plant growth substances produced by *Azospirillum brasilense* and their effect on the growth of pearl millet (*Pennisetum americanum* L.). Appl Environ Microbiol 37:1016–1024

Tomson AE, Zhmakova NA, Makarova NL, Naumova GV (2010) On free phenol compounds content in peat. In: Szajdak LW, Karabanov AK (eds) Physical, chemical and biological processes in soils. Wydawnictwo Prodruk, Poznań, pp 509–514

Trasar-Cepeda C, Leiros C, Gil-Sortes F, Seona S (1998) Towards a biochemical quality index for soils: an expression relating several biological and biochemical properties. Biol Fertil Soils 26:100–106

United States Environmental Protection Agency, Office of Prevention, Pesticides and Toxic Substances (H-7508W) EPA-738-F-92-001. August 1992, R.E.D. FACTS, Indole-3-Butyric Acid. pp 1–4

Verdonck O, De Vleeschauwer D, Penninck R (1983) Cocofibre dust a new growing medium for plants in the tropics. Acta Hortic 133:215–220

Yang SJ, Du ZY, Yu Y, Zhang ZL, Sun XY, Xing SJ (2011) Effects of root pruning on physicochemical characteristics and biological properties of winter jujube rhizosphere soil. Plant Soil Environ 11:493–498

Zhen Ma Z, Ge L, Lee ASY, Yong JWH, Tan SN, Ong ES (2008) Simultaneous analysis of different classes of phytohormones in coconut (*Cocos nucifera* L.) water using high-performance liquid chromatography and liquid chromatography - tandem mass spectrometry after solid-phase extraction. Anal Chim Acta 610:274–281

Printed by Printforce, the Netherlands